CHANGES OF ATMOSPHERIC CHEMISTRY AND EFFECTS ON FOREST ECOSYSTEMS

Nutrients in Ecosystems

VOLUME 3

Changes of Atmospheric Chemistry and Effects on Forest Ecosystems

A Roof Experiment without a Roof

Edited by

REINHARD F. HÜTTL

Chair of Soil Protection and Recultivation,
Brandenburg University of Technology at Cottbus, Germany

and

KLAUS BELLMANN

Berlin, Germany

Springer-Science+Business Media, B.V.

Library of Congress Cataloging-in-Publication Data:

Changes of atmospheric chemistry and effects on forest ecosystems : a roof experiment without a roof / edited by Reinhard F. Hüttl and Klaus Bellmann.
p. cm. — (Nutrients in ecosystems ; v. 3)

1. Scots pine—Ecophysiology—Germany (East) 2. Scots pine—Effect of air pollution on—Germany (East) 3. Scots pine—Effect of atmospheric deposition on—Germany (East) 4. Forest ecology—Germany (East) I. Hüttl, R. F. II. Bellmann, Klaus. III. Series.
QK494.5.P66C448 1999
585'.2–dc21 99-25199

ISBN 978-90-481-5224-7 ISBN 978-94-015-9022-8 (eBook)
DOI 10.1007/978-94-015-9022-8

Printed on acid-free paper

Contents

List of authors

Dr Klaus Bellmann, Potsdam Institute for Climate Impact Research, PO Box 601203, 14412 Potsdam, Germany

Dr Charlotte Bergmann, Chair of Recultivation and Soil Protection, Technical University of Brandenburg, 03013 Cottbus, Germany

Dr Gert E Dudel, Institute of General Ecology and Environmental Protection, 01735 Tharandt, Germany

Dr Hans-Peter Ende, Center for Agricultural Landscape and Land Use Research, 15374 Müncheberg, Germany

Dr Markus Erhard, Potsdam Institute for Climate Impact Research Telegrafenberg, 14473 Potsdam, Germany

Dr Rüdiger Grote, Chair of Forest Growth, Technical University Munich, Am Hochanger 13, 85354 Freising, Germany

Dr Reinhard F Hüttl, Chair of Soil Protection and Recultivation, Technical University of Brandenburg, 03013 Cottbus, Germany

Dr Lüttschwager, Center for Agricultural Landscape and Land Use Research, 15374 Müncheberg, Germany

Dr B Münzenberger, Institute of Microbial Ecology and Soil Biology, Centre for Agricultural Landscape and Land Use Research Müncheberg, Dr Zinn Weg 18, 16225 Eberswalde, Germany

Dr U Neumann, Institute of Forest Growth and Forest Computer Science, Tharandt Dresden University of Technology, Germany

Dr S Rust, Chair of Recultivation and Soil Protection, Technical University of Cottbus, 03013 Cottbus, Germany

Dr Horst Schulz, Department of Chemical Ecotoxicology, Centre for Environmental Research Leipzig, Permoserstr. 15, 04318 Leipzig, Germany

Dr Florian Strubelt, Centre for Agricultural Landscape and Landuse Research, Research Station, 16225 Eberswalde, Germany

Dr M Weisdorfer, Brandenburg Technical University of Cottbus, Chair of Soil Protection and Recultivation, PO Box 101344, 03013 Cottbus, Germany

1
Introduction to the SANA-project (SANA: regeneration of the atmosphere above the new states of Germany – effects on forest ecosystems)

K. BELLMANN and R. GROTE

1. The history of air pollution and deposition in East Germany

1.1. *Emission*

Former East Germany was one of the countries in Europe which had extremely high emissions of pollutants. Energy production was mainly based on the use of soft coal. Power plants as well as industry lacked the necessary equipment for pollutant reduction. Thus in 1989, East Germany emitted in total 5.3 Mt of SO_2 and 2.1 Mt of alkaline dusts, representing an average of 48.8 kg km ^{2} and 19.0 kg km^{-2}, respectively (Finkbeiner *et al.*, 1993). The emission per capita was 18-fold higher than in West Germany.

Point sources of energy and industrial production, emitting more than 100.000 t SO_2 a^{-1}, were concentrated in the southern parts of the country, which during meterological inversion situations led to an extraordinary high load of deposition in this region and to peak concentrations of between 2800 µg m^{-3} (Bitterfeld) and 4500 µg m^{-3} (Leipzig) SO_2 (Figure 1). Between 1970 and 1989, one year before German reunification, the emissions of SO_2, NO_X and NH_3 increased considerably (by 27, 42 and 24%, respectively). Alkaline dust emission, however, decreased by 16% due to the installation of dust filters into power plants (Table 1).

1.2. *Deposition*

The highest concentrations of air pollutants, particularly SO_2, occurred close to the industrial centre and decreased towards the north-east in accordance with the prevailing wind direction, which is from the south-west. Thus the Dübener Heide, a forested region of about 400 km^2 situated to the east of the main centre of industry, has been exposed to a steep deposition gradient (Figure 2). In the mid 1980s, the annual SO_2 concentration on the western border of the Dübener Heide reached approximately 150 µg m^{-3}, while in the east only half this concentration prevailed (app. 70 µg m^{-3}). Background concentrations were reached in the lowlands of north-east Germany, (37 µg m^{-3}, average of the years 1979 to 1990) at Neuglobsow. Unfortunately, no information about historical O_3, NO_X and NH_3 concentrations are available.

R.F. Hüttl and K. Bellmann, Changes of Atmospheric Chemistry and Effects on Forest Ecosystems, 1–15.

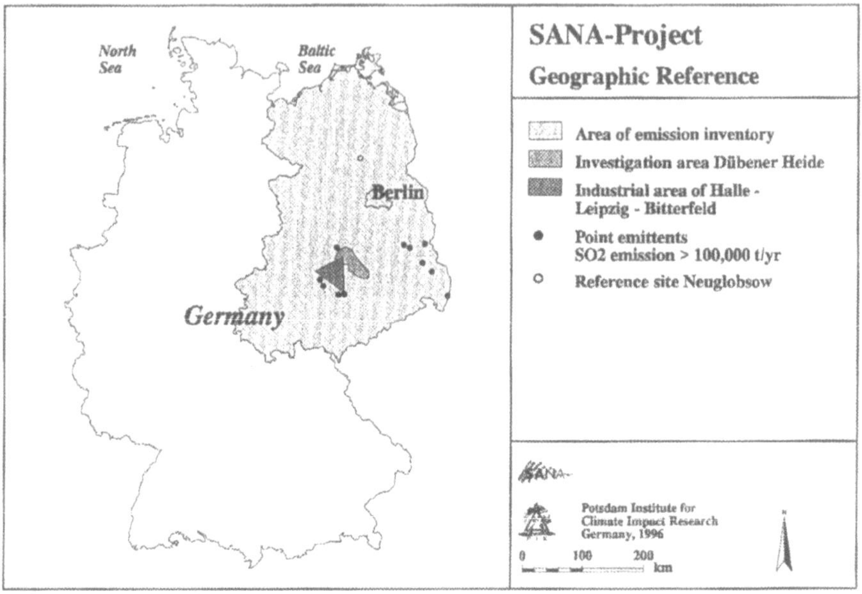

Figure 1. Location of the investigation area

Table 1. Emissions in the former GDR (East Germany) and projection for 1996

	1970		1989		1996	
	Mt a^{-1}	% of 1989	Mt a^{-1}	%	Mt a^{-1}	% of 1989
SO_2	4328	79	5506	100	1211	22
NO_X	0478	71	0677	100	0461	68
NH_3	0263	81	0325	100	0120	36
Dust	2498	119	2092	100	0188	9

(Data from Inst. of Energetic, Stuttgart)

Consequently, the deposition of sulphur and basic cations followed this gradient of deposition (Table 2 a+b), resulting, for example, in a pH value of precipitation of about 6.2 in the west, but 4.6 in the east in 1970 (Table 2c). A slightly different gradient was detected for nitrogen deposition, which decreased generally from the north to the south and from the borders of the region to the centre. This is mainly due to the distribution of agricultural production sites, which emitted large amounts of ammonium (Table 2d).

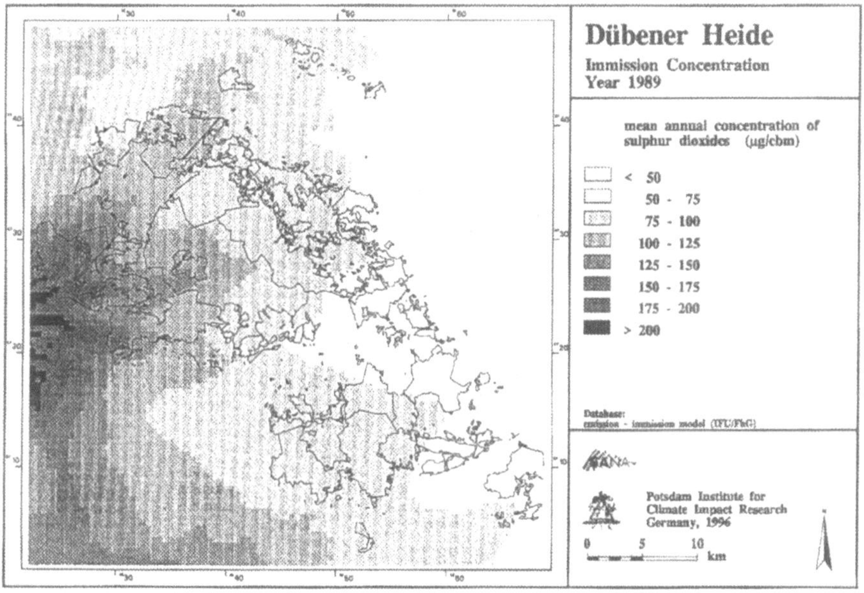

Figure 2. Mean annual SO_2 concentration in 1989 (data from Inst. for Atmospheric Environmental Research, Garmisch-Partenkirchen, see also Erhard and Flechsig, this volume)

1.3. Ecosystem responses

Since forest plantations, particularly Scots pine (*Pinus sylvestris* L.) stands, played an important economic role in the former GDR, investigations of pine forests in the Dübener Heide, including soils and ground vegetation, are available from the late 1950s onwards.

Soil responses

With regard to soil response, two types of development can be distinguished: i) alkalinization and re-acidification, and ii) nitrogen accumulation. Due to the deposition of alkaline dust with large amounts of basic cations, the pH value of the upper soil horizons increased until the mid 1960s despite the deposition of acid SO_4^{2-} and NO_3^-. At this time, only 26 % (8.500 ha) of the total forested area in the Dübener Heide showed a pH value below 4.2. With increasing installation of dust filters in lignite fired power plants, the deposition of basic cations (i.e. acid neutralization capacity) started to decrease, whereas acid deposition remained more or less unchanged or even increased until the mid 1980s. Thus, a re-acidification of the upper soil layers began to take place. In 1989, 51% (16.700 ha) of the forest in the Dübener Heide was found to have a pH value below 4.2. This re-acidification is an ongoing process, despite the decrease of acid deposition in recent years (see Konopatzky and Freyer, this volume).

Table 2a. Deposition development of sulphur at the western and eastern boarder of the Dübener Heide as well as at background sites north of Berlin (after Erhard and Flechsig, this volume)

	West kg SO_4-S ha^{-1} a^{-1}	%	East kg SO_4-S ha^{-1} a^{-1}	%	North kg SO_4-S ha^{-1} a^{-1}	%
1970	43	100	20	100	11	100
1985/89	55	127	28	140	14	127
1993/95	11	26	10	50	12	109
1970		100		46		26
1985/89		100		51		25
1993/95		100		91		109

Table 2b. Deposition development of calcium and magnesium at the western and eastern borders of the Dübener Heide as well as at background sites north of Berlin (after Erhard and Flechsig, this volume)

	West kg (Ca+Mg) ha^{-1} a^{-1}	%	East kg (Ca+Mg) ha^{-1} a^{-1}	%	North kg (Ca+Mg) ha^{-1} a^{-1}	%
1970	77	100	30	100	6	100
1985/89	65	84	23	77	6	100
1993/95	11	14	8	27	6	100
1970		100		38		8
1985/89		100		35		9
1993/95		100		72		54

Table 2c. Deposition development of the precipitation pH at the western and eastern borders of the Dübener Heide as well as at background sites north of Berlin (after Erhard and Flechsig, this volume)

	West precipitation pH	%	East precipitation pH	%	North precipitation pH	%
1970	6.2	100	5.1	100	(4.6*)	100
1985/89	5.6	90	4.6	90	4.3	93
1993/95	4.4	71	4.2	82	4.4	96
1970		100		82		74*
1985/89		100		82		76
1993/95		100		95		100

*estimated

Table 2d. Deposition of nitrogen at the western and eastern borders of the Dübener Heide and of background sites north of Berlin (after Erhard and Flechsig, this volume)

	West kg (tot. N) ha^{-1} a^{-1}	%	East kg (tot. N) ha^{-1} a^{-1}	%	North kg (tot. N) ha^{-1} a^{-1}	%
1970	13	100	13	100	9	100
1985/89	16	123	14	108	10	111
1993/95	13	100	12	92	16	178
1970		100		100		69
1985/89		100		88		63
1993/95		100		92		123

Although pH values are still too high to damage the vegetation directly, there is an increasing danger that potentially toxic (heavy) metals (Al, Fe, Mn), that have been deposited together with the alkaline dust during recent decades, may become more and more available for plant uptake (Heinsdorf *et al.*, 1990).

Like many other regions in Germany, the Dübener Heide was also exposed to considerable nitrogen deposition, originating from industrial processes and agricultural production, particularly through manure fertilisation from animal farming. Additionally, large forest areas were fertilised with urea by plane. This kind of fertilisation was applied from the early 1970s until the mid 1980s, generally at total amounts of 300 kg N ha^{-1}, distributed over three to four years. Especially in damaged forests, this procedure was repeated three to four times, leading to an additional input of up to 1000 kg N ha^{-1}. So far, no net loss of nitrogen (due to increased percolation) has been detected even at these sites. Thus, total nitrogen storage increased (in average over all forests in the Dübener Heide) by 518 kg N ha^{-1} between 1967 and 1989. During the same period the C/N ratios of the humus layer decreased from 31 to 24, which led to a change in humus types. Raw humus forms decreased from about 71 to 18%, whereas raw humus-type moder increased from 18 to 68% (see Konopatzky and Freyer, this volume).

Tree responses

Needle yellowing, relative needle losses and necroses of pine trees within the Dübener Heide area were investigated as early as 1965 by Lux (Lux, 1965), who distinguished between different degrees of damage (Figure 3). These levels of damage generally matched the regional distribution of SO_2 concentration (compare with Figure 2). The correlation of the spatial distribution of direct foliage damage with the distribution of the SO_2 concentration is supported by the repetition of the investigations in 1974 (Lux and Stein, 1975) as well as by recent results (Table 3).

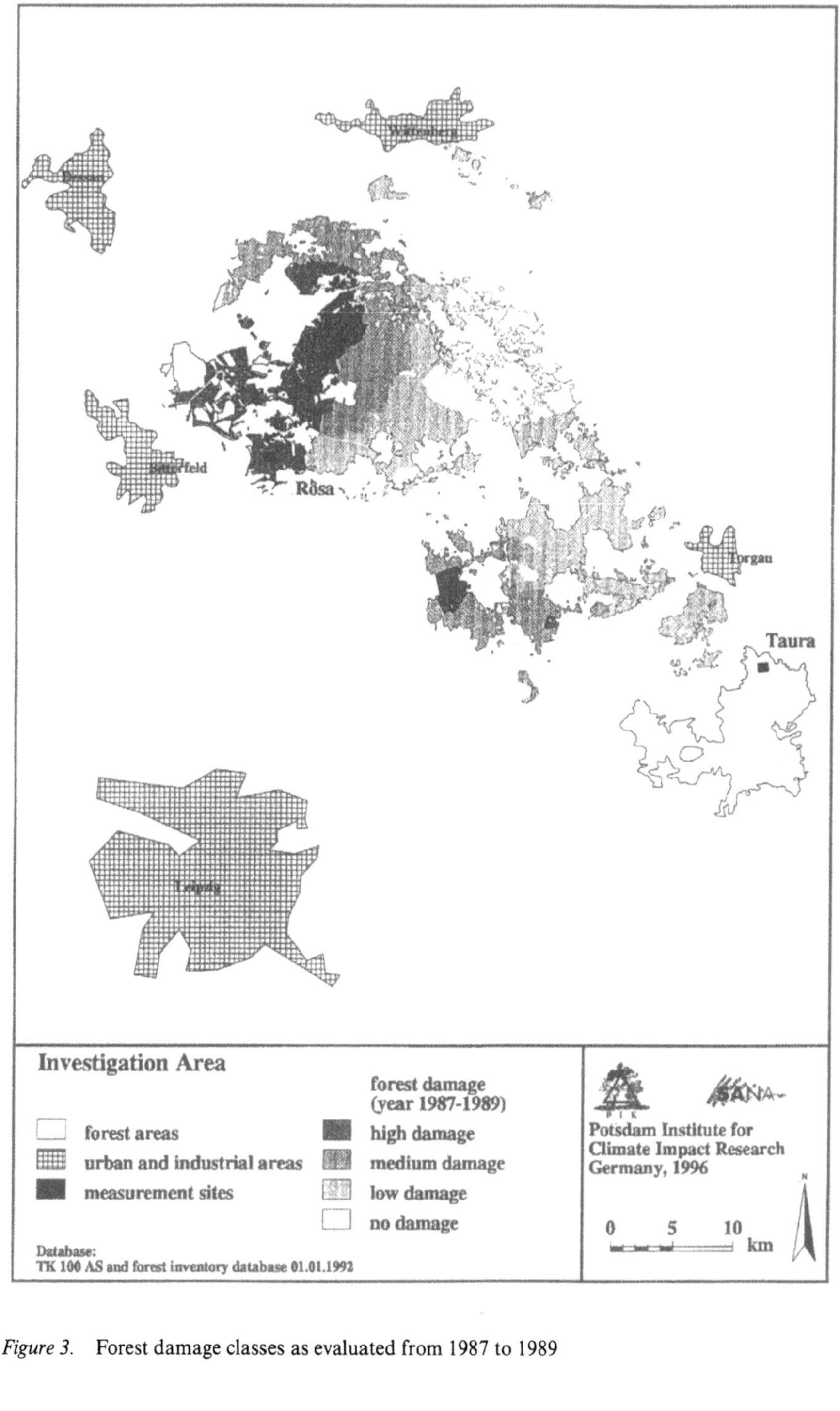

Figure 3. Forest damage classes as evaluated from 1987 to 1989

Table 3. Development of foliage longevity and annual diameter increase (statistically cleared from weather impacts) at the western borders of the Dübener Heide and areas north of Berlin (background site)

	Relative foliage longevity (%)*		Radial growth index (mm a^{-1})**	
	Dübener Heide, west	Background	Dübener Heide, west	Background
1961/63	55	100	0.9	1.05
1984/89	56	94	0.8	0.85
1990/95	67	75	1.33	0.75

*(after Gluch, in Anonymus, 1997), ** (after Neumann, this volume)

Clearly visible symptoms of foliage damage may reduce the photosynthetically active surface and, thus, decrease the assimilate gain of the trees. In addition, direct damage to the photosynthetic apparatus (e.g. Kropff, 1989; Meng *et al.*, 1994) and the surface structure of the needles (Huttunen, 1983; Hüttl, 1997; Manninen and Huttunen, 1995) is known to occur at high SO_2 concentrations. This impact may result in a further decrease of assimilation capacity as well as in an increased nutrient loss via crown leaching. Furthermore, the functioning of stomata, and thus, the water use efficiency, is negatively affected under these conditions (Meng and Arp, 1994).

The overall response of trees to SO_2 pollution, however, is difficult to estimate, because not only the direct but also the indirect effects have to be considered, which may mitigate or increase the negative direct impact. For example, light conditions are improved in a sparsely foliated canopy and thus, the photosynthetic production per foliage area may be increased. Although an affected stomata functioning may lead to further damage of the photosynthetic apparatus, nitrogen uptake through stomata can be increased, which may improve the nutrition of the trees (McLeod *et al.*, 1990). If stomata are still effective, however, co-occurring drought stress is able to reduce the uptake of SO_2 to a certain degree (Tesche *et al.*, 1989). The drought stress, on the other hand, is possibly affected by competition from ground vegetation (Hofmann *et al.*, 1990), as well as the nitrogen content of the soil, which influences the fine root distribution and can potentially impact fine roots and mycorrhiza (Heinsdorf, 1976; Ritter, 1990). Finally, the ability of the trees to compensate for tissue damage through changes in allocation has to be taken into account (Mooney *et al.*, 1988). These considerations demonstrate clearly that the dominant processes in the whole ecosystem, including soil and ground vegetation, and their relationships have to be identified, if the overall effect under changing conditions is to be evaluated.

Stand responses

In response to the continuous influence of pollution, both the productivity and stability of forests are likely to decrease (see e.g. Ulrich and Pankrath, 1983). In the Dübener Heide, a substantial decrease of stemwood production and an increased tree mortality, resulting in a sparser stand density, was observed (see Table 3). To compensate for this decrease in productivity, the plantations were fertilised with nitrogen to counterbalance the negative photosynthetic effects of SO_2 (Heinsdorf, 1978). Indeed, a substantial increase in wood production was successfully brought about, particularly in stands on poor sites (Heinsdorf, 1986; Niefnecker, 1982; Niefnecker, 1985; see Table 4). However, stemwood production decreased again when the nitrogen concentrations in the needles – that increased with increasing nitrogen availability – exceeded a value of approximately 2.1% (Hofmann *et al.*, 1990; Krauß *et al.*, 1986).

Table 4. Relative responses of diameter growth and relative foliage mortality on nitrogen fertilisation and SO_2 air pollution

	Diameter growth		Foliage losses	
	Polluted sites	Non-polluted sites	Polluted sites	Non-polluted sites
Fertilised	105	140	80	85
Non-fertilised	75	100	145	100

Changes in ground vegetation

In response to the increasing alkalinity and nitrogen availability, changes in the composition of ground vegetation were observed from the early 1960s onwards, particularly at formerly nutrient poor sites (Bergmann and Flöhr, 1988; Lux, 1964; Tölle and Hofmann, 1970). Generally, a higher species diversity with an increasing proportion of nitrophilic species was found (see also Konopatzky and Freyer, this volume). With the ongoing increase in nitrogen availability, during the 1980s more and more broad-leaved species developed in the understorey. This development was supported by an increased availability of light, due to SO_2-induced needle losses and an increased stem mortality. The dominant grass species in most pine plantations in 1989 were found to be *Calamagrostis epigejos* (app. 1.4 t DW ha^{-1} in the west and 1.9 t dw ha^{-1} in the east) and *Avenella flexuosa*. Typical representatives of woody plant species in these ecosystems were *Padus serotina* and *Sambucus nigra*.

Since (1) the understorey accumulates nutrients (app. 15 kg N ha^{-1}, above ground) and since (2) the daily transpiration rate of the understorey can reach the same magnitude as the transpiration from a sparsely foliated pine canopy, there has been discussion about whether the competition may affect the tree physiology and/or the stem growth significantly (e.g. Hofmann *et al.*, 1990). However, it has not been possible to show any clear results so far.

2. Expected developments

In 1990, after the German reunification, it became apparent that the industrial structure of the former GDR would change drastically. This was forced by the general breakdown of industry in the test region, and supported by the introduction of more sophisticated emission reduction technologies at the main point sources (see Table 1 for projections). Additionally, it was expected that further restrictions of polluting emissions would be rapidly demanded from policy.

As far as nitrogen emissions from non-industry sources (i.e. from traffic) were concerned, the situation was more complicated. On the one hand, emissions from agriculture (mainly NH_3) were expected to decrease, either due to economically forced changes or because of stricter legal regulations. On the other hand it became apparent that increased car traffic would lead to increased emissions of nitrogen in the form of NO_X. Thus, the regional pattern of total nitrogen deposition would change considerably. An additional factor was that forest fertilisation had been discontinued since the mid 1980s.

In this situation, it was not known how forest ecosystems would respond to the severe environmental changes in the region of the Dübener Heide, but it was assumed that the relevant developments would be of considerable importance for forest production and ecosystem stability. An understanding of the underlying processes was required not only to assist the policy of Germany, but also to provide information for pollution abatement strategies in the neighbouring countries of Poland and the Czech Republic. For scientific research, it was a unique opportunity to study developments of forests under gradients of deposition as well as under different initial conditions. The task addressed was thus similar to that addressed by various roof projects in different European countries (Visser *et al.*, 1994), but the investigation could be executed with much less effort and much more realistic conditions (roof experiment without roof) and on a much broader scale.

3. Former integrated assessments of forest decline and production

The analysis of available investigation results showed that forest growth and stand development depend on a number of influences that are more or less interrelated on different scales in time and space. Therefore, special attention was given to the development of computer models that integrate from the physiological or patch level up to the regional scale, are sensitive to air pollution and deposition and can be evaluated by means of experimental data.

In 1990, three different approaches for regional assessments of forest development were available that were sensitive to air pollution stress:

- The Regional Risk Model (Mäkelä, 1989), which is based on forest damage data of Norway spruce in the Ore Mountains, Germany, and is used to calculate risk levels for European forests (Alcamo *et al.*, 1985).

- The Forest Dieback Model (Lenz and Schall, 1987), which is based on investigations into Norway spruce in the German Fichtel-Mountains, calculates tree vitality as a response to direct and indirect deposition impacts.

- The Environmental Prediction and Decision Support Model (PEMU) (Bellmann *et al.*, 1988) is based on empirical tree growth functions for Scots pine, which are modified according to a multitude of stress conditions. It has been evaluated for areas in north-east Germany and has been used for estimations of forest production on a European scale (Nilsson *et al.*, 1992).

None of these approaches aims to provide a holistic picture of the forest ecosystem. They are, in fact, highly empirical and thus cannot easily be transferred from one region or species to another. Furthermore, a negative or positive feedback can only be considered within the range of conditions used for developing the dose-response functions, a factor which has been particularly critical for the first two models, as they were developed under conditions of an acidic environment with poor nutrient supply.

On the other hand, many investigations all over the world have led to an increased understanding of ecosystem processes, resulting in new models that represent tree growth based on a mechanistic approach requiring only initial conditions and external driving variables (e.g. Bossel and Schäfer, 1989; Mohren, 1987; Running and Coughlan, 1988). Those models had been designed to simulate full carbon balances of forest plantations. However, they were not suitable for long-term assessments because they had been evaluated with high-resolution data of physiological measurements. Stand processes, like tree mortality, had been based on empirical relations or were not been included at all in the models, and soil processes had been generally underrepresented. Furthermore, the initialisation requirements of the models are difficult to satisfy, which generally prevents application on a regional scale.

4. Tasks

As outlined above, severe changes of the environmental conditions in the Dübener Heide were expected, but the response of forest ecosystem development to these changes were largely unknown. It was obvious, however, that the possible impacts would not be the same over the whole area, but had to be differentiated according to the different regional patterns of SO_2 and nitrogen deposition, and different initial conditions. Thus, three tasks were identified, which could be used in a stepwise assessment of the regional forest development:

- Analysis of basic physical and eco-physiological processes at selected test sites, which had been exposed to different deposition loads, in order to obtain and to deepen the understanding of responses and feedback mechanisms within the ecosystem. Measurements should be accompanied by the construction of a new physiologically-based model, which includes the most important state variables and fluxes of carbon, nitrogen and water between atmosphere, pedosphere and biosphere. Based on these balance estimates, the model should be used to describe stand development processes mechanistically as a function of stem carbon.

- Collection of information about soil and forest conditions on a regional scale. This information includes past as well as current stand and soil conditions to explore cumulative effects on forest growth. To relate the quantified responses to the past environmental conditions, regional differentiated climate and deposition data are necessary and management information at each site to be assessed is required. All regional information should be implemented in a geographical information system (GIS) and should serve for initialisation and evaluation of the stand growth model, which could thus be applied on a regional scale.

- Development of deposition scenarios and regional impact assessment. Starting with the current conditions of forest soils and forest stands, possible future developments under different deposition scenarios should be assessed using the evaluated forest growth model. On the basis of the results, possible abatement strategies and management options should be discussed.

As models should play the part of a control in the project, they should not only be related to the collection of data at the intensively investigated site, but should also coordinate and interrelate the activities of the measuring groups.

The resulting model should be suitable to represent small-scale measurements as well as long-term forest developments, recorded on a regional scale. Thus, the model must be sensitive to daily climate and deposition input data, but must also account for forest management effects including fertilisation. Furthermore, the demand on initialisation data must be in accordance with the soil and stand data that are available within the region.

5. Project implementation

5.1. The framework of SANA

The ecological impact research programme itself is part of a larger research framework, which was suggested shortly before the German reunification by researchers from both parts of the country. The first sub-projects were launched in 1991 and the main investigations were executed during 1993 until the end of 1995, supported by the Federal Ministry for Research and Technology and coordinated by the Fraunhofer Institute for Atmospheric Environment Research in Garmisch-Partenkirchen. The whole framework is divided into the following programmes:

- Pollutant emission inventory, co-ordinated by the Institute for Energetic, Stuttgart.The project includes analysis and modelling of (daily) emission from point and non-point sources in the former GDR since 1970.

- Distribution and transport of pollutants, chemical transformation and modelling of regional pollution development, co-ordinated by the Fraunhofer Institute for Atmospheric Environment Research, Garmisch-Partenkirchen

- Ecological impacts, co-ordinated by the Potsdam Institute for Climate Impact Research in Potsdam and the Institute for Forest Ecology, Eberswalde at the Centre for Agricultural Landscape and Land Use Research, Müncheberg.

As outlined above for the ecological impact research, each of these programmes covers a number of sub-projects necessary to fulfil its specific task within the framework. Like the sub-projects of each programme, the programmes themselves are related to the other in a hierarchical order. Thus, it was possible to assess the whole chain of air pollution impact efficiently.

5.2. The research programme on ecological impacts

According to the tasks identified above, the specific sub-projects are determined firstly by the dominant gaps in knowledge, secondly by the modelling demand on initialisation data, and thirdly by the demand to summarise results and to formulate them on a regional scale (Table 5). A number of sub-projects investigated processes in the soil, humus, ground vegetation and trees at specific sites, which were selected along a gradient of air pollution:

The first site (Rösa) is located on the western border of the Dübener Heide, close to the former industrial centres. It represents a typical high-pollution site, which was exposed to extreme SO_2 concentrations but also received high amounts of alkaline dusts as well as nitrogen from fertilisation. The second site at Taura was exposed to medium pollution load, less dust deposition and relatively small amounts of nitrogen. It is located in the east of the target area. The third site at Neuglobsow was used as a background site, because it is located further north and was relatively unaffected by the industrial emissions. The results from these measuring groups served for process identification and parameterisation as well as short-term evaluation of the forest growth model.

A specific project was implemented for regional soil monitoring to support the modelling activities with initialisation data for regional assessments. This included a re-measuring of old soil samples taken in the same region in 1970, as well as the gathering of new data. Forest inventory data were supplied by the

Table 5. Sub-projects within the SANA programme for ecological impact research

Number	Task
1.1	Foliage dynamics and nutrition
1.2	Regional soil dynamics under pine forest ecosystems
1.3	Element transport and transformation within the soil
1.4	Transpiration and hydraulic conductance of pine forests
1.5	Carbon and nitrogen dynamics in the humus layer and litterfall
1.6	Canopy photosynthesis and respiration
1.7	Diameter and stem volume growth dynamics
1.8	SO_2 induced biochemical stress indicators
1.9	Dynamics of fine roots and mycorrhiza
1.10	Lichen occurrence as indicators for SO_2 deposition
2.1	Ecosystem modelling
2.2	Regional modelling

State Forest Service and digitised by the regional modelling group, and deposition estimates were taken from the other programmes within the SANA framework. Weather data were directly available from weather recording stations close to the Dübener Heide. The modelling activities are divided into two groups, which are concerned with forest growth modelling on the one hand and regional computations on the other.

6. Concluding remarks

During the past few years several integrated research projects have been completed, which were concerned with pollution impacts on forests. The programme on 'Response of Plants to Interacting Stresses' (ROPIS) in the United States and the 'Dutch Priority Programme on Acidification' in the Netherlands are prominent examples of such projects. As in SANA, field investigations were executed which served to construct complex forest growth models, which in turn were used to analyse past and possible future impacts of pollutants and other stress factors on forest growth (Heij *et al.*, 1991; Weinstein *et al.*, 1991). More recently, focussing on the impact of climatic changes, such models were also used to assess the impact on a regional scale, simulating net primary production or stemwood growth in selected (representative) forests subject to atmospheric deposition loads along a gradient of conditions (Aber *et al.*, 1995; Bowes and Sedjo, 1993; Running, 1994).

There is no single case, however, where environmental conditions changed on a regional scale as drastically as in Eastern Germany, resulting in ecosystem changes that can actually be observed from year to year. Therefore, the ecological part of SANA can be considered as a 'roof experiment without a roof'. Furthermore, the availability of long-term climatic and deposit information together with forest inventory data and research results from several decades provides an extensive data set for model evaluation. These unique circumstances made it possible for the first time to assess a dynamic forest development, which is based on the smallest homogeneous units of space, on a regional scale.

7. References

Aber JD, Ollinger SV, Federer CA et al. 1995. Predicting the effects of climate change on water yield and forest production in the northeastern United States. Clim. Res. 5, 207–222.

Alcamo J, Hordijk L, Kämäri J, Kauppi P, Posch M, Runca E. 1985. Integrated analysis of acidification in Europe. J. Env. Managment. 21, 47–61.

Anonymus, 1997. SANA – Wissenschaftliches Begleitprogramm zur Sanierung der Atmosphäre über den neuen Bundesländern, BMBF, Bonn.

Bellmann K, Lasch P, Hofmann G, Anders S, Schulz H. 1988. The PEMU Forest-Impact-Model FORSTK, Pine Stand Decline and Wood Supply Model. In: Systems Analysis and Simulation. Eds. A Sydow, SG Tzaestas, R Vichnevetsky. Akademie Verlag, Berlin, Germany.

Bergmann H-J, Flöhr W. 1988. Zur Wirkung von Fremdstoffen in den Wäldern der DDR unter besonderer Berücksichtigung einer Veränderung der Bodenflora. Soz. Forstwirtschaft. 38(6), 164–166.

Bossel H, Schäfer H. 1989. Generic simulation model of forest growth, carbon and nitrogen dynamics, and application to tropical acacia and European spruce. Ecol. Modell. 48, 221–265.

Bowes MD, Sedjo RA. 1993. Impacts and responses to climate change in forests of the MINK region. Climatic Change. 24(1–2), 63–82.

Finkbeiner A, Friedrich R, Grimm O. 1993. Modellierung der Emissionen ausgewaehlter umweltrelevanter Schadstoffe und Bestimmung ihrer zeitlichen Trends. University of Stuttgart, Stuttgart, Germany.

Heij GJ, De Vries W, Posthumus AC, Mohren GMJ. 1991. Effects of air pollution and acid deposition on forests and forest soils. In: Acidification Research in The Netherlands. Final report of the Dutch Priority Programme on Acidification. Studies in Environmental Science. Eds. GJ Heij, T Schneider. pp. 97–137. Elsevier, Amsterdam.

Heinsdorf D. 1976. Feinwurzelentwicklung in Kiefernbestockungen unterschiedlichen Alters nach N-Düngung. Beitr. Forstwirtschaft. 4, 199–204.

Heinsdorf D. 1978. Zweckmäßige N-Düngefolgen in jungen Kiefernbestockungen. Beitr. Forstwirtschaft. 1, 33–39.

Heinsdorf D. 1986. Trockensubstanzproduktion in gedüngten und ungedüngten 14- bis 15jährigen Kieferndickungen auf Kippböden. Beitr. Forstwirtschaft. 20(1), 14–21.

Hofmann G, Heinsdorf D, Krauß H-H. 1990. Wirkung atmogener Stickstoffeinträge auf Produktivität und Stabilität von Kiefern-Forstökosystemen. Beitr. Forstwirtschaft. 24(2), 59–73.

Huttunen S, Laine K. 1983. Effects of air-borne pollutants on the surface wax structure of *Pinus sylvestris* needles. Ann. Bot. Fennici. 20, 79–86.

Krauß HH, Heinsdorf D, Hippeli P, Tölle H. 1986. Untersuchungen zu Ernährung und Wachstum wirtschaftlich wichtiger Nadelbaumarten im Tiefland der DDR. Beitr. Forstwirtschaft. 20(4), 156–164.

Kropf MJ. 1989. Modelling short-term effects of sulphur dioxide. 1. A model for the flux of SO_2 into leaves and effects on leaf photosynthesis. Neth. J. Pl. Path. 95, 195–213.

Lenz R, Schall P. 1987. Darstellung waldschadensrelevanter Ökosystembeziehungen als Grundlage von dynamischen Modellen und Hypothesensimulationen am Beispiel der Stickstoffhypothese, 17. Jahrestagung der Gesellschaft für Ökologie, Göttingen. pp. 633–641.

Lux H. 1964. Beitrag zur Kenntnis des Einflusses der Industrieexhalationen auf die Bodenvegetation in Kiefernforsten (Dübener Heide). Archiv für Forstwesen. 13(11), 1215–1223.

Lux H. 1965. Die großräumige Abgrenzung von Rauchschadenszonen im Einflußbereich des Industriegebietes um Bitterfeld. Wiss. Z. Techn. Univers. Dresden. 14(2), 433–442.

Lux H, Stein G. 1975. Ergebnisse der Wiederaufnahme des Weiserflächennetzes in den StFB der Dübener Heide und Schlußfolgerungen für die weitere forstwirtschaftliche Bewirtschaftung, TU Dresden, Tharandt.

Mäkelä A. 1989. A regional model for risk to forests by direct impacts of Sulfur. Syst. Anal. Model, Simul. 6(6), 439–450.

Manninen S, Huttunen S. 1995. Scots pine needles as bionindicators of sulpur deposition. Can. J. For. Res. 25, 1559–1569.

McLeod AR, Holland MR, Shaw PJA, Sutherland PM, Darrall NM, Skeffington RA. 1990. Enhancement of nitrogen deposition to forest trees exposed to SO_2. Nature. 347(6290), 227–279.

Meng FR, Arp PA. 1994. Modelling photosynthetic responses of a spruce canopy to SO_2 exposure. For. Ecol. Manage. 67, 69–85.

Meng F-R, Cox RM, Arp PA. 1994. Fumigating mature spruce branches with SO_2: effects on net photosynthesis and stomatal conductance. Can. J. For. Res. 24, 1464–1471.

Mohren GMJ. 1987. Simulation of forest growth, applied to Douglas Fir stands in the Netherlands. Agricultural University, Wageningen, The Netherlands, 184 pp.

Mooney HA, Küppers M, Koch G, Gorham J, Chu C, Winner WE. 1988. Compensating effects to growth of carbon partitioning changes in response to SO_2-induced photosynthetic reduction in radish. Oecologia. 75, 502–506.

Niefnecker W. 1982. Volumenmehrzuwachs nach aviotechnischer Großflächendüngung in Kiefernreinbeständen. Soz. Forstwirtschaft. 32(8), 233–235.

Niefnecker W. 1985. Ertragskundliche Untersuchungsergebnisse zur stabilisierenden Wirkung der mineralischen Großflächendüngung in immisionsgeschädigten Kiefernwaldgebieten. Beitr. Forstwirtschaft. 19(3), 120–124.

Nilsson S, Sallnäs O, Dunker P. 1992. Future Forest Resources of Western and Eastern Europe. The Parthenon Publishing Group, IIASA, Laxenburg, Austria.

Ritter G. 1990. Zur Wirkung von Stickstoffeinträgen auf Feinwurzelsystem und Mykorrhizabildung in Kiefernbeständen. Beitr. Forstwirtschaft. 24(3), 100–104.

Running SW. 1994. Testing Forest-BGC ecosystem process simulations across a climatic gradient in Oregon. Ecological Applications. 4(2), 238–247.

Running SW, Coughlan JC. 1988. A general model of forest ecosystem processes for regional applications. I. Hydrologic balance, canopy gas exchange and primary production processes. Ecol. Modell. 42, 125–154.

Tesche M, Feiler S, Michael G, Rant H, Bellmann C. 1989. Physiologische Reaktionen der Fichte (*Picea abies*) auf komplexen SO_2- und Trockenstreß. Teil 1: Reaktionen auf gleichzeitiges Einwirken von SO_2 und Trockenheit. Eur. J. For. Path. 19, 281–292.

Tölle H, Hofmann G. 1970. Beziehungen zwischen Bodenvegetation, Ernährung und Wachstum mittelalter Kiefernbestände im nordostdeutschen Tiefland. Archiv für Forstwesen. 19(4), 385–400.

Ulrich B, Pankrath J. 1983. Effects of cumulations of air pollutants in forest ecosystems. Reidel Publishing, Dortrecht.

Visser PHB, Beier C, Rasmussen L et al. 1994. Biological response of five forest ecosystems in the EXMAN project to input changes of water, nutrients and atmosperic loads. For. Ecol. Manage. 68, 15–29.

Weinstein DA, Beloin RM, Yanai RD, 1991. Modeling changes in red spruce carbon balance and allocation in response to interacting ozone and nutrient stresses. Tree Physiol. 9, 127–146.

2
Site and stand description

R.F. HÜTTL and K. BELLMANN

To facilitate interpretation of the experimental results, a short description of the test sites along a historic gradient of air pollutant deposition is provided (Figure 1).

1. Location

The site **Rösa** is located close to the centre of the heavy industry belt around Bitterfeld and received its main deposition load from this region. The experimental area is located about 10 km east of Bitterfeld at the western edge of the Dübener heath. Previously, this site was exposed to very high levels of alkaline dust desposits and high acid input rates. Within the deposition gradient Rösa represents the site with the largest atmogenic deposition load.

The site **Taura** is located north-east of Leipzig. Because of its distance from Bitterfeld, it represents a deposition gradient to the site Rösa. On the other hand, this site is influenced by the emissions of Leipzig/Halle and, thus, received a moderate atmogenic deposition load.

The site **Neuglobsow** is located in the north of Brandenburg, about 3 km south of the Stechlin lake. Considering its remote deposition regime, it serves as a background site.

2. Soils, stands and emission dynamics

The sites Rösa and Taura are located in the loess-free pleistocene region between the rivers Elbe and Mulde in the area of Saale glaciation related morainic debris complexes of the Dübener and Dahlener heath. Specifically, Rösa is located in the area of fluvio-glacial sand deposits of the frontal moraine area of the Dübener heath, whereas the site Taura is located in the bottom morainic deposits of the Dahlener heath. The site Neuglobsow is located in the area of the Weichsel moraine, which – compared to the Saale glaciation – is the younger moraine. At all three sites, parent material for soil development is glacial outwash sand. The wide distribution of these sandy substrates in the north-eastern lowlands and their relatively homogeneous thickness provide the basis for adequate similarity of all three test sites related to particle size,

R.F. Hüttl and K. Bellmann, Changes of Atmospheric Chemistry and Effects on Forest Ecosystems, 17–20.

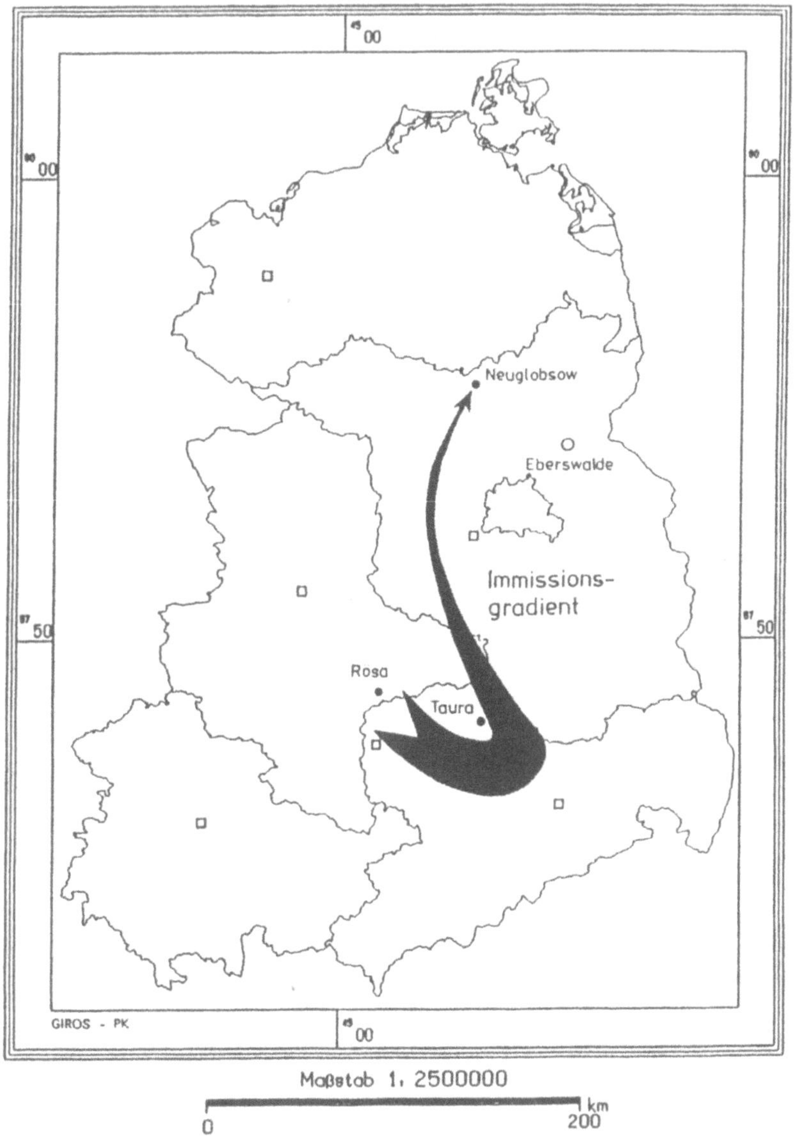

Figure 1. Geographical location along the deposition gradient of the test sites Rösa (heavy deposition load), Taura (moderate deposition load) and Neuglobsow (background site) in the north-eastern lowlands of Germany

Table 1. Selected site and stand parameters of the three experimental Scots pine sites including management and historical aspects

	Location		
Site parameters	Rösa	Taura	Neuglobsow
Annual precipitation (mm)[1]	566	565	586
Mean annual temperature (°C)[1]	8.87	8.87	8.18
Soil type	Spodi-dystric Cambisols, sand	Cambic Podsols, silty sand	Dystric Cambisols, sand
Humus form	Mor/moder	Moder	Moder
Thickness (cm)	6/4	5[2]	4[2]
pH(KCl)	4.1/4.5	3.4	3.0
Base saturation (%)[3]	57/85	37	35
C (%)	34/22	39	44
N (%)	1.7/1.7	1.5	1.6
Dry mass (Mg ha^{-1})	106/99	65	55
Ecosystem type	Calamagrostio-Cultopinetum sylvestris	Avenello-Cultopinetum sylvestris	Myrtillo-Cultopinetum sylvestris
Tree species	*Pinus sylvestris* (L.)	*Pinus sylvestris* (L.)	*Pinus sylvestris* (L.)
Stand age (1995) (yrs)	61	45	65
Number of stems (n ha^{-1})	935	853	1043
Basal area (m^2 ha^{-1})	34	28	36
Diameter at breast height (cm)	20.7	20.6	21.0
Mean tree height (m)	16.0	18.0	20.1
Anthropogenic influences	Litter raking until ca. 19000	Litter raking until ca. 1900	Former agricultural utilization
Fertilization	1970–1972, 1976-1979,1983-1985 N-fertilization; 100 kg N ha^{-1} a^{-1}	Application of urea (quantity unknown)	1973/74 N-fertilization (quantity unknown)
Previous stand	Scots pine	Scots pine	Afforestation of former arable land

[1]Mean values: Rösa (station Wittenberg) and Taura (station Torgau) 1964–1990; station Neuglobsow 1996–1990

[2]In Taura Oh-material was usually mixed with the upper mineral soil and not sampled seperately; in Neuglobsow Oh-material could not be distinguished as a horizon

[3]Values measured in August 1994, pH in 1:5 1 *N* K_4Cl-extracts, exchangeable bases in 0.5 *N* NH_4Cl-extracts

Table 2. Annual mean values ($\mu g\ m^{-3}$) of various air pollutants for the time period 1988 to 1993 as documented in measuring stations close to the respective test sites (numbers in parenthesis are approximative values)

Site	Measuring station	1988	1991	1992	1993
			SO_2		
Rösa	Greppin	143	104	73	62
Taura	Melpnitz	(70)	(50)	(40)	36
Neuglobsow	Neuglobsow	18	–	9	11
			NO_x		
Rösa	Greppin	(18)	20	23	28
Taura	Melpnitz	(20)	(20)	(20)	(20)
Neuglobsow	Neuglobsow	(10)	(11)	12	10
			O_3		
Rösa	Greppin	(45)	50	55	58
Taura	Melpnitz	(38)	(38)	(38)	39
Neuglobsow	Neuglobsow	(45)	(50)	54	52

composition, and other physical soil parameters, a context which is particularly important when trying to compare the water regime of the rooted solum of forest ecosystems. From these sandy substrates, acid brown earth (distric cambisols) varying in degree of podsolization (podsols/spodisols) have developed (Table 1). Also the humus forms are comparable at all three sites. However, the thickness of the organic matter layers differs, being largest at the site Rösa.

Table 1 also provides information about further site parameters as well as about characteristics of the 40–65-yr-old Scots pine (*Pinus sylvestris* L.) stands; data on site history are also presented, such as stand management, forest floor vegetation and ecosystem type. The former application of mineral nitrogen (Table 1) aimed to compensate for forest damage related to the negative impact of air pollution.

Table 2 details the development of the SO_2-, NO_x-, and O_3-concentrations in the atmosphere during the years 1988, 1991, 1992, and 1993. These data prove the steep historic deposition gradient of air pollutants between the three test sites.

3
Nutrient element distribution in the above-ground biomass of Scots pine stands

H.-P. ENDE and R.F. HÜTTL

1. Introduction

The development of the trees' canopy and the nutrient distribution within the single compartments of the canopy are important factors for the calculation of nutrient balances. Repeated observation provided, these parameters are also reliable indicators for the health status of forest ecosystems. For the three forest ecosystems investigated in the SANA program, the nutrient element distribution in the canopies has been investigated by annually repeated needle analysis. In addition, the quantity and distribution of above-ground biomass has been assessed after a whole-tree harvest at the end of the experimental period. The results have served as important input variables for the pine ecosystem model FORSANA (cf. Grote *et al.*, this volume).

2. Materials and methods

2.1. Sampling

In October 1993, 1994 and 1995, respectively, needle samples were withdrawn from the upper crown of 15 dominant trees of each experimental stand. The samples were separated by needle age classes and – after estimating the degree of needle tip necroses (ocular, five classes; Ende *et al.*, 1995) – divided into two parts, one of which was utilized for the biochemical analyses (Schulz *et al.*, this volume). Additionally, at a whole tree harvest in August 1995 needle samples were selected separately by whorls and needle age classes, respectively; these samples were utilized for direct LAI (leaf area index) assessment with an image analysis system (Lüttschwager *et al.*, this volume) and biomass measurements. The analytical program described below accounted for about 7500 single analyses.

2.2. Laboratory analyses

By weighing 100 needle pairs before and after oven-drying, the respective fresh and dry weights of 1000 needles were calculated.

R.F. Hüttl and K. Bellmann, Changes of Atmospheric Chemistry and Effects on Forest Ecosystems, 21–35.

The needle samples were dried at max. 70°C for at least 48 h to ensure sulfur compounds could still be fractionated. Potential weight differences to oven-dry weight have been quantified. For total carbon, nitrogen, and sulfur estimations, 200 mg pulverized needle material were analyzed in a Vario EL (Elementar, Germany) CNS analyzer specially equipped with an NDIR-BINOS for more precise reproduction of total-S contents in needle material compared to standard CNS analyzers. Another proportion of the needle material was digested with HNO_3 conc. and HCl conc., consecutively, in a microwave digestor MLS 1200 in PTFE pressure cups at max. 400 W and was then filtered. Phosphorus was estimated in a phosphate-molybdate-complex (blue) from the digestion solution by spectrophotometry (Kontron Uvikon 930). Magnesium, calcium, manganese, iron, and zinc contents were estimated from the digestion solution by flame AAS (Perkin-Elmer 4100) according to Perkin-Elmer prescriptions. Copper, cadmium, and lead contents were estimated from the digestion solution by graphite-tube AAS (Perkin-Elmer 3030 ZEEMAN) according to Perkin-Elmer prescriptions.

2.3. *Quality control*

The chemical analyses were referenced using the certified pine needle standard 1575 of the U.S. Dept. of Commerce, National Bureau of Standards, as well as a number of internal pine needle standards.

3. Results

The investigations in the nutrient element distribution in the pine needles as well as in the biomass distribution derived from whole-tree harvests demonstrate a pronounced differentiation of the three experimental sites.

3.1. *Nitrogen*

The course of the N-contents between 1991 and 1995 (Figure 1) indicates a seasonally extremely high N-supply at the Rösa experimental site. The range of sufficient N supply of *Pinus sylvestris* as defined by Bergmann (1993), 14 to 17 mg g^{-1} d.w., was exceeded by up to 40% in Oct. 1993. N contents between 18 and 21 mg g^{-1} d.w. as observed at Taura and Rösa, but not at Neuglobsow, have been classified as optimal for pine stands of the northeastern lowlands of Germany, whereas initial damage due to high N-input have been reported at N-contents exceeding 23.5 mg g^{-1} d.w. (Hippeli, 1991; Hofmann and Krauß, 1988; Krauß and Heinsdorf, 1983). The N-contents of the Neuglobsow pine stand are lowest, but sufficient (low-ranged according to Krauß and Heinsdorf, 1983) and take a very balanced course during the time period investigated. The high N-contents at Rösa are also reflected by an increased needle dry weight (Table 1).

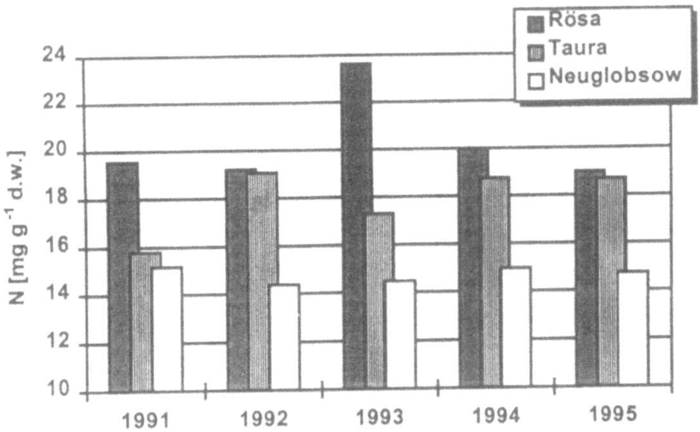

Figure 1. Temporal development of the N contents of Scots pine needles (current needles) from the three experimental sites. Time of sampling: October; data from 1991 and 1992: Schulz (personal communication)

Table 1. Temporal development of the dry weight (g) of 1000 needles (current needles) from the three experimental sites. Time of sampling: October

	Rösa	Taura	Neuglobsow
1993	29.42	26.44	18.33
1994	25.72	26.07	22.83
1995	23.21	23.07	21.15

3.2. Phosphorus

The P contents at the Rösa experimental site are remarkably low (Figure 2). The range of sufficient supply defined by Bergmann (1993), 1.4–3.0 mg g^{-1} d.w., was not attained by any of the needle samples analyzed. However, this does not mean acute P deficiency. According to Hofmann and Krauß (1988), P contents below 1.19 mg g^{-1} d.w. pine foliage represent average conditions of glacial sand sites of the northeastern lowlands of Germany. Besides, no P deficiency symptoms have been observed in the experimental areas during the time of investigation. Another indication for insufficient P supply is given by the course the single needle age classes took (Figure 3): the regular decrease of the P-contents with needle age led to a value below 1.0 mg g^{-1} d.w. at Rösa in the 2-year-old needle age class. Imbalanced P supply can also be diagnosed by the N:P ratio (Figure 4). While at Neuglobsow and Taura this value oscillates around 10, it clearly exceeds 16 at Rösa due to the high N as well as low P contents there.

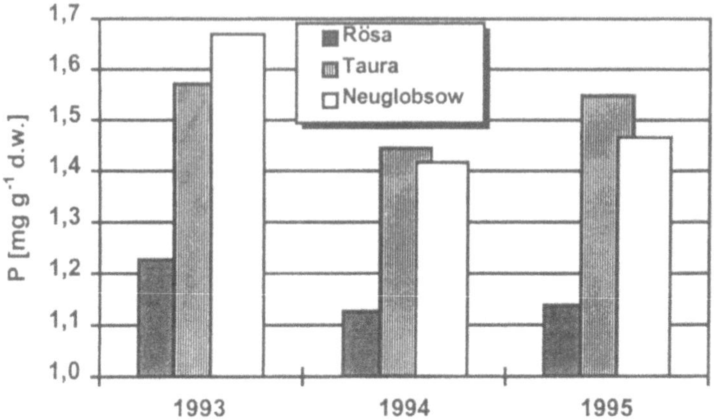

Figure 2. Temporal development of the P contents of Scots pine needles (current needles) from the three experimental sites. Time of sampling: October

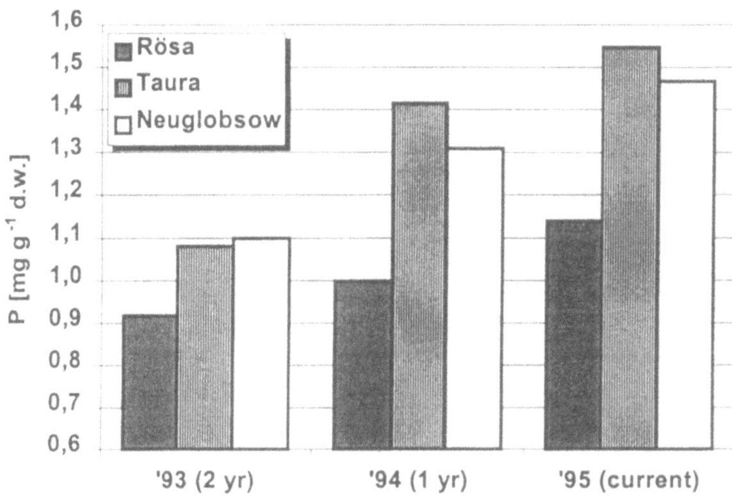

Figure 3. P contents of the three needle age classes of Scots pine trees from the three experimental sites. Time of sampling: October, 1995

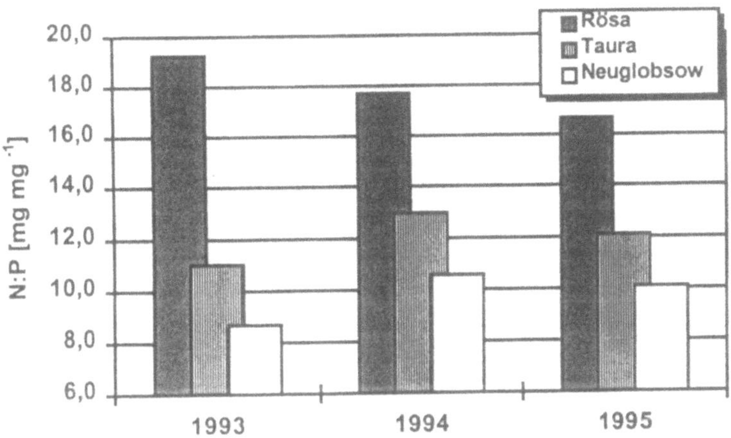

Figure 4. Temporal development of the ratio N contents to P contents of Scots pine needles from the three experimental sites. Time of sampling: October, current needles

3.3. *Sulfur*

During the time of observation, the S contents are on a higher level at Rösa and Taura compared to Neuglobsow (Figure 5). According to Hofmann und Krauß (1988), at Neuglobsow (1.9 mg g^{-1} d.w.) average S contents were not exceeded, whereas at Taura and Rösa the S contents indicated anthropogeneous influence. The fluctuations of the S contents are comparatively high at all three sites.

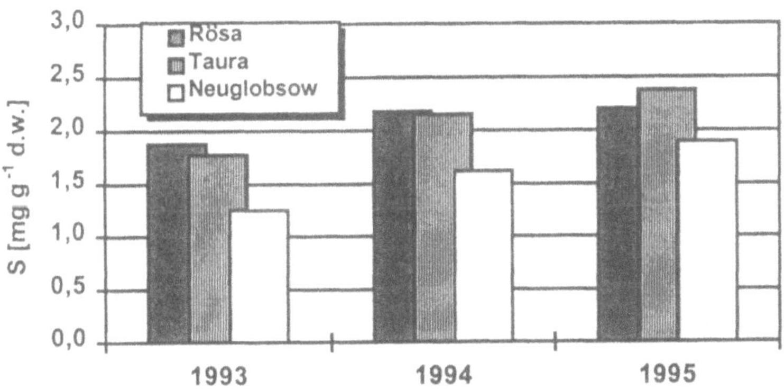

Figure 5. Temporal development of the S contents of Scots pine needles (current needles) from the three experimental sites. Time of sampling: October

3.4. Magnesium

From 1993 to 1995 there was little differentiation in the Mg contents (Figure 6) of the current needles between the sites and there weree within a range of sufficient supply (Bergmann, 1993: 1.0–2.0 mg Mg g^{-1} d.w.; Hofmann and Krauß, 1988: 0.6–0.9 mg Mg g^{-1} d.w.).

With regard to the high N supply demonstrated above, the N:Mg ratio has to be taken into consideration, as a pronounced imbalance between these elements may indicate severe nutritional disturbances even at absolute Mg values classified as sufficient (cf. Hüttl, 1991). Related to dry matter contents, N:Mg ratios between 25 and 30 as found at Rösa (Figure 7) indicate nutritional imbalances (Ende *et al.*, 1995). Nevertheless, the values decreased down to October, 1995, below critical levels.

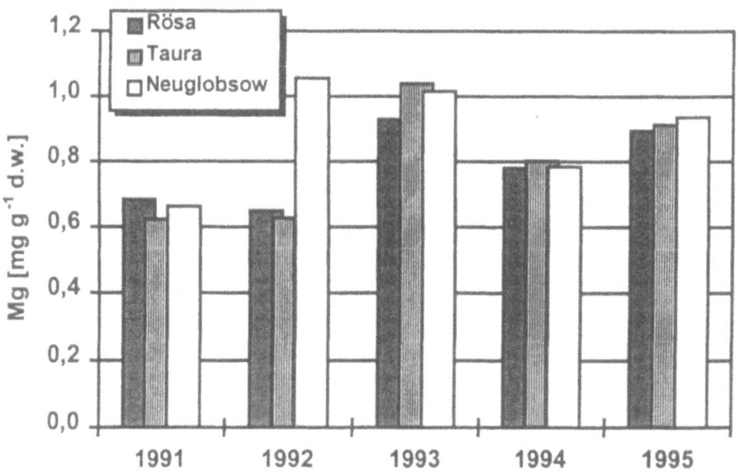

Figure 6. Temporal development of the Mg contents of Scots pine needles (current needles) from the three experimental sites. Time of sampling: October; data from 1991 and 1992: Schulz (personal communication)

3.5. Potassium

Between 1993 and 1995 the K contents (Figure 8) followed a course within a range of sufficient supply (Bergmann, 1993: 4.0–8.0 mg K g^{-1} d.w.; Hofmann and Krauß, 1988: 3.9–4.9 mg K g^{-1} d.w.) at Neuglobsow; at the other sites the K contents were on a higher average level and exhibited pronounced oscillations, typical for this very mobile element.

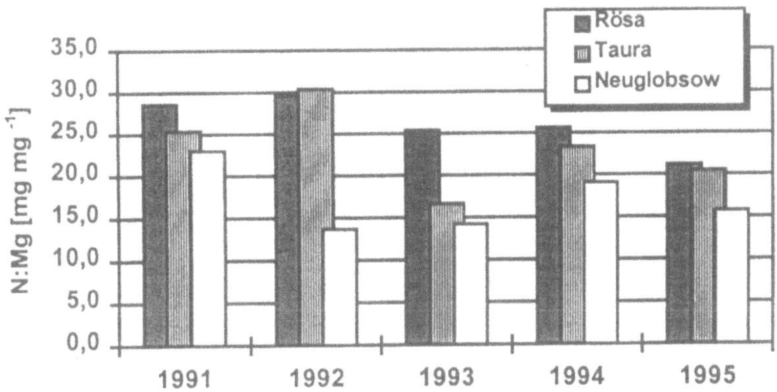

Figure 7. Temporal development of the ratio of N content to Mg content of Scots pine needles (current needles) from the three experimental sites. Time of sampling: October; data from 1991 and 1992: Schulz (personal communication)

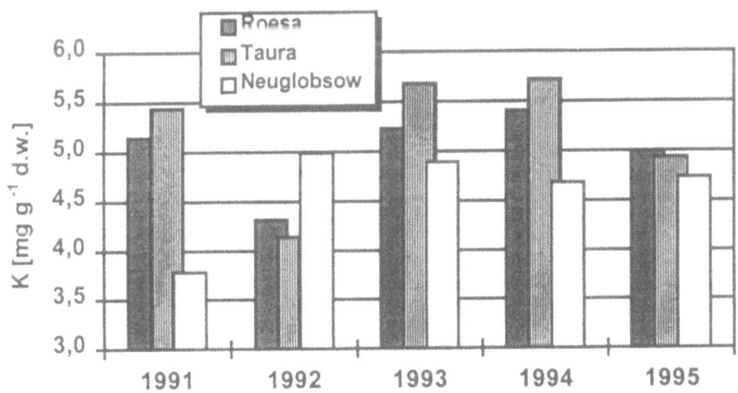

Figure 8. Temporal development of the K contents of Scots pine needles (current needles) from the three experimental sites. Time of sampling: October; data from 1991 and 1992: Schulz (personal communication)

3.6. *Calcium*

Between 1993 and 1995, the Ca contents (Figure 9) at Neuglobsow followed a course within a range of sufficient supply (Bergmann, 1993: 2.5–6.0 mg Ca g^{-1} d.w.; Hofmann and Krauß, 1988: 2.1–3.1 mg Ca g^{-1} d.w.) with the exception of 1993; at the other sites higher values were obtained. More obvious than with K, the relatively higher dust load of the sites within the industrialized areas are detectable. Higher foliage contents can be explained by the low mobility of Ca in plants that leads to an accumulation of Ca in older biomass compartments.

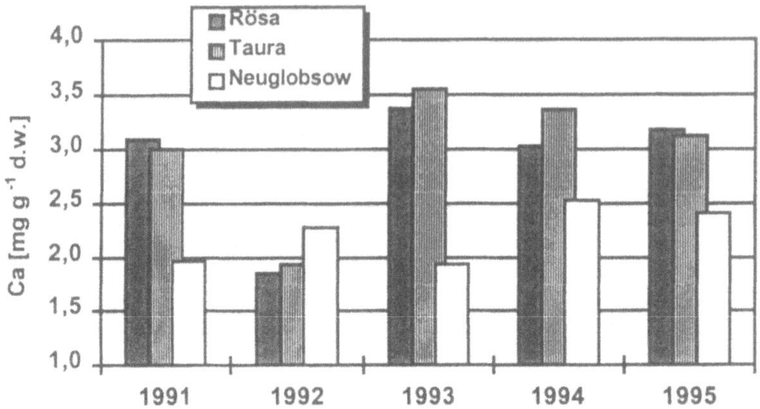

Figure 9. Temporal development of the Ca contents of Scots pine needles (current needles) from the three experimental sites. Time of sampling: October; data from 1991 and 1992: Schulz (personal communication)

3.7. *Manganese*

The variations in the Mn contents of needles (Figure 10) were within a range typical for this element (100 to 1000 cf. Fiedler and Höhne, 1985). The uptake of Mn is highly dependent on the redox conditions in the soil determined by pH and soil O_2 concentrations; this can be clearly demonstrated at Rösa. Varying soil moisture conditions may be responsible for the course of the Mn contents at Neuglobsow.

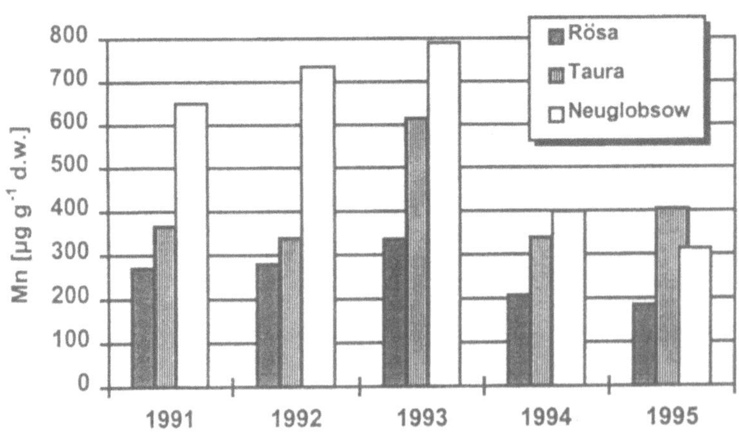

Figure 10. Temporal development of the Mn contents of Scots pine needles (current needles) from the three experimental sites. Time of sampling: October; data from 1991 and 1992: Schulz (personal communication)

3.8. Zinc

There was little differentation in the Zn contents of needles (Figure 11) between the three sites; they did not indicate an insufficient supply (deficiency level: lower than 20 g Zn g^{-1} d.w.; Bergmann, 1993).

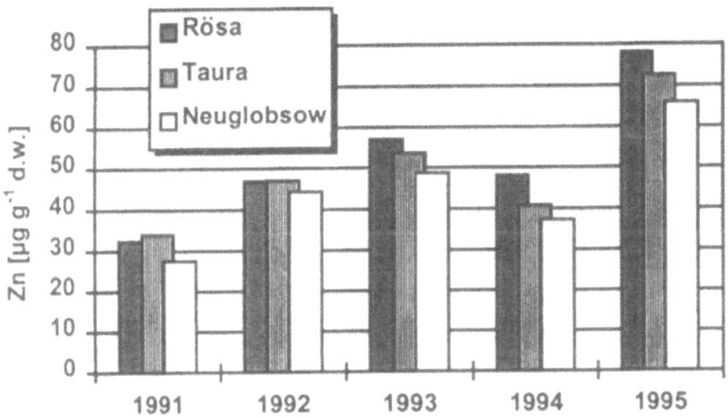

Figure 11. Temporal development of the Zn contents of Scots pine needles (current needles) from the three experimental sites. Time of sampling: October; data from 1991 and 1992: Schulz (personal communication)

3.9. Copper

The Cu contents of needles (Figure 12) were sufficient at all three sites. According to Hofmann and Krauß (1988) average Cu contents are 3.0–4.8 g Cu g^{-1} d.w. in current needles; deficiencies occur below 1.9 g Cu g^{-1} d.w.

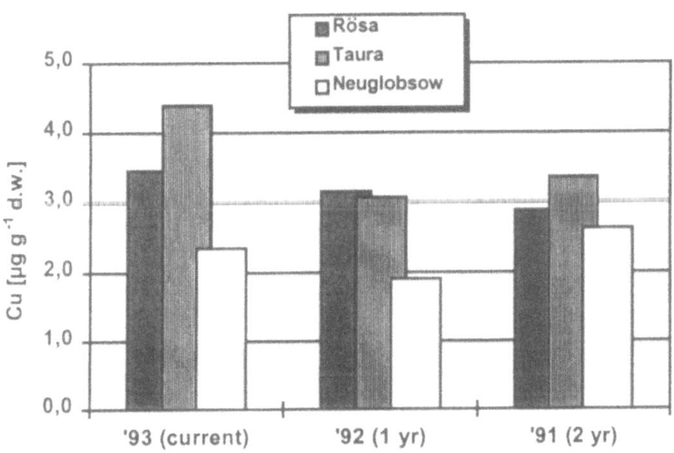

Figure 12. Cu contents of the three needle age classes of Scots pine trees from the three experimental sites. Time of sampling: October, 1993

3.10. Cadmium, lead

The analytical data presented here were obtained from unwashed needle samples so they do not allow for a differentiation between heavy metals deposited at the outer surface of the needles and heavy metals absorbed. The Pb as well as the Cd contents (Figure 13) do not indicate extremely high uptake or deposition rates of these elements. Nevertheless, the pronounced increase in Pb content in the older needle age classes of Rösa and Taura is an indicator of the relatively short distance to potential sources like industry and traffic.

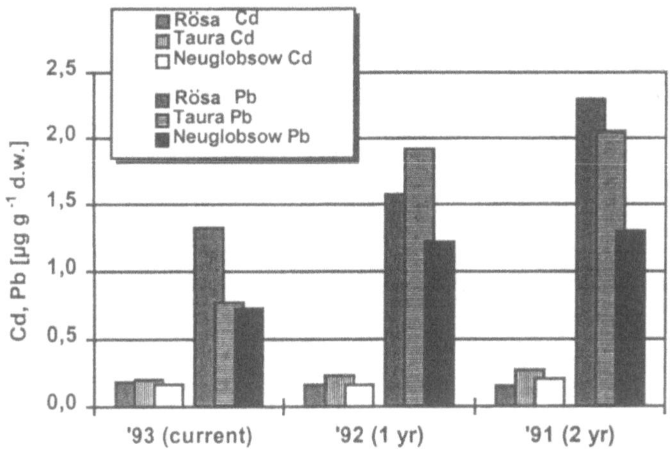

Figure 13. Cd contents (left hand groups of columns) and Pb contents (right hand groups of columns) of the three needle age classes of Scots pine trees from the three experimental sites. Time of sampling: October, 1993

3.11. Above-ground biomass compartments

The estimation of needle tip necroses mirrors the quantitative differentiation of the deposition history of the three sites (Rösa > Taura > Neuglobsow; Figure 14). During the time of observation, the tip necroses regularly increased from the youngest to the eldest needle age class.

At the whole tree harvest in 1995, crown dimensions and crown biomass were not correlated (Table 2). The crown biomass of Neuglobsow is lower than that of the other sites. The highest branch biomass was measured at Taura, the highest needle biomass at Rösa (Table 2). The estimated annual needle yield of 1995 was correlated with the diameter at breast height of the trees (Figure 15).

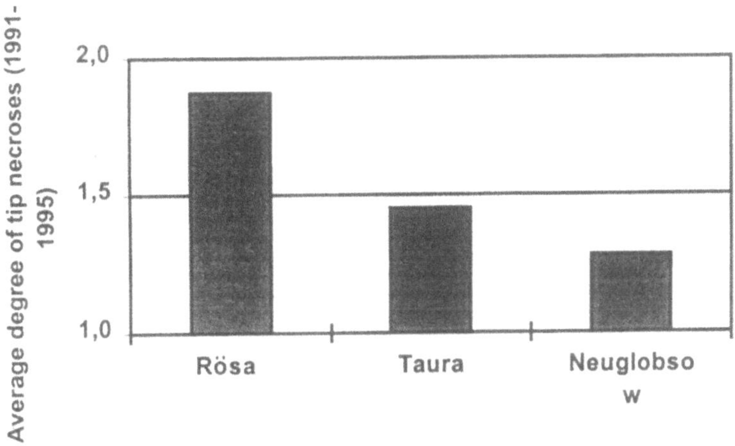

Figure 14. Degree of needle tip necroses of Scots pine needles at the three experimental sites. Mean values from samples 1991–1995, time of sampling: October

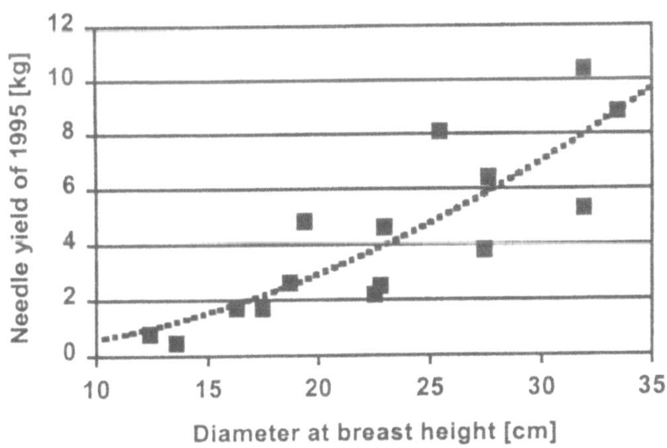

Figure 15. Estimated needle yield of 1995 in relation to diameter at breast height of Scots pine

Table 2. Canopy dimensions and canopy biomass of single Scots pine trees (weighted mean over all tree classes)

	Rösa	Taura	Neuglobsow
Length of crown (m)	6.84	7.58	8.10
Canopy projection area (m^2)	13.98	15.50	15.98
Biomass of branches (kg)	18.30	22.70	9.97
Biomass of needles (kg)	11.53	9.54	6.17

4. Discussion

Since the structural changes in the industrial production of eastern Germany from 1989 onwards, the atmospheric deposition of alkaline dust and SO_2 has been reduced drastically. Investigations of the atmospheric chemistry above the 'new states' of Germany by Marquardt and Brüggemann (1995) indicate that the dust components of emissions have been reduced more effectively than the SO_2 emissions. Even though nutritional investigations as presented here do not allow for conclusions in terms of actual deposition rates, they do demonstrate the differentiation of the sites investigated with regard to historical loads and actual supply. In comparison with findings of the other project parts of the SANA program, there are indications for ecosystem-internal processes that can be seen as reactions to the changed input situation as regards quality and quantity.

The reduction of the NH_4-N-deposition in the 'new states' of Germany expected as a consequence of the breakdown of industrial cattle farming stands against an increase of the NO_3-N deposition (assumably due to strongly increased traffic). In 1994, the annual total-N input (throughfall) at Rösa accounted for approx. 20 kg N ha^{-1} (Schaaf and Hüttl, 1995). In addition, the soil N stores at this site are comparatively high also as a consequence of repeated N fertilization before 1989 (see introduction). Optimal conditions for mineralization at Rösa (Fischer *et al.*, 1995), supported by favorable weather conditions during late summer 1993, had led to an extremely high N supply in the autumn of 1993, proven by needle N contents and needle dry weights (Table 1). This had also produced a remarkable shift in the N:Mg ratio (Figure 8) at Rösa in 1993. At the end of the investigative period, the absolute N contents at Taura were in the same range as at Rösa, but in terms of the N:P ratio (Figure 4) as well as the Mg contents (Figures 7 and 8) Taura is the site with the more balanced nutrition. At Neuglobsow, there is no indication for imbalanced nutrition due to N supply: The absolute N contents of current needles (Figure 1), the needle dry weight (Table 1) as well as the development of the N:Mg ratio (Figure 8) are typical for sufficient but sub-optimal supply. The fine root intensity and mycorrhiza frequency at this site (Münzenberger *et al.*, 1996, this volume) also demonstrate that Neuglobsow is far from being an N-saturated system.

The explanation for the low P supply of the Rösa stand certainly has several aspects. First of all, under the pH regime found in the rooting zone (around pH 6.0; Weisdorfer *et al.*, 1996, this volume) phosphate uptake may be suppressed by the competing uptake of nitrate ions which – at least temporarily – are at very high concentrations compared to Taura and Neuglobsow. An indication for this assumption is that the N:P ratio of the needle contents from 1993 – 1995 was highest at Rösa (Figure 4). Secondly, the fine root intensity and mycorrhiza frequency (Münzenberger *et al.*, 1995, this volume) were lowest at Rösa. Thirdly, a high pH value and high Ca concentrations as found in the rooting zone at Rösa counteract the solubility of calcium phosphates that may have evolved during longer dry periods such as the late summer of 1993. The phosphate concentrations measured in the soil solution are very low and do not indicate significant differences between the three sites (Weisdorfer, 1995, personal communication).

According to Schaaf and Hüttl (1995) the annual S input of 1994 was twice as high at Rösa as at Neuglobsow. Additionally, the S stores in the mineral soil of Rösa are historically at the highest level of the three sites (around 2000 kg ha^{-1}; Weisdorfer, 1995, personal communication), and a very high annual sulfate release from the mineral soil at 100 cm depth was found in 1994 (Weisdorfer *et al.* 1995, this volume). This explains the higher S_t contents in the needles of this site compared with Neuglobsow.

Malcolm and Garforth (1977) found statistical significance for a relation between elevated S supply (in this case due to a nearby SO_2 emitter) and total-S (S_t) contents of needles if 1.5 mg S g^{-1} d.w. were exceeded. Based on the frequently quoted fixed relation of organic S to N compounds (S_{org}:N_{org}; for pine: 0.03 on atomic weight basis; cf. Kelly and Lambert, 1972), they calculated the potential proportion of inorganic S compounds (S_{inorg}). A number of authors (overview by Dijkshoorn and van Wijk, 1967) demonstrated that elevated S_{inorg} is a very good indicator of an excessive S supply to forest trees. The sulfur sources may be dry and/or wet deposition (overview by Dijkshoorn and van Wijk, 1967), sulfate fertilizer (cf. Ende, 1991) or high S release from the soil (this study).

When calculating S_{inorg} from the N_t and S_t contents estimated in all October samples from all three sites using S_{org}:N_{org} = 0.03, a hypothetical relation between S_t and S_{inorg} results (Figure 16) that is indifferent between 1.4 and 2.0 mg S_t g^{-1} d.w. At a level below 1.4 the inorganic S fraction is below 0.3 mg S_{inorg} g^{-1} d.w. (less than 25% of S_t); all data in this range belong to the Neuglobsow site. At S_t contents above 2.0, the inorganic S fraction clearly increases up to more than 1.0 mg S_{inorg} g^{-1} d.w. or 45% of S_t (these date belong to Taura and Rösa). Under the assumptions described above this means that during 1994 and 1995 the pines at Taura and Rösa had taken up high amounts of excessive S, a large proportion of which was not transformed into organic compounds (proteins). Detailed investigations in the S compounds of the needles have been carried out by Schulz *et al.* (this volume) confirming the complexity of the S and N metabolism. Further investigations are needed in this field to clarify the role of soil-borne sulfate in the soil-plant relationship.

The Ca soil stores at Rösa, also extremely high as a consequence of massive dust deposition (Figure 10), supply sufficient cations accompanying the SO_4^{2-} leached from the upper soil. Thus a strongly accelerating soil acidification is not to be expected for the time being. At Taura, in 1994, high S-amounts were also leached from the upper soil horizons (which is also documented by needle analyses data in Figure 5). As the dust-borne soils stores of basic cations are much lower at Taura than at Rösa, a proceeding cation (Mg, K) depletion of the mineral soil has to be expected if the SO_4^{2-} leaching continued at the level measured at this site. At this time of observation, however, there is no evidence for such a cation depletion (Figures 6, 8, 9). In this context the pronounced oscillation of the N:Mg ratio at this site is conspicuous (Figure 7).

The course the trace elements took during the time of observation renders no indication for disharmonic or deficient supply. Only the Cu supply of the Neuglobsow experimental site temporarily approached the deficiency threshold level. The Cd and Pb contents of the (unwashed) needles indicate a higher needle age dependent accumulation of these elements near the industrialized areas (Rösa, Taura). The absolute needle contents of these elements, however, do not exceed critical levels. The course the trace elements took during the time of observation renders no indication for disharmonic or deficient supply. Only the Cu supply of the Neuglobsow experimental site temporarily approached the deficiency threshold level. The Cd and Pb contents of the (unwashed) needles indicate a higher needle age dependent accumulation of these elements near the industrialized areas (Rösa, Taura). The absolute needle contents of these elements, however, do not exceed critical levels.

5. Conclusions

Six years after the structural changes in the industrial production of Eastern Germany and the resulting reduction of emissions, the development of the nutritional status of the ecosystems still mirrored the different deposition situation (historically, present) of the pine stands compared at least in the case of N, S, K, Ca and the heavy metals investigated. Six years after the structural changes in the industrial production of Eastern Germany and the resulting reduction of emissions, the development of the nutritional status of the ecosystems still mirrored the different deposition situation (historically, present) of the compared pine stands at least in the case of N, S, K, Ca and the heavy metals investigated.

Acknowledgements

The scientific advice of Dr W. Gluch, Halle, as well as the technical assistance of Mrs A. Müsebeck and Mrs R. Riedelsheimer (both ZALF-Institute of Forest Ecology, Eberswalde) are gratefully acknowledged.

References

Bergmann W. 1993. Ernährungsstörungen bei Kulturpflanzen, 3rd ed. Fischer, 835 p.

Dijkshoorn W, van Wijk AL. 1967 The sulphur requirements of plants as evidenced by the sulphur-nitrogen ratio in the organic matter – a review of published data. Plant Soil. 26, 129–157.

Ende H-P. 1991. Wirkungen von Mineraldünger in Buchen- und Fichtenbeständen des Grundgebirgs-Schwarzwaldes. Freiburger Bodenkundl. Abh. 27, 98.

Ende H-P, Gluch W, Hüttl RF. 1995. Ernährungskundliche und morphologische Untersuchungen im Kronenraum von *Pinus sylvestris* L., Atmosphärensanierung und Waldökosysteme. Eds RF Hüttl, K Bellmann, W Seiler W. Umweltwissenschaften Vol. 4. Blottner, Taunusstein 112–128.

Fiedler HJ, Höhne H. 1985. Mengen- und Spurenelementgehalt der Eibennadeln in Abhängigkeit von biologischen und standortkundlichen Faktoren, Mikronährstofforschung, ed Inst. f. Pflanzenernährung, Jena: Publ. Inst. Pflanzenern. Jena, AdL, DDR 29–31.

Fischer T, Bergmann C, Hüttl RF. 1995. Auswirkungen sich zeitlich ändern der Schadstoffdepositionen auf Prozesse des Kohlenstoff- und Stickstoffumsatzes im Boden, Atmosphärensanierung und Waldökosysteme, ed Hüttl R F Bellmann K Seiler W: Umweltwissenschaften Vol. 4. Blottner, Taunusstein 144–160.

Hippeli P. 1991. Düngung der Kiefer im nordostdeutschen Tiefland, Ber. aus Forschg. u. Entw., Forschungsanst. f. Forst- u. Holzwirtsch. Eberswalde Vol. 25, 35–43.

Hofmann G, Krauß H-H. 1988. Die Ausscheidung von Ernährungsstufen für die Baumarten Kiefer und Buche auf der Grundlage von Nadel- und Blattanalysen und Anwendungsmöglichkeiten in der Überwachung des ökologischen Waldzustandes, Sozialistische Forstwirtschaft. 38, 272–273.

Hüttl RF. 1991. Die Nährelementversorgung geschädigter Wälder in Europa und Nordamerika, Freiburger Bodenkundl. Abh. 28, 440 p.

Hüttl RF, Bellmann K, Seiler W. 1995 Atmosphärensanierung und Waldökosysteme. UmweltWissenschaften, Vol. 4. Blottner, Taunusstein

Kelly J, Lambert MJ. 1972. The relationship between sulphur and nitrogen in the foliage of *Pinus radiata*. Plant Soil. 37, 395–407.

Krauß H-H, Heinsdorf D. 1983. Untersuchungen über den Einfluß der Emission N-haltiger Abprodukte des VEB Schweinezucht und -mast Eberswalde auf die umliegenden Waldbestände, report (unpubl.), Forschungsanst. f. Forst- u. Holzwirtsch. Eberswalde.

Malcolm DC, Garforth MF. 1977. The sulphur:nitrogen ratio of conifer foliage in relation to atmospheric pollution with sulphur dioxide. Plant Soil. 47, 89–102.

Marquardt W, Brüggemann E. 1995. Quantitative und qualitative Relationen zwischen Emissionsveränderungen in der ehem. DDR und der Naßdeposition nach einem Ferntransport im Wolkenniveau, Annalen der Meteorologie. 31, 155–156.

Münzenberger B, Lehfeldt J, Schmincke B, Strubelt F, Hüttl RF. 1995. Reaktionen mykorrhizierter und nicht mykorrhizierter Kiefernfeinwurzeln auf sich ändernde Schadstoffdepositionen in verschiedenen Kiefernwaldökosystemen in den neuen Bundesländern, Atmosphärensanierung und Waldökosysteme, ed Hüttl R F Bellmann K Seiler W: Umweltwissenschaften Vol. 4. Blottner, Taunusstein 161–179.

Schaaf W, Hüttl RF. 1995. SANA E: Ökosystemare Untersuchungen in drei Kiefernbeständen unterschiedlicher atmosphärischer Belastung im Nordostdeutschen Tiefland, Annalen der Meteorologie. 31, 149–150.

Weisdorfer M, Schaaf W, Hüttl RF. 1995. Auswirkungen sich zeitlich ändernder Schadstoffdepositionen auf Stofftransport und -umsetzung im Boden, Atmosphärensanierung und Waldökosysteme. Eds RF Hüttl, K Bellmann, W Seiler. Umweltwissenschaften Vol. 4. Blottner, Taunusstein 56–74.

4
Responses of sulphur- and nitrogen-containing compounds in Scots pine needles

H. SCHULZ, G. HUHN and S. HÄRTLING

1. Introduction

Monitoring strategies are essential to determine changes or trends in forest ecosystems or populations of trees. These strategies involve repeated measurements over time of selected chemical and biological variables. The information can be of critical importance to hypothesis formulation, hypothesis testing, ecological prediction and ecotoxicological risk assessment (Smith, 1990).

In recent years a variety of standard measurements have been developed that are useful in tree health assessments. The sulphur/nitrogen status of higher plants and their relationships have been suggested as an important indication level of factors affecting tree health (Dijkshoorn and Van Wijk, 1967; Kelly and Lambert, 1972; Malcolm and Garforth, 1977). In addition to the sulphur and nitrogen supply obtained from the soil, dry and wet depositions of sulphur dioxide (SO_2), sulphate sulphur (SO_4^{2-}-S), nitrogen oxides ($NO_x = NO + NO_2$) and ammonia (NH_3) were found to affect the needle sulphur and/or nitrogen content (Huttunen *et al.*, 1985; Heinsdorf, 1993; Innes, 1995; Meng *et al.*, 1995). Therefore, numerous attempts have been made to determine different sulphur and nitrogen fractions as well as their ratios as indices of SO_2 and NO_x depositions in forests and their effects on conifer needles (Gasch *et al.*, 1988; Legge *et al.*, 1988; Kaiser *et al.*, 1993; Manninen, 1995; Polle *et al.*, 1994; Härtling and Schulz, 1995; Schulz *et al.*, 1995; Huhn and Schulz, 1996).

Against the background of air pollution changes in eastern Germany since reunification the purpose of our investigations in this part of the SANA-E programme was to estimate long-term effects of main airborne pollutants on Scots pines (*Pinus sylvestris* L.) and monitor effects of revitalization. In a field study along a deposition gradient needles from young and adult pine stands were collected at three test areas, which differed in air pollution of SO_2 and NO_x. Various biomarkers of S- and N-metabolization as well as visible damage symptoms were used in order to investigate changes of multiple deposition on pine stands and the consequences in the vitality of pine trees. In particular, amounts of sulphur, sulphate, nitrogen, non-protein nitrogen, soluble protein nitrogen, thiols (cysteine, glutathione), amino acids (glutamine, arginine) as well as needle necrosis were measured to assess stress to which the Scots pine trees were exposed during the period 1991 to 1995. In comparison to adult pine

R.F. Hüttl and K. Bellmann, Changes of Atmospheric Chemistry and Effects on Forest Ecosystems, 37–63.

stands, the young pine stands show comparable responses to airborne pollutants (Schulz *et al.*, 1995). This paper reports the results obtained with adult pine stands.

2. Materials and methods

Field sites and sampling conditions

Pine needles were collected from adult Scots pine stands (*Pinus sylvestris* L.) growing at the sites Neuglobsow, Taura and Rösa. The sites are described in the general introduction and site description chapters.

At each area 5 test sites were selected in which 15 trees were randomly chosen. Sampling of the needles always took place in October, between 10th and 20th, 1991–1995; harvest time was between 10 a.m. and 3 p.m. From each tree one twig was cut in the sun-crown using a mobile elevator platform. Only the current year's needles were collected for (bio)chemical analysis. For biochemical analysis mix samples were prepared by mixing equal amounts of collected needles from 15 branches. The mixed needles were immediately deep-frozen in liquid nitrogen and stored at –80°C until biochemical analysis. For chemical analysis branches of the first needle age class were placed in plastic bags and transported in a refrigerator to the laboratory. For sample processing, the needles were cleaned with distilled water, pooled to a mixed sample, one hour shock-dried at 110°C and then dried at 60°C for six days. Dried needles were ground to a powder using a ultracentrifugal mill ZM1 with a 0.5 mm sieve (Retsch, Hanau, Germany).

Soil samples were taken in October 1993 and 1995 from the humic horizon (O_F/O_H) by using a sharp-edged soil cutting frame, adjacent to the trees selected for harvest. The individual samples were combined in order to obtain one mixed sample from each plot per site. The fresh samples were passed through a 2 mm sieve, freeze-dried, milled and stored for further chemical analysis.

For the determination of sulphate-sulphur, nitrate-nitrogen, ammonium-nitrogen and calcium in throughfall of investigated pine stands, bark samples were collected in October 1991–1995. The samples were dried at 30°C, milled and stored for further chemical analysis as described by Huhn *et al.* (1995). The throughfall rates were calculated according to Schulz *et al.* (1997).

Needle necrose

Needle necroses for the first and second age class were described by judging needle colour (chloroses) and visible damage syptoms (necroses) according to the following key: 1 = green needles, no signs of chlorophyll loss in the needle tips, 1.5 = green needles with partially chlorotic tips, 2 = all needles with chlorotic tips, 2.5 = some needles with 1–5 mm necrotic tips, 3 = all needles with 1–5 mm necrotic tips, 3.5 = some needles with 5–10 mm necrotic tips, 4 = all needles with 5–10 mm necrotic tips. Needle necroses for each class were summed so that mean values are based on totals of 15 twigs per test site.

2.1. *Analytical procedures*

Total sulphur (S) and organic sulphur (S_{org})
The total S content of needles and soils was determined using a Smat-5500 sulphur analyser (Ströhlein, Germany) by combustion of a 100 mg sample in an oxygen stream by 1350°C and subsequent IR-detection of the released SO_2. For calibration a coal standard (1115% S) from the National Institute of Standards and Technology was used. The organic sulphur content was calculated as the difference between total sulphur and sulphate sulphur. All measurements were checked with a reference material (internal Norway spruce needle standard) from Landesamt für Umweltschutz (Baden-Württemberg).

Sulphate sulphur, nitrate nitrogen, phosphate (SO_4^{2-}-S, NO_3^--N, PO_4^{3-}-P)
The anion contents of needles, barks and soils were measured using a Dionex ion chromatograph (Dionex GmbH, Idstein, Germany). 500 mg dried and milled sample material were extracted with 25 ml deionized water on a shaker for 45 min. After centrifugation, the supernatant was membrane filtered. A 100 ml sample was separated on a Dionex AS 12A column. The eluent consists of a carbonate/hydrogencarbonate solution.

Total nitrogen (N), non-protein nitrogen (NPN), protein nitrogen (PN)
Total nitrogen of needles and soils was determined by using the micro-Kjeldahl method. 250 mg needle powder or 2 g milled humus layer were digested with sulphuric acid and Wieninger catalyst mixture. For measuring the non-protein nitrogen content of needles, 500 mg needle powder was treated with 10% trichloroacetic acid (20 ml) in a boiling water bath (10 min). After protein precipitation overnight, the filtrate was used for Kjeldahl-analysis. The nitrogen content of protein was calculated as the difference between total nitrogen and non-protein nitrogen. All procedures are described by Huhn and Schulz, 1996.

Ammonium nitrogen (NH_4^+-N)
Measurements of the ammonium nitrogen content of bark and soils were carried out by water steam distillation of bark extracts (12.5 g bark/50 ml 1% K_2SO_4, w/w) with 32% (w/w) sodium hydroxide solution into a flask filled with 60 ml of 0.5 mol/L boric acid solution and back titration to the initial pH-value by 0.05 mol/L hydrochloric acid.

Element analysis (magnesium, calcium, potassium)
0.5 g of dried and ground needle samples were digested with 2.5 ml conc. nitric acid at 150°C and 1.2 MPa in a Microwave Digester (CEM MDS 2000, CEM Corporation, Matthews, USA). The element contents of the digestion solution and the 0.5 mol/L ammonium chloride soil extracts (calcium) were analyzed by ICP-AES (Jobin Yvon JY 24, Longjumeau Cedex, France).

Soluble protein
The content of soluble protein in extracts of needle acetone dry powder (25 mg dry powder in 1 ml of 20 mmol/L KH_2PO_4/K_2HPO_4, pH 7.5) was determined according to the methods given by Schulz (1981) and Bradford (1976).

Glutathione
Frozen pine needles were pulverized in liquid nitrogen using a micro dismembrator (Braun, Melsungen, Germany). 0.2 g of needle powder were homogenized for 45 s in 7 ml 0.1 mol/L HCl and 1 mmol/L EDTA at 4°C. After centrifugation (10 min at 26 500*g*) 0.4 ml of the supernatant were mixed with 0.6 ml CHES buffer (pH 9.3) and reduced with 0.1 ml 3 mmol/L DTT at ambient temperature for 1 h. 495 ml of this mixture was derivatized with 20 ml 15 mmol/L monobromobimane at ambient temperature in the dark for 15 min. The determination of the fluorescence derivatives were carried out using HPLC according to Härtling and Schulz (1995).

Glutamine, arginine
0.7 g of the pulverized needles were extracted (cutter rod 10N, Ultra Turrax T 25, IKA-Labortechnik, Staufen, Germany) for 1 min in 7 ml 4% (w/w) sulfosalicylic acid. The extract was centrifugated for 5 min at 4°C and 26 500*g*. 4 ml of the supernatant were pipetted into a 5 ml volumetric vessel, then 100 ml of 30% (w/w) NaOH were added to adjust the pH to about 3, and it was filled up with sulfosalicylic acid of pH 3. The crude needle extract was used for derivatization and HPLC separation.

The derivatization procedure of amino acids with 9-fluorenylmethyl chloroformate and the HPLC conditions was carried out as described by Huhn and Schulz (1996).

Superoxide dimutase (SOD)
The SOD activity was assayed as the sum of isozymes SOD1 and SOD3 after enzyme extraction of acetone dry powder with 20 mmol/L KH_2PO_4/EDTA, pH 8.0 and separating them on a 7% polyacrylamide gel according to Schulz (1983).

Total chlorophyll
Pulverized needles were extracted with 80% acetone. The chlorophyll content was calculated using the extinction coefficients by Lichtenthaler and Wellburn (1983).

Reference parameter
The dry weights for all analytical procedures were taken after drying the pulverized needles for 96 h at 80°C.

2.2. *Statistics*

At least 3 aliquots of individual needle, resp. bark and soil samples were measured. Tables were used to calculate site means (n = 5, standard error). Mann–Whitney U-Test was used for testing pairwise differences between sites. Details are given in the table legends. Data which were not different at $p < 0.05$

are labelled with the same letter in the table. Significant differences are indicated by different letters. Further data analysis was carried out to determine correlations between needle, bark and soil parameters on the basis of 15 individuals per year.

3. Results

3.1. Pollution impacts

Air concentrations of SO_2, NO_x and O_3 recorded at meteorological stations located next to the sites, are given in Table 1. The concentrations of these air pollutants reflect the different deposition load of the sites and the temporal variations in the period from 1988–1995. Since 1990 the situation has changed. At all sites the annual mean of SO_2 concentration has decreased significantly. Especially in Rösa SO_2 decreased from 143 to 25 μg m^{-3}, but the differences among the sites persist. O_3 and NO_x do not show as large differences among the sites. In 1995 the annual mean of NO_x at Rösa (22 μg m^{-3}) was about twice as high as at Neuglobsow (9 μg m^{-3}). In contrast to SO_2, the annual means of NO_x are increasing in temporal course at Rösa. Compared to NO_x, the annual mean values of O_3 concentrations at all sites are relatively high, but differed less and show no temporal changes during the last few years. Amounts of atmospheric inputs of S and N compounds as well as Ca^{2+}, which are obtained by throughfall rates on the basis of pine bark analyses (Schulz *et al.*, 1996), are given in Figure 1. Compared to 1988, the throughfall rates of SO_4^{2-}-S, NH_4^+-N

Table 1. Mean annual concentrations of SO_2, NO_x and O_3 in mg m^{-3} measured at meteorological stations (Neuglobsow, Melpitz and Pouch) close to the sites

Test area	88	91	92	93	94	95
				SO_2		
Neuglobsow*	11	–	9	11	7	7
Taura*	–	–	–	37	23	19
Rösa**	143	106	72	61	32	25
				NO_x		
Neuglobsow*	–	–	12	10	9	10
Taura*	–	–	–	–	14	15
Rösa**	–	13	16	21	22	21
				O_3		
Neuglobsow*	44	–	53	52	56	51
Taura*	–	–	–	40	50	50
Rösa**	–	61	66	69	64	51

*Stations Neuglobsow and Melpitz (Taura) of the Umweltbundesamt (distance approximately 5 km),

**Station Pouch (Rösa) of the Landesamt für Umweltschutz Sachsen-Anhalt (distance approximately 3 km)

a

b

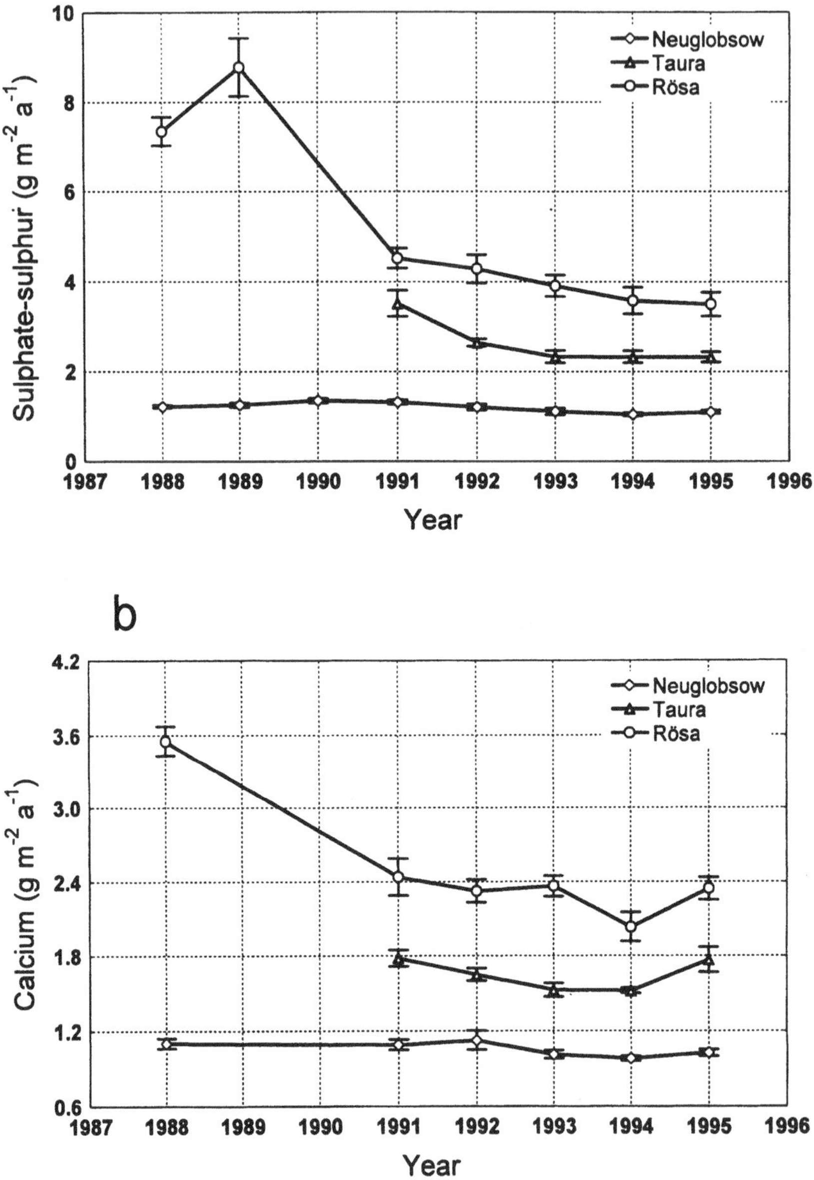
Neuglobsow
Taura
Rösa
Sulphate-sulphur (g m^{-2} a^{-1})
10
8
6
4
2
0
1987
1988
1989
1990
1991
1992
1993
1994
1995
1996
Year
Neuglobsow
Taura
Rösa
Calcium (g m^{-2} a^{-1})
4.2
3.6
3.0
2.4
1.8
1.2
0.6
1987
1988
1989
1990
1991
1992
1993
1994
1995
1996
Year

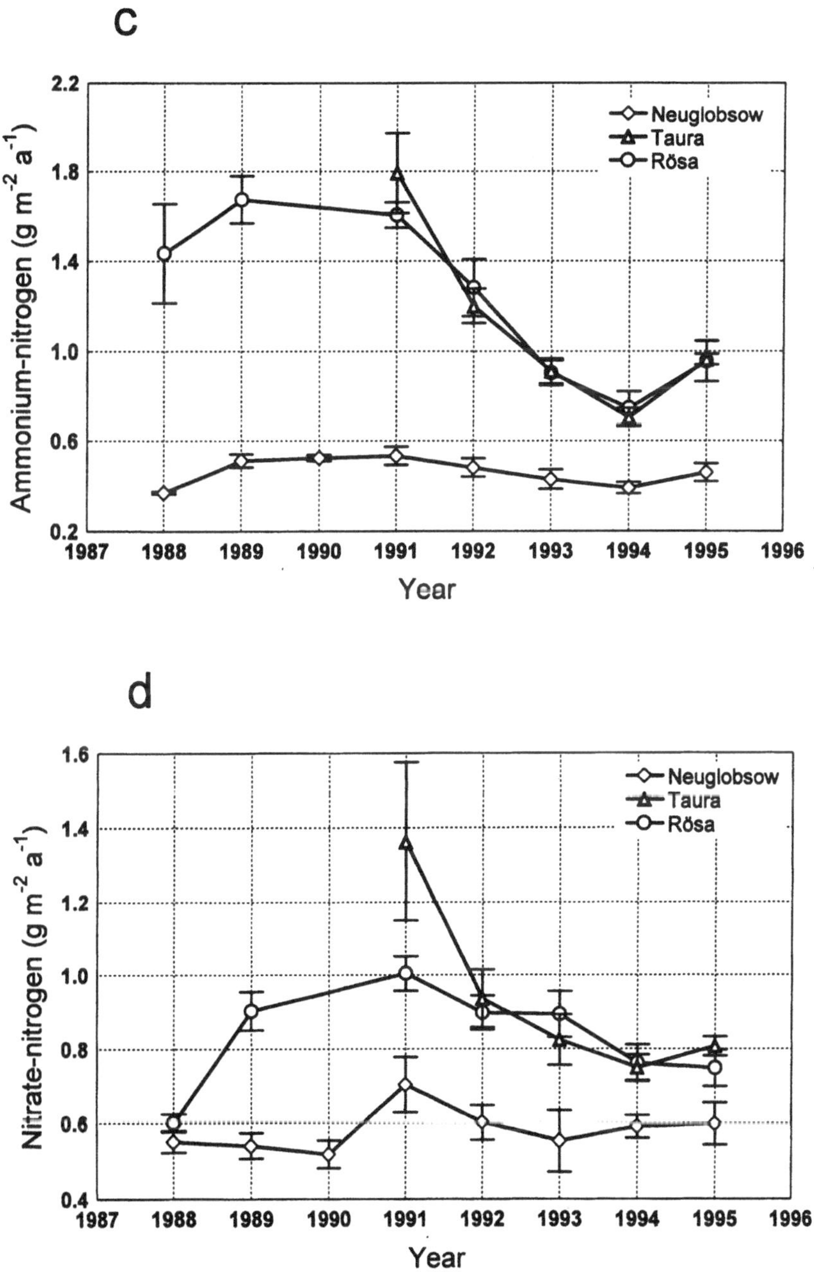

Figure 1. Yearly mean throughfall rates (TFR) of sulphate sulphur (a), calcium (b), ammonium nitrogen (c), and nitrate nitrogen (d) at the three field sites between 1989 and 1995. The TFR of dry and wet depositions were derived on the basis of bark analysis according to Schulz *et al.* (1996)

and Ca^{2+} strongly decreased at the sites Rösa and Taura (except Ca in 1995), while the inputs at Neuglobsow show only minor changes. Unlike SO_2, the decrease in throughfall rates of nitrate nitrogen is much smaller at all sites. In 1995 the throughfall rates of nitrate nitrogen were above those of 1988.

3.2. Correlations between needle parameters, pollutant concentrations and throughfall rates

In a first correlation analysis the sulphur and nitrogen fractions, including SO_4^{2-}-S and NPN of current year's needles, were correlated with the pollution impacts. Highly significant correlations were found between yearly mean SO_2 concentrations of the ambient air and the SO_4^{2-}-S contents of the pine needles (Figure 2) as well as between NO_x concentrations and the NPN contents of the needles (Figure 3). Remarkably high correlation coefficients were also observed for needle SO_4-S contents vs. throughfall rates of SO_4^{2-}-S ($r^2 = 0.91$) or needle NPN contents vs. throughfall rates of nitrate ($r^2 = 0.90$).

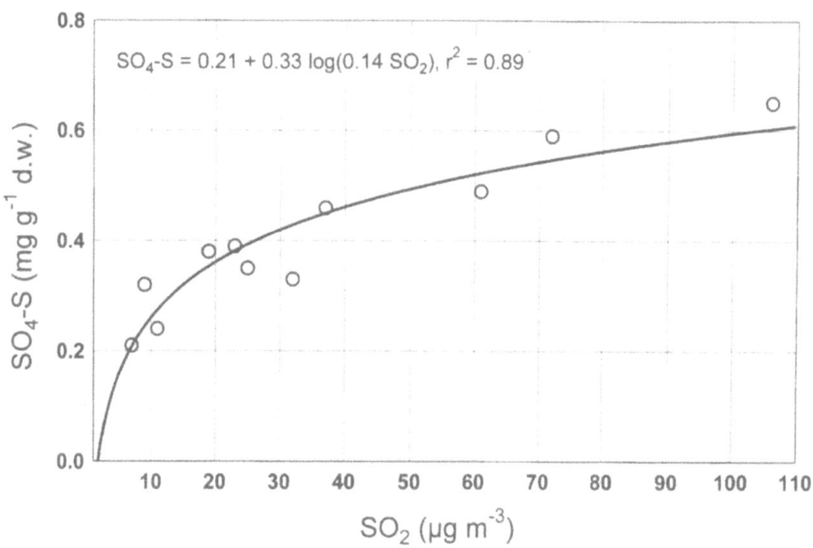

Figure 2. Relationship for mean sulphate sulphur (SO_4^{2-}-S) contents in current year's pine needles vs. yearly means of sulphur dioxide (SO_2) concentration at the three field sites

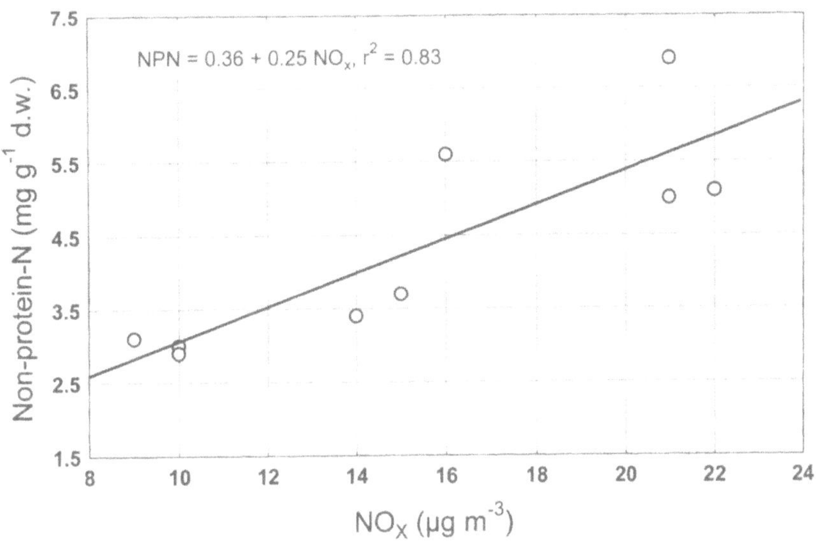

Figure 3. Relationship for mean non-protein nitrogen (NPN) contents in current year's pine needles vs. yearly means of nitrogen oxide (NO_x) concentrations at the three field sites

3.3. Spatial and temporal variations

The data in Table 2 show that contents of SO_4^{2-}-S in the current year's needles from the higher SO_2-polluted sites Taura and Rösa are almost 1.5–2 times higher than in needles from Neuglobsow. The same relations were also obtained for total sulphur, even with smaller differences between the sites and smaller temporal variations. These observations were made during the whole investigation period from 1991–1995. Since 1993, the SO_4^{2-}-S content in the current year's needles has decreased significantly at all sites. The different sulphur exposition affected the contents of non-protein sulphur forms in pine needles too. Clear site-dependent differences (Neuglobsow < Taura < Rösa) were found for the soluble thiols cysteine and glutathione, but their contents indicate no temporal variations at any of the sites. The organic sulphur fraction shows neither clear site-dependent nor temporal variations. Thus, the mean S_{org}/SO_4^{2-} ratios of needles from pine stands are significantly smaller at the high polluted sites Rösa and Taura compared to the low polluted site Neuglobsow. There are clear changes with time.

Like the different nitrogen loads at the three sites the contents of the estimated nitrogen fractions also varied in the current year's needles (Table 3). In comparison with Neuglobsow, the needles from Taura and Rösa contained

Table 2. Contents of total sulphur, sulphate sulphur, organic sulphur, sulphate sulphur/organic sulphur ratio, glutathione and cysteine (mean values ± standard errors) of the current year's pine needles. Significance was determined according to the Mann–Whitney U-test. Different small letters (a, b) within the rows indicate significant differences between the test sites and different large letters (A, B) within the columns indicate significant temporal differences at $p < 0.05$ (reference year 1991 or 1992)

Year	Neuglobsow	Taura	Rösa
		Total sulphur	
1991	$1.35 \pm 0.03^{a,A}$	$1.72 \pm 0.04^{b,A}$	$1.86 \pm 0.05^{b,A}$
1992	$1.32 \pm 0.01^{a,A}$	$1.88 \pm 0.03^{b,B}$	$1.92 \pm 0.04^{b,A}$
1993	$1.25 \pm 0.05^{a,A}$	$1.67 \pm 0.05^{b,A}$	$1.85 \pm 0.06^{b,A}$
1994	$1.20 \pm 0.02^{a,B}$	$1.67 \pm 0.04^{b,A}$	$1.63 \pm 0.02^{b,B}$
1995	$1.09 \pm 0.05^{a,B}$	$1.66 \pm 0.04^{b,A}$	$1.60 \pm 0.03^{b,B}$
		Sulphate sulphur	
1991	$0.33 \pm 0.02^{a,A}$	$0.59 \pm 0.01^{b,A}$	$0.65 \pm 0.02^{b,A}$
1992	$0.32 \pm 0.02^{a,A}$	$0.59 \pm 0.02^{b,A}$	$0.59 \pm 0.03^{b,A}$
1993	$0.24 \pm 0.02^{a,B}$	$0.46 \pm 0.02^{b,B}$	$0.49 \pm 0.04^{b,A}$
1994	$0.21 \pm 0.01^{a,B}$	$0.39 \pm 0.02^{b,B}$	$0.33 \pm 0.02^{b,B}$
1995	$0.21 \pm 0.02^{a,B}$	$0.38 \pm 0.01^{b,B}$	$0.35 \pm 0.02^{b,B}$
		Organic sulphur	
1991	$1.02 \pm 0.02^{a,A}$	$1.12 \pm 0.03^{b,A}$	$1.21 \pm 0.04^{b,A}$
1992	$0.99 \pm 0.01^{a,A}$	$1.29 \pm 0.03^{b,B}$	$1.33 \pm 0.02^{b,B}$
1993	$1.01 \pm 0.03^{a,A}$	$1.20 \pm 0.03^{b,A}$	$1.37 \pm 0.02^{c,B}$
1994	$0.99 \pm 0.02^{a,A}$	$1.28 \pm 0.03^{b,B}$	$1.30 \pm 0.02^{b,A}$
1995	$0.88 \pm 0.03^{a,B}$	$1.28 \pm 0.03^{b,B}$	$1.25 \pm 0.02^{b,A}$
		Sulphate sulphur/organic sulphur	
1991	$0.33 \pm 0.02^{a,A}$	$0.53 \pm 0.01^{b,A}$	$0.54 \pm 0.03^{b,A}$
1992	$0.33 \pm 0.02^{a,A}$	$0.46 \pm 0.02^{b,B}$	$0.44 \pm 0.02^{b,B}$
1993	$0.24 \pm 0.01^{a,B}$	$0.38 \pm 0.02^{b,B}$	$0.36 \pm 0.03^{b,B}$
1994	$0.21 \pm 0.01^{a,B}$	$0.30 \pm 0.02^{b,B}$	$0.26 \pm 0.02^{a,b,B}$
1995	$0.23 \pm 0.02^{a,B}$	$0.30 \pm 0.01^{b,B}$	$0.28 \pm 0.01^{a,b,B}$
		Glutathione	
1992	$329 \pm 25^{a,A}$	$484 \pm 23^{b,A}$	$504 \pm 23^{b,A}$
1993	$314 \pm 12^{a,A}$	$469 \pm 12^{b,A}$	$441 \pm 24^{b,A}$
1994	$339 \pm 10^{a,A}$	$547 \pm 9^{b,A}$	$488 \pm 27^{b,A}$
1995	$284 \pm 18^{a,A}$	$444 \pm 15^{b,A}$	$431 \pm 13^{b,B}$
		Cysteine	
1992	$7.96 \pm 0.33^{a,A}$	$12.58 \pm 0.49^{b,A}$	$10.09 \pm 0.24^{c,A}$
1993	$6.34 \pm 0.12^{a,B}$	$8.90 \pm 0.31^{b,B}$	$9.77 \pm 0.63^{b,A}$
1994	$7.88 \pm 0.25^{a,A}$	$9.41 \pm 0.17^{b,B}$	$9.20 \pm 0.37^{b,A}$
1995	$7.17 \pm 0.37^{a,A}$	$10.84 \pm 0.52^{b,A}$	$11.06 \pm 0.41^{b,A}$

Table 3. Contents of total nitrogen, non-protein nitrogen, protein nitrogen, soluble protein, non-protein nitrogen/ protein nitrogen ratio, glutamine and arginine (mean values ± standard errors) of the current year's pine needles. Significance was determined according to the Mann–Whitney U-test. Different small letters (a, b) within the rows indicate significant differences between the test sites and different large letters (A, B) within the columns indicate significant temporal differences at $p<0.05$ (reference year 1991 or 1993)

Year	Neuglobsow	Taura	Rösa
		Total nitrogen (mg g^{-1} d.w.)	
1991	$15.2 \pm 0.3^{a,A}$	$16.1 \pm 0.4^{a,A}$	$19.2 \pm 1.2^{b,A}$
1992	$13.9 \pm 0.3^{a,B}$	$16.5 \pm 0.2^{b,A}$	$18.8 \pm 1.0^{b,A}$
1993	$14.6 \pm 0.4^{a,A}$	$16.9 \pm 0.2^{b,A}$	$21.4 \pm 1.1^{c,A}$
1994	$14.8 \pm 0.5^{a,A}$	$17.9 \pm 0.3^{b,B}$	$19.8 \pm 0.6^{c,A}$
1995	$14.0 \pm 0.6^{a,A}$	$17.8 \pm 0.1^{b,B}$	$19.2 \pm 0.5^{b,A}$
		Non-protein nitrogen (mg g^{-1}d.w.)	
1991	$2.9 \pm 0.1^{a,A}$	$3.0 \pm 0.1^{a,A}$	$5.7 \pm 0.9^{b,A}$
1992	$2.3 \pm 0.1^{a,B}$	$3.0 \pm 0.1^{b,A}$	$5.6 \pm 0.9^{c,A}$
1993	$3.0 \pm 0.1^{a,A}$	$3.6 \pm 0.2^{b,A}$	$6.9 \pm 1.1^{c,A}$
1994	$3.1 \pm 0.1^{a,A}$	$3.4 \pm 0.1^{a,B}$	$5.1 \pm 0.6^{b,A}$
1995	$2.9 \pm 0.1^{a,A}$	$3.7 \pm 0.1^{b,B}$	$5.0 \pm 0.5^{b,A}$
		Protein nitrogen (mg g^{-1} d.w.)	
1991	$12.3 \pm 0.3^{a,A}$	$13.0 \pm 0.3^{a,A}$	$13.4 \pm 0.3^{a,A}$
1992	$11.6 \pm 0.3^{a,A}$	$13.5 \pm 0.1^{b,A}$	$13.1 \pm 0.2^{b,A}$
1993	$11.6 \pm 0.4^{a,A}$	$13.3 \pm 0.3^{b,A}$	$14.5 \pm 0.2^{c,A}$
1994	$11.7 \pm 0.5^{a,A}$	$14.5 \pm 0.3^{b,B}$	$14.7 \pm 0.1^{b,B}$
1995	$11.1 \pm 0.6^{a,A}$	$14.1 \pm 0.1^{b,A}$	$14.2 \pm 0.3^{b,A}$
		Soluble protein (mg g^{-1} d.w.)	
1991	$1.3 \pm 0.2^{a,A}$	$4.5 \pm 0.8^{b,A}$	$7.2 \pm 0.9^{b,A}$
1992	$3.2 \pm 0.6^{a,B}$	$9.0 \pm 1.7^{b,A}$	$8.2 \pm 1.2^{b,A}$
1993	$2.2 \pm 0.3^{a,A}$	$3.4 \pm 0.6^{a,A}$	$4.9 \pm 1.3^{a,A}$
1994	$5.0 \pm 0.6^{a,B}$	$8.0 \pm 0.9^{b,B}$	$8.0 \pm 0.8^{a,b,A}$
1995	$2.5 \pm 0.2^{a,B}$	$6.4 \pm 1.1^{b,A}$	$5.6 \pm 0.7^{b,A}$
		Non-protein nitrogen/protein nitrogen	
1991	$0.24 \pm 0.01^{a,A}$	$0.23 \pm 0.01^{a,A}$	$0.42 \pm 0.07^{b,A}$
1992	$0.20 \pm 0.01^{a,B}$	$0.23 \pm 0.01^{a,A}$	$0.43 \pm 0.06^{b,A}$
1993	$0.26 \pm 0.01^{a,A}$	$0.27 \pm 0.02^{a,b,A}$	$0.47 \pm 0.07^{b,A}$
1994	$0.27 \pm 0.01^{a,A}$	$0.23 \pm 0.01^{a,A}$	$0.35 \pm 0.05^{a,A}$
1995	$0.26 \pm 0.01^{a,A}$	$0.26 \pm 0.01^{a,B}$	$0.35 \pm 0.04^{a,A}$
		Glutamine (mg g^{-1} d.w.)	
1993	$104 \pm 13^{a,A}$	$272 \pm 29^{b,A}$	$558 \pm 22^{c,A}$
1994	$192 \pm 28^{a,B}$	$314 \pm 28^{b,A}$	$433 \pm 30^{c,B}$
1995	$75 \pm 11^{a,A}$	$127 \pm 13^{b,B}$	$248 \pm 29^{c,B}$
		Arginine (mg g^{-1} d.w.)	
1993	$96 \pm 11^{a,A}$	$456 \pm 204^{b,A}$	$9287 \pm 3175^{c,A}$
1994	$29 \pm 9^{a,B}$	$256 \pm 178^{b,A}$	$4636 \pm 1793^{c,A}$
1995	$59 \pm 16^{a,A}$	$170 \pm 58^{a,b,A}$	$4075 \pm 1617^{b,A}$

higher amounts of total nitrogen, NPN, protein nitrogen and soluble protein. The nitrogen fractions indicate no clear temporal trend. Changes with increased contents of nitrogen and NPN has been measured in needles at Taura since 1993. The same site-dependent differences and temporal variations were observed for the NPN/PN-ratio. Clear effects indicate the amino acids glutamine and arginine, which are accumulated due to increased nitrogen loads at the sites. In comparison with the other sites, the mean values of glutamine and arginine in the needles at Rösa show a 5 and 100 fold increase in comparison to Neuglobsow, respectively. The arginine content is, however, highly related to the deposition load and shows the largest standard deviation of all indication parameters within the test sites.

Despite the dramatically reduced SO_2 deposition, the average of needle necroses as sum of current and previous pine needles is still unexpectedly high at Rösa (Table 4). In the investigation period from 1992–1995 at Taura, the needle necroses in fact showed a clear increase. The same trend was found at Neuglobsow, while no significant temporal changes were observed at Rösa. The current year's needles contained significantly higher total chlorophyll contents at the more polluted sites Taura and Rösa (Table 4). The total chlorophyll content of the needles has increased significantly at all sites since 1992.

Table 4. Needle necroses as sum (mean values ± standard errors) of previous and current year's pine needles and total chlorophyll content of current year's pine needles. Significance was determined according to the Mann–Whitney U-test. Different small letters (a, b) within the rows indicate significant differences between the test sites and different large letters (A, B) within the columns indicate significant temporal differences at $p < 0.05$ (reference year 1991 or 1992)

Year	Neuglobsow	Taura	Rösa
		Needle necroses	
1991	$3.17 \pm 0.13^{a,A}$	$3.34 \pm 0.09^{a,A}$	$4.46 \pm 0.16^{b,A}$
1992	$3.07 \pm 0.09^{a,A}$	$3.10 \pm 0.02^{a,B}$	$3.92 \pm 0.07^{b,B}$
1993	$3.75 \pm 0.14^{a,A}$	$3.85 \pm 0.06^{a,B}$	$4.99 \pm 0.10^{b,B}$
1994	$3.40 \pm 0.10^{a,A}$	$3.93 \pm 0.08^{b,B}$	$5.08 \pm 0.05^{c,B}$
1995	$3.79 \pm 0.05^{a,B}$	$3.89 \pm 0.16^{a,B}$	$4.35 \pm 0.09^{b,A}$
		Total chlorophyll (mg g^{-1} d.w.)	
1992	$2.09 \pm 0.12^{a,A}$	$2.62 \pm 0.10^{b,A}$	$2.53 \pm 0.05^{b,A}$
1993	$1.89 \pm 0.03^{a,A}$	$2.30 \pm 0.07^{b,B}$	$2.31 \pm 0.05^{b,B}$
1994	$2.39 \pm 0.06^{a,A}$	$2.80 \pm 0.08^{b,A}$	$2.86 \pm 0.06^{b,B}$
1995	$2.28 \pm 0.07^{a,A}$	$2.86 \pm 0.09^{b,A}$	$2.97 \pm 0.07^{b,B}$

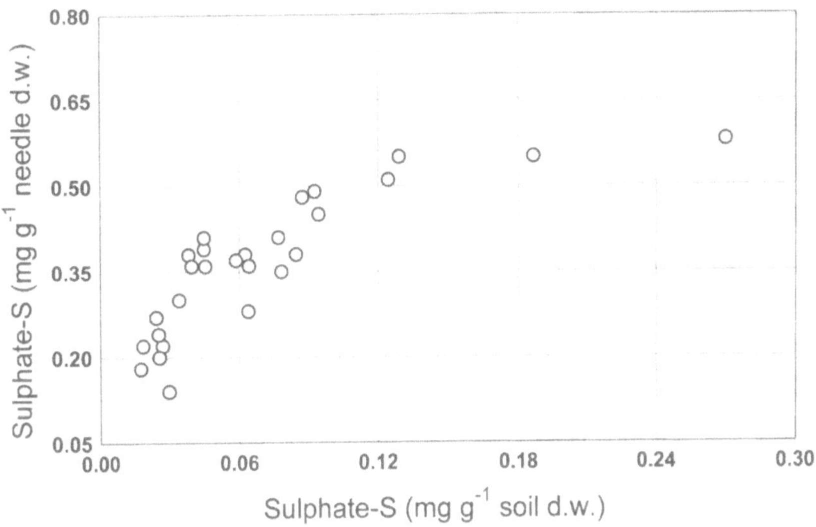

Figure 4. Correlation between the sulphate sulphur contents in current year's pine needles and in humus layers at the three field sites in 1993 and 1995

4. Discussion

Comparison of the measured sulphur dioxide (SO_2) and nitrogen oxide (NO_x = $NO + NO_2$) concentrations in the ambient air near Scots pine forests located in (N)euglobsow, (T)aura and (R)ösa (Table 1) with the determined SO_4^{2-}-S (Table 2, Figure 2) and NPN (Table 3, Figure 3) contents in current year's needles confirmed the accuracy of the presented results of this long-term study under field conditions between 1991–1995. SO_2-dependent SO_4^{2-}-S accumulation was observed repeatedly after exposure of Norway spruce (Gasch *et al.*, 1988, Kaiser *et al.*, 1993) as well as to Scots pine (Malcolm and Garforth, 1977; Manninen, 1995) to SO_2. With regard to NO_x , direct uptake and subsequent reduction in spruce needles has to be taken into account as a major path of assimilation (Schulze, 1989). Schulz *et al.* (1996) observed highly significant correlations between NO_x pollution and the protein nitrogen content of spruce needles in the field. Fumigation experiments with ^{15}N-nitrogen dioxide have shown that amino compounds are synthesized from the absorbed nitrogen dioxide in spruce needles (Nussbaum *et al.* 1993). In accordance with results reported by Schaaf *et al.* (1996) and measurements of natural isotope variations ($\delta^{15}N$ values) of ammonium nitrogen in pine barks and NPN in needles, we established that the accumulation of nitrogen in pine needles is a specific effect of NO_x. Hence, the contents of SO_4^{2-}-S and NPN in the current year's pine needles indicate significant spatial and temporal variations of air pollution after fundamental changes of the emission conditions in the eastern Germany five years after reunification.

The proved site-dependent variations (N < T < R) in the SO_4^{2-}-S content are the result of significant differences in SO_2 pollution between the three sites. On the other hand, the decreased ambient SO_2 concentrations were clearly indicated by reduced amounts of SO_4^{2-}-S in the needles of adult pine stands between 1991–1995 at all sites (Table 2). Similar spatial and temporal changes were observed in needles of young pine stands (Schulz *et al.*, 1995). Still, the question remained open whether observed SO_4^{2-}-S variations in pine needles were the result of the impact of SO_2 taken up via stomata, or was it possibly an indirect effect of different SO_4^{2-}-S deposition dissolved in throughfall (Figure 1a) which affected significantly different SO_4^{2-}-S contents in the humus layer (Table 5). Between 1993–1995 the SO_4^{2-}-S contents in the humus layer of the highly SO_2-polluted sites Taura and Rösa showed a significant decrease. At all sites, the organic sulphur fraction represents more than 90% of total sulphur. It

Table 5. Contents of different sulphur and nitrogen fractions as well as pH values (mean values ± standard errors) of the humus layer O_f/O_h-horizon of adult pine stands. Significance was determined according to the Mann–Whitney U-test. Different small letters (a, b) within the rows indicate significant differences between the test sites and different large letters (A, B) within the columns indicate significant temporal differences at $p < 0.05$ (reference year 1993)

Year	Neuglobsow	Taura	Rösa
		Total sulphur (mg g^{-1} d.w.)	
1993	$1.83 \pm 0.19^{a,A}$	$2.91 \pm 0.22^{b,A}$	$3.94 \pm 0.18^{c,A}$
1995	$1.90 \pm 0.10^{a,A}$	$2.66 \pm 0.22^{b,A}$	$3.13 \pm 0.27^{b,A}$
		Sulphate sulphur (mg g^{-1} d.w.)	
1993	$0.032 \pm 0.003^{a,A}$	$0.090 \pm 0.008^{b,A}$	$0.146 \pm 0.034^{b,A}$
1995	$0.023 \pm 0.002^{a,A}$	$0.043 \pm 0.004^{b,B}$	$0.062 \pm 0.007^{c,B}$
		Total nitrogen (mg g^{-1} d.w.)	
1993	$10.7 \pm 1.2^{a,A}$	$13.8 \pm 0.7^{a,b,A}$	$14.6 \pm 0.4^{b,A}$
1995	$11.8 \pm 0.5^{a,A}$	$12.8 \pm 0.7^{a,A}$	$11.3 \pm 1.1^{a,A}$
		Nitrate nitrogen (mg g^{-1}d.w.)	
1993	$0.0014 \pm 0.0002^{a,A}$	$0.0141 \pm 0.0051^{b,A}$	$0.0377 \pm 0.0094^{b,A}$
1995	$0.0007 \pm 0.0001^{a,B}$	$0.0310 \pm 0.0074^{b,A}$	$0.0175 \pm 0.0033^{b,B}$
		Ammonium nitrogen (mg g^{-1} d.w.)	
1993	$0.88 \pm 0.09^{a,A}$	$1.41 \pm 0.14^{b,A}$	$1.47 \pm 0.04^{b,A}$
1995	$0.89 \pm 0.06^{a,A}$	$1.10 \pm 0.13^{a,A}$	$1.01 \pm 0.09^{a,B}$
		pH (0.1 *N* KCl)	
1993	$2.92 \pm 0.02^{a,A}$	$3.47 \pm 0.05^{b,A}$	$4.24 \pm 0.09^{c,A}$
1995	$2.98 \pm 0.05^{a,A}$	$3.46 \pm 0.03^{b,A}$	$4.56 \pm 0.10^{c,A}$

appears that the total sulphur content varies as a function of microbial ecology and sulphate flux in deeper soil horizons (Vannier *et al.*, 1993), which are strongly affected by sulphate inputs with the throughfall (Schaaf *et al.*, 1996). In Figures 2 and 4 it is shown that both sulphur sources are assumed to be relevant in the field. If an accumulation of sulphur in pine needles is defined as an increase in needle SO_4^{2-}-S to over 200 g g^{-1} d.w. (Manninen, 1995), then the SO_4^{2-}-S contents in the current year's needles mirror the effective uptake from the atmosphere. This conclusion is in agreement with the data of Slovik *et al.*, (1995), which suggest that only the background SO_4^{2-}-S (100–200 g g^{-1}) in approx. Two-month-old needles is soil-dependent, whereas the rate of SO_4^{2-}-S accumulation in ageing needles (approximately six months old) of Norway spruce is readily explained by stomatal SO_2 uptake. According to (Malcolm and Garforth, 1977), an SO_4^{2-}-S content in pine needles of greater than 200 g g^{-1} in Scots pine is considered to be high and indicate uptake of SO_2 or inability of the tree to assimilate it effectively into organic compounds. The needle SO_4^{2-}-S content is indicative of exposure to sulphur bulk deposition. Wet deposition is thought to increase sulphate only slightly (100–200 mg g^{-1}) whereas sulphate of SO_2 may result in a many times higher increase (Slovik, 1996). The data in Table 2 show that oxidative SO_2 detoxification (SO_4^{2-} S formation) dominates in Scots pine needles at all sites. Thus, there is no demand to postulate a soil-dependent sulphate accumulation component for the three sites in this field study.

The absolute amount of SO_4^{2-}-S accumulated in the needles depends on the SO_2 concentration applied and the sulphur nutritional status (De Kok, 1990). Intracellular SO_4^{2-}-S may enter a metabolical storage pool, i.e. the vacuole and a metabolically active pool in the cytoplasm (Rennenberg and Polle, 1994). A small but distinct content of the cytoplasmic sulphate is reduced to cysteine and glutathione (De Kok, 1990). Transport of sulphur to the root may proceed in reduced and oxidized form, i.e. glutathione and sulphate, which are mobile in the phloem (Rennenberg and Herschbach, 1995). The glutathione contents in pine needles have remained unchanged at all sites since 1991 (Table 2). There is a relationship between glutathione and SO_4^{2-}-S which follows an optimum curve (Figure 5). After an initial increase of glutathione content a decrease follows at about 0.37 mg SO_4^{2-}-S g^{-1} d.w. This might suggest that SO_4^{2-}-S contents in excess of 0.37 mg g^{-1} d.w. probably affected the assimilatory sulphate reduction in pine needles. According to Kindermann *et al.* (1995) pines could release surplus sulphur (approximately 10%) in the form of H_2S. On the other hand glutathione is transported to the roots and regulates the uptake of sulphate from the soil (Herschbach and Rennenberg, 1994). Generally, the threshold of 0.37 mg SO_4^{2-}-S g^{-1} d.w. also indicates the formation of needle damage in the form of tip necroses (Schulz *et al.*,1996).

Like the needle SO_4^{2-}-S contents, the ratio of SO_4^{2-}-S/S_{org} is also a valid indication of SO_2 deposition. Unlike Manninen (1995) we found a significant relationship between the SO_4^{2-}-S/S_{org} ratio and the total sulphur content in pine needles (SO_4^{2-}-S /S_{org} = –0.064 + 0.247 S; r^2 = 0.75).

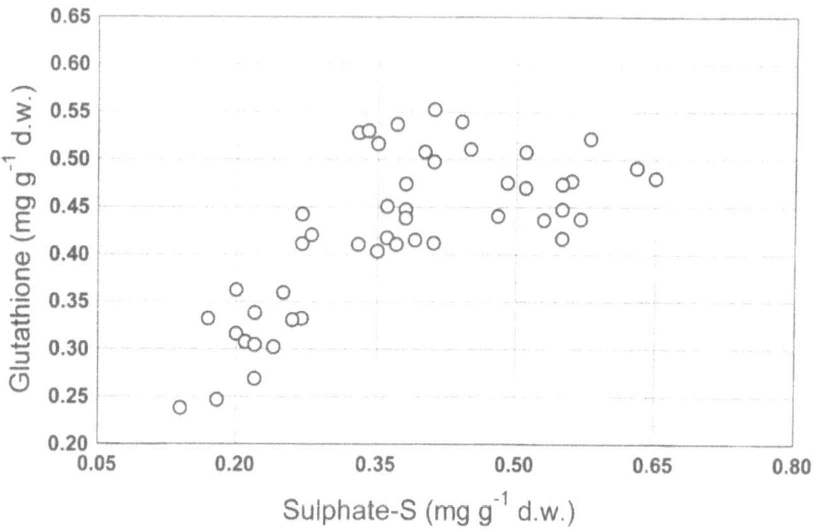

Figure 5. Correlation between the glutathione and sulphate sulphur (SO_4^{2-}-S) contents in current year's pine needles at the three field sites. The relationship shows a threshold at approximately 0.37 mg SO_4^{2-}-S g^{-1} d.w. as a result of assimilatory sulphate reduction (probable release of H_2S according to Kinderman *et al.*, 1995) or glutathione transport to sink organs

In contrast to SO_4^{2-}-S, the NPN fraction in the current year's needles (Table 3) is not clearly related to the changed nitrogen throughfall rates (Figure 1c,d) and to the annual mean ambient NO_x concentration at the three sites (Table 1). As is shown in Figure 6, it must be assumed that ammonium or nitrate uptake by the roots is influenced by nitrogen oxide in the ambient air of pine stands, because the ammonium contents in the humus layer and the throughfall rates of ammonium nitrogen show the same values at both sites (Table 5, Figure 1c). Thus it is not surprising that pine needles from Rösa with higher ambient NO_x concentration contain significantly higher levels of NPN and total nitrogen (Ende and Hüttl, these proceedings) than pine needles from Taura. Still, the relationship between NO_x absorption by the needles and total nitrogen uptake has not been elucidared, and at the time difficult to understand. In 1995 the NPN contents of the current year's needles at the same sites in Rösa were decreasing, although the mean annual NO_x concentrations have remained unchanged since 1993. Therefore it is also to conclude that the decreased NPN contents in needles from site Rösa were clearly affected by smaller ammonium contents in the humus layer (Table 5) as a result of decreased throughfall rates of ammonium and nitrate (Figure 1c,d). The uptake of ammonium in spruce and pine seedlings is reported to be considerably higher than that of nitrate (Ingestad, 1979; Lumme, 1994). Fumigation with nitrogen dioxide decreased nitrate uptake by the roots, when spruce seedlings were grown with nitrate plus ammonium (Muller *et al.*, 1996). Our field results are in agreement with these findings and investigations of other working groups in SANA (Ende and Hüttl; Schaaf and Hüttl, this volume). We found an increasing content of NPN in pine needles, when the ammonium content in the humus layer or the NO_x concentration was increasing at Rösa (Figure 6).

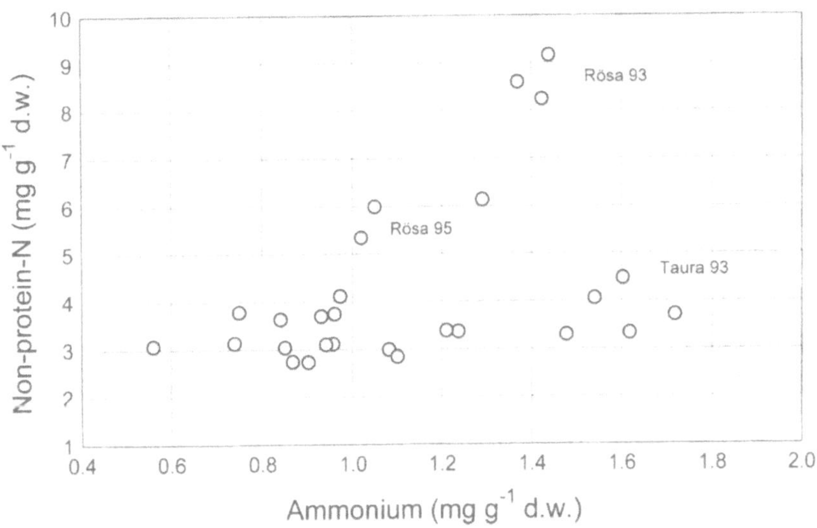

Figure 6. Correlation between the contents of non-protein nitrogen (NPN) in current year's pine needles and NaOH released ammonia in the humus layer at the three field sites. The NPN contents in plot at site Taura and Rösa differed as a result of significant differences in the mean NO_x concentration at both sites. In 1993, the NPN contents have been decreased at Rösa, because the throughfall rates of ammonium have been changed (Figure 1)

This increase is accompanied by accumulation of the amino acids glutamine and arginine in the NPN fraction (Table 3). In addition to glutamate, both nitrogen-containing compounds in needles from Rösa are dominant NPN forms, and these amino acids are usually found in Scots pines after nitrogen fertilization (Näsholm and Ericsson, 1990) or fumigation with atmospheric ammonia (Pérez-Soba *et al.*, 1995) as well as nitrogen dioxide (Nussbaum *et al.*, 1993). Therefore, the findings of spatial and temporal variations in NPN fractions in pine needles are indicative of accumulated glutamine and arginine pools in pine needles as a consequence of nitrogen uptake over the soil and/or air path. Glutamine as well as arginine play an important role as a storage form of nitrogen in Scots pines (Huhn and Schulz, 1996). Generally, amino acids can be cycled between leaves and roots to regulate the nitrogen demand (Touraine *et al.*, 1994; Aarnes *et al.*, 1995). Figure 7a,b and c show that glutamine, arginine and soluble protein are highly correlated with the NPN fraction in pine needles. It is assumed that at approximately 4 mg NPN g^{-1} d.w. there is a

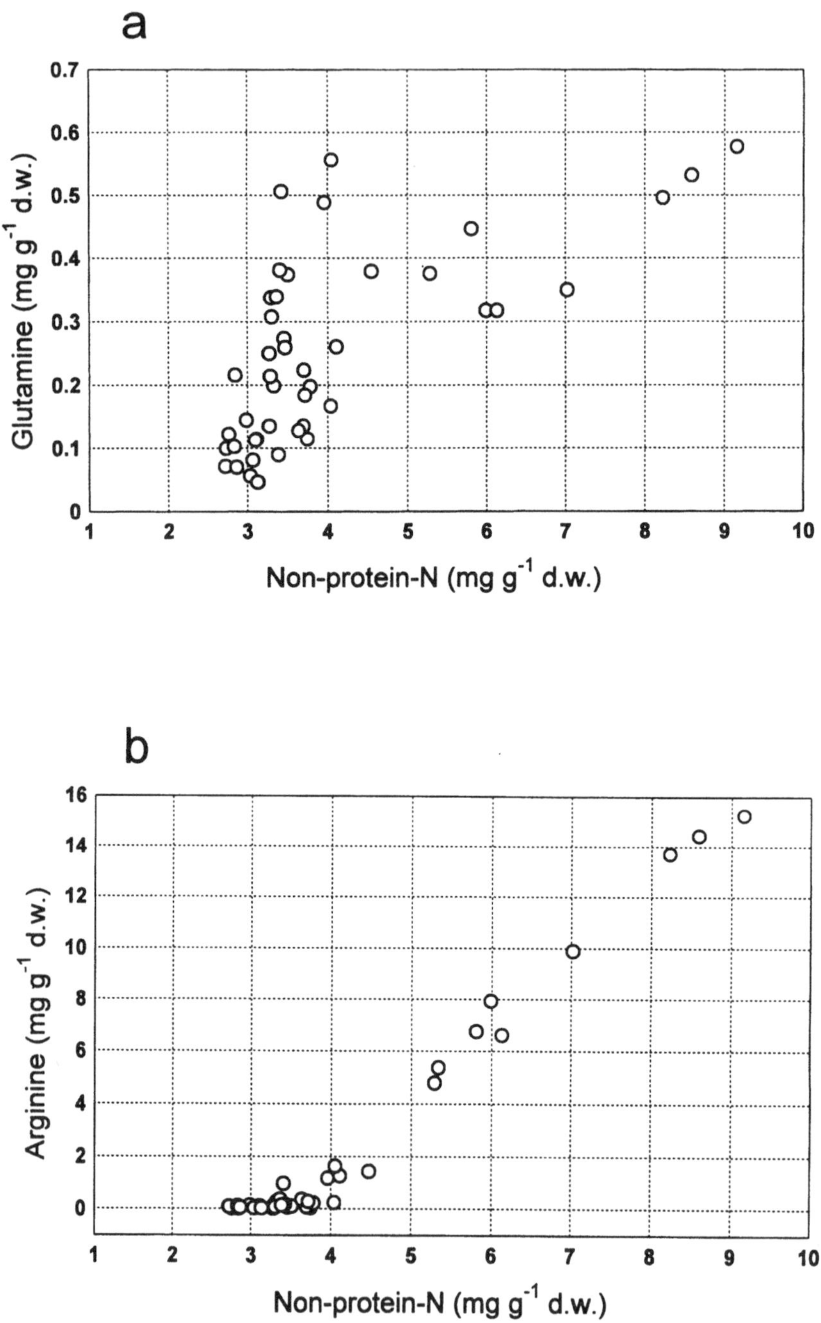
a
0.7
0.6
0.5
0.4
0.3
0.2
0.1
0
Glutamine (mg g-1 d.w.)
1
2
3
4
5
6
7
8
9
10
Non-protein-N (mg g-1 d.w.)
b
16
14
12
10
8
6
4
2
0
Arginine (mg g-1 d.w.)
1
2
3
4
5
6
7
8
9
10
Non-protein-N (mg g-1 d.w.)

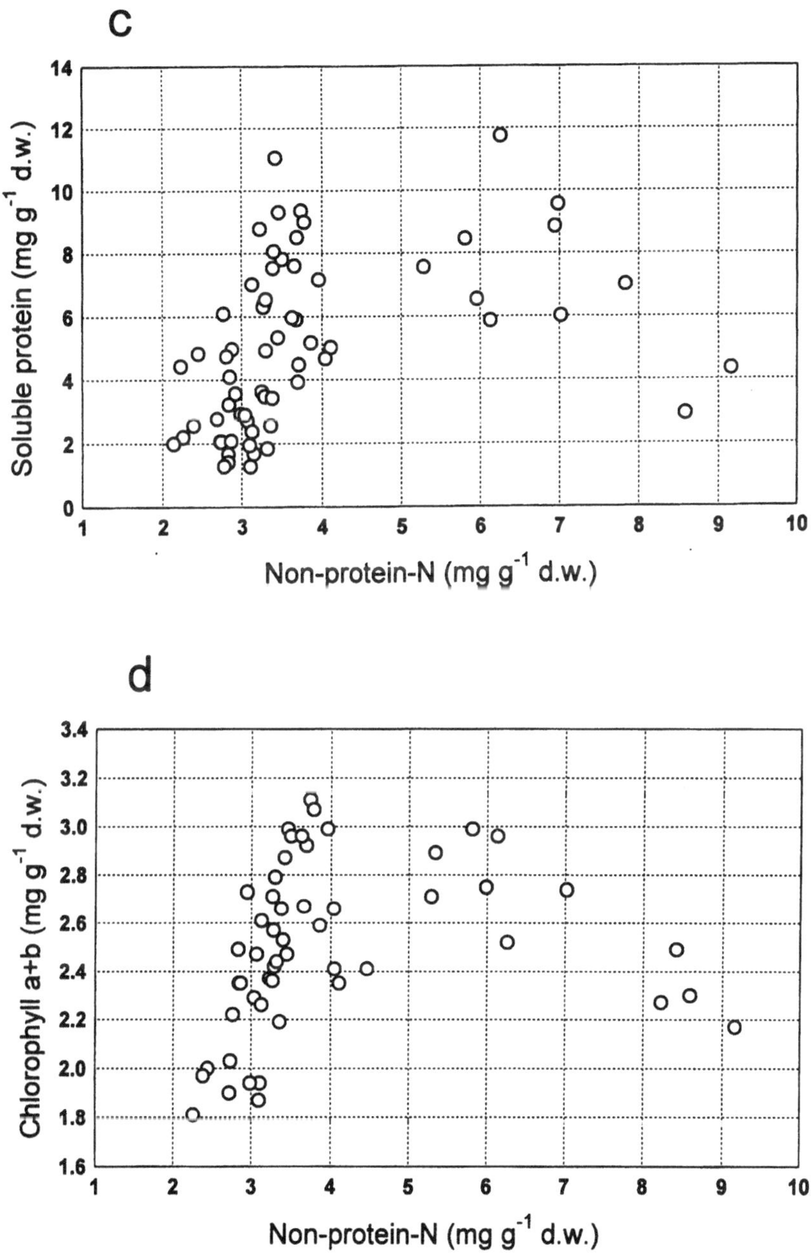

Figure 7. Correlation between the glutamine (a), arginine (b), soluble protein (c) as well as total chlorophyll (d) with non-protein nitrogen (NPN) contents in current year's pine needles from the three field sites. The relationships show a threshold at 4 mg NPN g^{-1} d.w.

threshold in the synthesis or storage of both amino acids as well as soluble protein. At this threshold the capacity for the synthesis of glutamine and the storage of soluble protein, too, is saturated (Figure 7c). In this case, the pine needles at Taura reached a total nitrogen content which did not exceed 18 mg g^{-1} d.w. (Table 3), a level considered optimal for growth (Heij and Schneider, 1991; Heinsdorf, 1993). Further nitrogen uptake via roots or needles must now be channelled or detoxified by storage in arginine, instead of being assimilated into the protein nitrogen fraction. Hofmann *et al.* (1990) found growth decrease in pine stands with needle concentrations at about 18 mg N g^{-1} d.w. The declined contents of soluble protein was possibly due to transport of amino acids in other parts of trees (Millard, 1988; Muller *et al.*, 1996). This observation is in agreement with the above described glutathione decrease by enhanced sulphate accumulation in pine needles. We suppose such storage and translocation mechanism for surplus nitrogen in pine needles at Rösa, where the NPN contents exceed the threshold during the whole field study. These results confirm the important role of glutamine and/or arginine in Scots pine as form of nitrogen transport and storage form (Pietilainen and Lähdesmäki, 1986; Nommik *et al.*, 1994). Arginine may be formed to remove excess ammonium which is not needed for growth (Näsholm and Ericsson, 1990). Beside arginine has the lowest C/N ratio of 1.5 among the amino acids (Kim *et al.*, 1987), and this has a favourable effect on nitrogen mobilization from this pool (Kozlowski *et al.*, 1991) as well as on the energy required for nitrogen assimilation. In addition, arginine responds directly to nitrogen fertilization (Pérez-Soba and De Visser, 1994) and seems to dominate the NPN pool in pine affected by high nitrogen deposition (Van Dijk and Roelofs, 1988; Kainulainen *et al.*, 1993). Another interpretation is that deficiencies in mineral nutrients, especially phosphorus deficiencies, promote arginine accumulation (Rabe and Lovatt, 1986; Houdik and Roelofs, 1993). In fact, we did not find drastically different phosphate contents in the current year's needles from all sites between 1993 and 1995. The mean phosphate content in the humus layer is significantly lower at Rösa due to an increased presence of calcium and a high pH value (Table 6). Also the contents of other mineral nutrients in pine needles, such as magnesium and potassium show no drastically site-dependent differences, but in contrast to 1991, the contents of both nutrient elements are significantly lower in the previous as well as in the current year's needles in 1995 (Table 7). Generally, there are no relationships between the needle contents of arginine and phosphate or magnesium and potassium. Therefore, we deduce the site-dependent ($N<T<R$) arginine contents in pine needles are a specific response to different nitrogen depositions and not a result of different contents of other mineral nutrients in pine needles. The needle contents of arginine and glutamine at site Rösa significantly decreased as a result of a decline in throughfall rates of ammonium and nitrate since 1993 (Figure 1c,d). In contrast to these findings, the protein nitrogen fraction shows no temporal changes (Table 3). This suggests that a maximum content of needle protein was reached in the trees at all sites in 1995.

Table 6. Contents (mean values ± standard errors) of water soluble phosphate (current years needles and humus layer) and calcium (0.5 *M* NH_4Cl extracts of humus layer) as well as pH values (0.1 *M* KCl) at the field sites. Significance was determined according to the Mann–Whitney U-test. Different letters (a, b) within the columns indicate significant site dependent differences ($p < 0.05$)

	Needle Phosphate (mg g^{-1} d.w.)		
Test site	1993	1994	1995
Neuglobsow	1.60 ± 0.05^{a}	1.88 ± 0.05^{a}	1.75 ± 0.04^{a}
Taura	1.56 ± 0.13^{a}	$1.69 \pm 0.10^{a,b}$	1.55 ± 0.06^{a}
Rösa	1.62 ± 0.03^{a}	1.42 ± 0.05^{b}	1.65 ± 0.06^{a}

	Humus Layer				
	Phosphate (mg g^{-1} d.w.)		Calcium (mg g^{-1} d.w.)	pH (KCl)	
Test site	1993	1995	1993	1993	1995
Neuglobsow	0.090 ± 0.010^{a}	0.075 ± 0.004^{a}	1.25 ± 0.17^{a}	2.92 ± 0.02^{a}	2.98 ± 0.05^{a}
Taura	0.040 ± 0.006^{b}	0.035 ± 0.005^{b}	2.59 ± 0.24^{b}	3.47 ± 0.05^{b}	3.46 ± 0.03^{b}
Rösa	0.005 ± 0.001^{c}	0.003 ± 0.001^{b}	3.49 ± 0.39^{b}	4.24 ± 0.09^{c}	4.56 ± 0.10^{c}

Table 7. Contents (mean values ± standard errors) of nutrients as well as various sulphur/ratios (on a gram atom basis) in current and previous year's pine needles. Significance was determined according to the Mann–Whitney U-test. Different large letters (A, B) within the columns indicate significant temporal differences at $p < 0.05$ (reference year 1991)

		Magnesium (mg g^{-1} d.w.)		Potassium (mg g^{-1} d.w.)	
Test site	Year	Current	Previous	Current	Previous
Neuglobsow	1991	0.71 ± 0.03^{A}	0.74 ± 0.05^{A}	4.06 ± 0.15^{A}	4.86 ± 0.15^{A}
	1995	0.76 ± 0.02^{A}	0.52 ± 0.02^{B}	4.37 ± 0.13^{A}	4.08 ± 0.17^{B}
Taura	1991	0.62 ± 0.03^{A}	0.78 ± 0.03^{A}	5.08 ± 0.27^{A}	5.19 ± 0.28^{A}
	1995	0.74 ± 0.03^{A}	0.58 ± 0.06^{B}	4.72 ± 0.10^{A}	4.09 ± 0.33^{B}
Rösa	1991	0.64 ± 0.03^{A}	0.77 ± 0.01^{A}	4.75 ± 0.11^{A}	4.40 ± 0.16^{A}
	1995	0.80 ± 0.04^{B}	0.55 ± 0.02^{B}	5.35 ± 0.14^{B}	4.06 ± 0.28^{A}

		SO_4^{2-}-S/NPN		S/N	
Test Site	Year	Current	Previous	Current	Previous
Neuglobsow	1991	0.050 ± 0.003^{A}	0.052 ± 0.006^{A}	0.039 ± 0.001^{A}	0.040 ± 0.001^{A}
	1995	0.031 ± 0.003^{B}	0.027 ± 0.003^{B}	0.034 ± 0.001^{B}	0.041 ± 0.001^{A}
Taura	1991	0.086 ± 0.004^{A}	0.071 ± 0.011^{A}	0.047 ± 0.002^{A}	0.049 ± 0.001^{A}
	1995	0.045 ± 0.002^{B}	0.056 ± 0.004^{A}	0.041 ± 0.001^{A}	0.047 ± 0.002^{A}
Rösa	1991	0.055 ± 0.010^{A}	0.047 ± 0.007^{A}	0.043 ± 0.002^{A}	0.040 ± 0.003^{A}
	1995	0.032 ± 0.004^{A}	0.023 ± 0.003^{B}	0.037 ± 0.001^{B}	0.035 ± 0.002^{A}

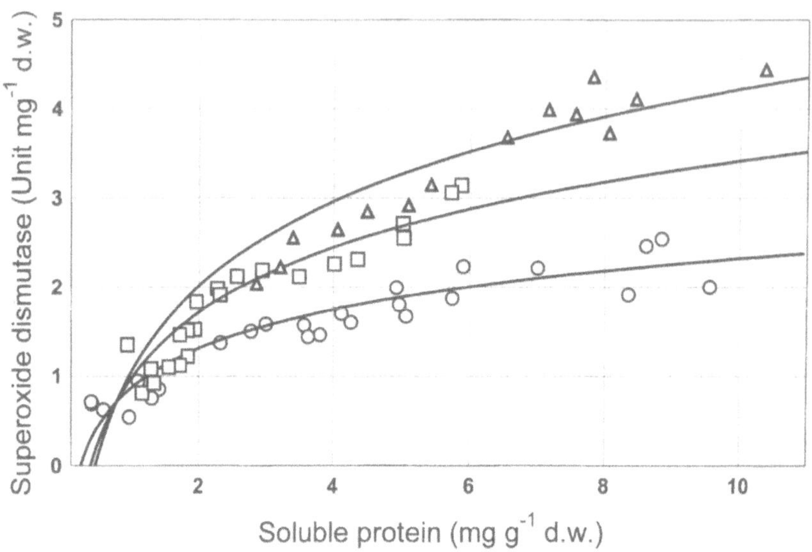

Figure 8. Correlations between superoxide dismutase (SOD) activities and soluble protein contents in current year's pine needles from the three field sites in 1991, 1993 and 1994. The relationships indicate significantly increased activities of SOD at the polluted sites Taura and Rösa (mean protein contents >5 mg g^{-1} d.w.) since 1991. In opposite to results reported by Polle *et al.* (1994), the temporal variations of SOD activities increase as well as the specific SOD activities, because the contents of soluble protein show only small temporal changes (Table 3)

Nevertheless, the needles from adult pine stands show damage symptoms at Rösa as well as in Taura and less in Neuglobsow, although the yearly means of SO_2 concentration show a clear decreasing trend at all sites (Table 1). We found the same effect in the young pine stands at all sites (Schulz *et al.*, 1995). There is no definite, plausible explanation for the still almost unchanged needle necroses in the investigation period between 1991 and 1995. On the other hand the total chlorophyll content continuously increased since 1991 (Table 4). For the present, the in Figure 7d proved relationship between chlorophyll and NPN contents in the current year's needles are consistent with results reported by Pérez-Soba *et al.* (1994). The relationship in Figure 7d shows that the chlorophyll content increased in proportional with the nitrogen uptake in pine needles, but at high NPN contents (>4 mg NPN g^{-1} d.w.), the chlorophyll content decreases with an increase in needle damage symptoms. This observed relationship is in agreement with temporal variations in chlorophyll contents in pine needles from Rösa. Since 1994 the chlorophyll content increases with slightly decreased NPN contents (Tables 3 and 4). On the other hand, it is well known that in SO_4^{2-}-S accumulated spruce needles the counterions potassium and magnesium are increased (Kaiser *et al.*, 1993). According to Slovik (1996), the stomatal uptake of SO_2 and NO_x induces an additional cation demand for sulphate neutralization in the cytoplasm and vacuoles. Therefore, the chronic SO_2 and NO_x pollution can cause mineral deficiency symptoms mainly in the previous year's pine needles, which may induce needle chloroses as well as necroses, when the supply of potassium and magnesium is insufficient in the humus layer or other soil horizons. Therefore, the decrease of magnesium and potassium in previous year's needles at all sites in 1995 should be seen in connection with the drastic reduction in alkaline dusts, mainly at the sites Taura and Rösa, and probably as an early indication of increased demand of both nutrient elements for actual neutralization processes, which can also explain the not significantly changed needle necroses in the investigation period between 1991 and 1995. Another way of interpretating the mainly observed tip necroses of the previous year's needles at Taura and Rösa is that the accumulation of SO_4^{2-}-S in pines is driven by phytotoxic oxygen radicals which cause oxidative stress (Asada and Takahashi, 1987) or induce imbalance in the sulphur/nitrogen status (Malcolm and Garforth, 1977). Shaw *et al.* (1993) suggest that the mean threshold SO_2 concentration for visible injury in Scots pine needles is in the range 6–8 nl l^{-1} (16–21 g m^{-3}). On the basis of this critical level of SO_2 for Scots pines it is not surprising that needles collected from Taura and Rösa indicated clear tip necroses. The mean SO_2 concentrations reached this threshold at both sites in 1995. According to results reported by Slovik (1996), SO_2 is the sole key cause of needle necrosis (Slovik *et al.*, 1992). We found enhanced contents of soluble protein associated with a substantial increase in the activity of superoxide dismutase, which indicates radical formation during SO_2 oxidation to sulphate (Figure 8). This could be a defence reaction of the organism to the multiple stress conditions (Table 1) and to the primary needle damage. On the other hand we observed an imbalance in the sulphur/nitrogen status of pine needles as a consequence of declined SO_4^{2-}-S, because NPN contents were unchanged. In addition, the previous year's needles show significantly decreased magnesium contents (Table 7). All these (bio)chemical markers indicate vitality loss and an early recognition of needle necroses in adult pine stands at Rösa, Taura and less in Neuglobsow, although the ambient SO_2 concentration at all sites has decreased since 1991.

In accordance with Slovik (1996), the decrease of magnesium and potassium in the previous year's needles at all sites is probably an indication of increased demand of both nutrient elements for actual neutralization processes. The ratios of (SO_4^{2} -S/NPN as well as S/N show spatial and temporal variations as a result of strongly decreased SO_4^{2-}-S contents in pine needles from all sites. In contrast with the normal S/N ratios (S/N = 0.030 described by Malcolm and Garforth, 1977), the values of current and previous needles indicate approx. 17% (Neuglobsow), 46% (Taura) and 21% (Rösa) excess SO_4^{2-}-S in 1995.

At all sites high concentrations of ozone were measured throughout the whole investigation period (Table 1). Therefore, it would seem also possible that especially ozone and probably organic air pollutants such as semivolatile organic compounds, might be indirectly involved in the irreversible damage to pine needles (Jung *et al.*, 1994; Plümacher and Schröder, 1994).

Based on the present results we conclude that SO_4^{2-}-S and NPN as well as the sulphur- and nitrogen-containing compounds glutathione, glutamine and arginine can be used as specific and highly sensitive biomarkers of sulphur or/ and nitrogen inputs over the air and soil path. In the case of needle necroses, however, SO_2 concentration and NO_x deposition should be considerered together with water supply (Lüttschwater *et al.*, this volume) and photosynthesis (Dudel *et al.*, this volume) which are important for modelling physiological processes (S. Bellmann *et al.*, this volume).

Acknowledgement

This study is supported by the Bundesministerium für Bildung, Wissenschaft, Forschung und Technologie (BMBF) Bonn, grant no.: P12/103764.

The authors thank the Umweltbundesamt and the Landesamt für Umweltschutz, Halle for providing data on air pollutants. We would like to thank Ms I. Geier, M. Herrmann, G. Pelzl and R. Rudloff for their excellent technical assistance.

References

Aarnes H, Eriksen AB, Southon TE. 1995. Metabolism of nitrate and ammonium in seedlings of Norway Spruce (*Picea abies*) measured by in vivo ^{14}N and ^{15}N NMR spectroscopy. Physiol Plant. 94, 384–390.

Asada K, Takahashi M. 1987. Production and scavenging of active oxygen in photosynthesis. In: Photoinhibition. Eds. D. Kyle, C. Osmond. pp. 222–287. Elsevier Scientific Publishers BV, Amsterdam.

Bradford MM. 1976. A rapid and sensitive method for the quantification of microgram quantities of protein utilizing the principle of protein-dye binding. Anal Biochem. 72, 248–254.

De Kok L. 1990. Sulfur metabolism in plants exposed to atmospheric sulfur. In: Sulfur Nutrition and Sulfur Assimilation. Eds. H. Rennenberg, C. Brunold, L. De Kok, I. Stulen. pp 111–131. SPB Academic Publishing BV, The Hague.

Dijkshoorn W, Van Wijk AL. 1967. The sulphur requirements of plants as evidenced by the sulphur-nitrogen ratio in the organic matter. A review of published data. Plant Soil. 26, 129–157.

Gasch G, Grünhage L, Jäger H-J, Wentzel K-F. 1988. Das Verhältnis der Schwefelfraktionen in Fichtennadeln als Indikator für Immissionsbelastungen durch Schwefeldioxid. Angew Bot. 62, 73–84.

Härtling S, Schulz H. 1995. Ascorbat- und Glutathiongehalt in verschiedenartig schadstoffbeeinflußten Nadeln von *Pinus sylvestris* L. Forstw Cbl. 114, 40–49.

Heij GT, Schneider T. 1991. Acidification research in The Netherlands. Environmental Science Series nr 46. pp 51–137. Elsevier, Amsterdam.

Heinsdorf D. 1993. The role of nitrogen in declining Scots pine forests (*Pinus sylvestris*) in the lowland of East Germany. Water Air Soil Pollut. 69, 21–35.

Herschbach C, Rennenberg H. 1994. Influence of glutathione (GSH) on net uptake of sulfate and sulfate transport in tobacco plants. J Exp Bot. 45, 1069–1076.

Hofmann G, Heinsdorf D, Krauß H-H. 1990. Wirkung atmogener Stickstoffeinträge auf Produktivität und Stabilität von Kiefern-Forstökosystemen. Beitr Forstwirtsch. 24, 59–73.

Houdijk ALFM, Roelofs JGM. 1993. The effects of atmospheric nitrogen deposition and soil chemistry on the nutritional status of *Pseudotsuga menziesii*, *Pinus nigra* and *Pinus sylvestris*. Environ Pollut. 80, 79–84.

Huhn G, Schulz H, Stärk H-J, Tölle R, Schüürmann G. Evaluation of regional heavy metal deposition by multivariate analysis of element contents in pine tree barks. Water Air Soil Pollut. 84, 367–383.

Huhn G, Schulz H. 1996. Contents of free amino acids in Scots pine needles from field sites with different levels of nitrogen deposition. New Phytol. In press.

Huttunen S, Laine K, Torvela H. 1985. Seasonal sulphur contents of pine needles as indices of air pollution. Ann Bot Fennici. 22, 343–359.

Ingestad T. 1979. Mineral nutrient requirements of *Pinus sylvestris* and *Picea abies* seedlings. Physiol Plant. 45, 373–380.

Innes JL. 1995. Influence of air pollution on the foliar nutrition of conifers in Great Britain. Environ Pollut. 88, 183–192.

Jung K, Rolle W, Schlee D, Tintemann H, Gnauk T, Schüürmann G. 1994. Ozone effects on nitrogen incorporation and superoxide dismutase activity in spruce seedlings (*Picea abies* L.) New Phytol. 128, 505–508.

Kaiser W, Dittrich A, Heber U. 1993. Sulfate concentrations in Norway spruce needles in relation to atmospheric SO_2: a comparison of trees from various forests in Germany with trees fumigated with SO_2 in growth chambers. Tree Physiol. 12, 1–13.

Kainulainen P, Staka H, Mustaniemi A, Holopainen JK, Oksanen J, 1993, Conifer aphids in air-polluted environment. II. Host plant quality. Environ Pollut. 80, 193–200.

Kelly J, Lambert MJ. 1972. The relationship between sulfur and nitrogen in the foliage of *Pinus radiata*. Plant Soil. 37, 395–407.

Kim YT, Glerum C, Stoddart J, Colombo SJ. 1987. Effect of fertilization and free amino acid concentrations in black spruce and jack pine containerized seedlings. Can J For Res. 17, 27–30.

Kindermann G, Hüve K, Slovik S, Lux H, Rennenberg H. 1995. Emission of hydrogen sulfide by twigs of conifers – a comparison of Norway spruce (*Picea abies* (L.) Karst), Scots pine (*Pinus sylvestris* L.) and Blue spruce (*Picea pungens* Engelm). Plant Soil. 169, 421–423.

Kozlowsky TT, Kramer PJ, Pallardy SG. 1991. The physiological ecology of woody plants. Academic Press, Inc., San Diego, California, USA.

Legge AH, Bogner JC, Krupa SV. 1988. Foliar sulphur dpecies in pine: a new indicator of a forest ecosystem under air pollution stress. Environ Pollut. 55, 15–27.

Lichtenthaler HK, Wellburn AR. 1983. Determinations of total carotinoids and chlorophylls a and b of leaf extracts in different solvents. Biochem Soc Trans. 603, 591–592.

Lumme I. 1994. Nitrogen uptake of Norway spruce (*Picea abies* Karst) seedlings from simulated wet deposition. Forest Ecol Manage. 63, 87–96.

Malcolm DC, Garforth MF. 1977. The sulphur:nitrogen ratio of conifer foliage in relation to atmospheric pollution with sulphur dioxide. Plant Soil. 47, 89–102.

Manninen S. 1995. Assessing the critical level of SO_2 for Scots pine (*Pinus sylvestris* L.) in northern Europe on the basis of needle sulphur fractions, sulphur/nitrogen ratios and needle damage. Dissertation, University of Oulu. Oulu.

Meng FR, Bourque CP-A, Belczewski RF, Whitney NJ, Arp PA. 1995. Foliage response of spruce trees to long-term low-grade sulfur dioxide deposition. Environ Pollut. 90, 143–152.

Millard P. 1988. The accumulation and storage of nitrogen by herbaceous plants. Plant Cell Environ. 11, 1–8.

Muller B, Touraine B, Rennenberg H. 1996. Interaction between atmospheric and pedospheric nitrogen nutrition in spruce (*Picea abies* L. Karst.) seedlings. Plant Cell Environ. In press.

Näsholm T, Ericsson A. 1990. Seasonal changes in amino acids, protein and total nitrogen in needles of fertilized Scots pine trees. Tree Physiol. 6, 267–281.

Nômmik H, Pluth DJ, Larsson K, Mahendrappa MK. 1994. Isotopic fractionation accompanying fertilizer nitrogen transformations in soil and trees of a Scots pine ecosystem. Plant Soil. 158, 169–182

Nussbaum S, Von Ballmoos P, Gfeller H et al. 1993. Incorporation of atmospheric $^{15}NO_2$-nitrogen into free amino acids by Norway spruce *Picea abies* (L.) Karst. Oecologia. 94, 408–414.
Pérez-Soba M, Stulen I, Van der Erden LJM. 1994. Effect of atmospheric ammonia on the nitrogen metabolism of Scots pine (*Pinus sylvestris*) needles. Physiol Plant. 90, 629–636.
Pérez-Soba M, De Visser PHB. 1994. Nitrogen metabolism of Douglas fir and Scots pine as affected by optimal nutrition and water supply under conditions of relativly high atmospheric nitrogen deposition. Trees. 9, 19–25.
P³rez-Soba M, Dueck TA, Puppi G, Kuiper PJD. 1995. Interactions of elevated CO_2, NH_3 and O_3 on mycorrhizal infections, gas exchange and N metabolism in saplings of Scots pine. 176, 107–116.
Pietiläinen P, Lähdesmäki P. 1986. Free amino acid and protein levels, and g-glutamyltransferase activity in *Pinus sylvestris* apical buds and shoots during the growing season. Scand J For Res. 1, 387–395.
Plümacher J, Schröder P. 1994. Accumulation and fate of C_1/C_2-chlorocarbons and trichloroacetic acid in spruce needles from an Austrian mountain site. Chemosphere. 29, 2467–2476.
Polle A, Eiblmeier M, Rennenberg H. 1994. Sulphate and antioxidants in needles of Scots pine (*Pinus sylvestris* L.) from three SO_2-polluted field sites in eastern Germany. New Phytol. 127, 571–577.
Rabe E, Lovatt CJ. 1986. Increased argenine biosynthesis during phosphorus deficiency. A response to the increased ammonia content of leaves. Plant Physiol. 81, 774–779.
Rennenberg H, Polle A. 1994. Metabolic consequences of atmospheric sulfur influx into plants. In: Plant Responses to the Gaseous Environment. Molecular, Metabolic and Physiological Aspects. Eds. RG Alscher, AR Wellburn. pp 165–180. Chapman and Hall, University Press, Cambridge.
Rennenberg H, Herschbach C. 1995. Sulfur nutrition of trees: a comparison of spruce (*Picea abies*) and beech (*Fagus sylvatica* L). Z Pflanz Bodenk. 158, 513–517.
Schaaf W, Hüttl RF, Weisdorfer M, Blechschmidt R, Schütze J. 1996. Auswirkungen sich zeitlich ändernder Schadstoffdepositionen auf Stofftransport und -umsetzung im Boden. In: SANA Wissenschaftliches Begleitprogramm zur Sanierung der Atmosphäre über den neuen Bundeländern, Abschlußbericht Teil IV. Ed. Fraunhofer-Institut für Atmosphärische Umweltforschung (IFU), Garmisch-Partenkirchen.
Schulz H. 1981. Enzymatisch-ökologische Untersuchungen an einigen Bodenpflanzen eines naturnahen Berg-Fichtenwaldes. Methodik der Probenahme, Probenaufbereitung und Enzymextraktion. Flora. 169, 135–149.
Schulz H. 1983. Aktivitätsbestimmung der Superoxid-Dismutase-isoenzyme von *Pinus sylvestris* Nadeln im Polyacrylamidgel. Biochem Physiol Pflanzen. 178, 249–261.
Schulz H, Huhn G, Jung K, Härtling S, Schüürmann G. 1995. Biochemical responses in needles of Scots pine (*Pinus sylvestris*) from air polluted field sites in eastern Germany. In: Air Pollution III. Volume 4: Observation and Simulation of Air Pollution: Results from SANA and EUMAC. Eds. A Ebel, N Moussiopoulos. pp 33–42. Computational Mechanics Publications, Southhampton, Boston.
Schulz H, Weidner M, Baur M et al. 1996. Recognition of air pollution stress on Norway spruce (*Picea abies* L.) on the basis of multivariate analysis of biochemical parameters: a model field study. Angew Bot. 70, 19–27.
Schulz H, Huhn G, Niehus B, Liebergeld G, Schüürmann G. 1997. Determination of throughfall rates on the basis of pine barks loads: results of a pilot field study. Air Waste Manage Assoc. 47, 510–516.
Schulze ED. 1989. Air pollution and forest decline in a spruce (*Picea abies*) forest. Science. 244, 776–783.
Shaw PJA, Holland MR, Darrall NM, McLeod AR. 1993. The occurrence of SO_2-related foliar symptoms on Scots pine (*Pinus sylvestris* L.) in an open-air forest fumigation experiment. New Phytol. 123, 143–152.
Slovik S, Heber U, Kaiser WM, Kindermann G, Körner C. 1992. Quantifizierung der physiologischen Kausalkette von SO_2-Immissionsschäden. (II) Ableitung von SO_2-Immis-sionsgrenzwerten für chronische Schäden an Fichten. Allg Forst Zeitschrift. 17, 913–920.

Slovik S, Siegmund A, Kindermann G, Riebeling R and Balázs Á. 1995, Stomatal SO_2 uptake and sulfate accumulation in needles of Norway spruce stands (*Picea abies*) in Central Europe. Plant Soil, 168–169, 405–419.

Slovik S. 1996. Chronic SO_2- and NO_x-pollution interferes with the K^+ and Mg^{2+} budget of Norway spruce trees. J Plant Physiol. 148, 276–286.

Smith WH. 1990. Air pollution and forests. Interaction between air contaminants and forest ecosystems. Springer-Verlag New York, Berlin, Heidelberg. 585 p.

Touraine B, Clarkson DT, Muller B. 1994. Regulation of NO_3^- uptake at the whole plant level. In: A Whole Plant Perspective on Carbon-nitrogen Interactions. Eds. J Roy, E. Garnie. pp 11–30. SPB Academic Publishing, The Hague.

Van Dijk HFG, Roelofs JGM. 1988. Effects of excessive ammonium deposition on the nutritional status and condition of pine needles. Physiol Plant. 73, 494–501.

Vannier C, Didon-Lescot J-F, Lelong F, Guillet B. 1993. Distribution of sulphur forms in soils from beech and spruce forests of Mont Lozére (France). Plant Soil. 154, 197–209.

5 Photosynthetic capacity, respiration and water use efficiency in Scots pine stands

G.E. DUDEL, U. FEDERBUSCH and A. SOLGER

1. Introduction

Air pollutants (O_3, SO_2, NO_x) may affect the metabolic activity of leaves or needles, and organelles of plants, directly or indirectly by soil acidification and nutrient imbalance in particular in conifers: Many of publications refer to this phenomenon (e.g. Ulrich and Matzner, 1983; Ulrich *et al.*, 1984; Lange *et al.*, 1985; Schulze *et al.*, 1989; Führer *et al.*, 1993; Heber *et al.*, 1994; other quotations ibid.). This were mainly short-term fumigation experiments, which could be used to detect causal biochemical mechanisms (Norby *et al.*, 1989 regarding NO_2; Pfanz and Heber, 1986 regarding SO_2; Wallin and Skärby, 1992 regarding ozone; Heber *et al.*, 1994 regarding interactions). In contrast to this, long-term investigations focusing on the pollutant concentrations as they normally occur in the field were conducted in rare cases only (Schulze *et al.*, 1987; Keller and Hässler, 1986; Führer *et al.*, 1993).

The functional effects of air pollutants on photosynthetic performance, transpiration, and biomass production have been ascertained primarily in young plants, grasses, herbs, and in individual exposure experiments. Contrary to this, the responses of the main tree species (Scots pine) stocking on comparatively dry, nutrient-poor, glacially coined sites in the east of Germany, and partly growing in the surroundings of large pollution sources, has not yet been the object of an entire ecosystem approach (Lux, 1965; Lux and Stein, 1977; Gluch, 1988; Heinsdorf, 1993).

In general only few assessments of the photosynthetic performance and transpiration of mature coniferous stands are available (e.g. Schulze *et al.*, 1987; Pfanz and Beyschlag, 1993 for spruce) and of Scots pine in particular (Künstle and Mitscherlich, 1975, 1976; Troeng and Lindner, 1982; Umweltbundesamt, 1990; Faensen-Thiebes *et al.*, 1989, 1991). They are completely absent as far as the drastically changing air pollution loads are concerned which were present in the East of Germany during the past few years. No attempt has been made so far to compare stands influenced by critical air pollution in the past with a control area of similar climatic conditions but with low air pollution past and present. Moreover, a quantification of metabolic capacities, varying with air pollution loads or site conditions, as related to several scales (needle, twig, crown section, stand) is missing. Also, the influence

R.F. Hüttl and K. Bellmann, Changes of Atmospheric Chemistry and Effects on Forest Ecosystems, 65–95.

of seasonal conditions on the physiological adaptation of gas exchange under different pollutant loads has not been the topic of investigation so far.

The SANA project (rehabilitation of the atmosphere above the New Länder of the Federal Republic of Germany) supplies complex information on the quantity and dynamics of air pollution along a large-scale pollution gradient and, especially, the soil-related site factors (Weisdoerfer *et al.*, this volume). Thus, conclusions about ecosystems become possible, which normally, because of their spatial and temporal dimensions are unsuitable for a short time experimental approach. Here, the attempt has been made to find causal explanations for air pollution effects, using a model-aided quantification of the stand-related gas exchange performance. They cover the interpretation of phenomena in the soil and nutrient or metabolic systems (Schulz *et al.*, this volume) as well as retrospective information (increment analyses: Wenk *et al.*, this volume). In particular, in pine forests of the north-east German lowland, that are characterized by a different air pollution history (Leipzig-Halle-Bitterfeld area: Taura and Rösa; as well as Neuglobsow in the north-east of Berlin) and drastically declining loads, the following topics have been dealt with:

- Is there still a negative impact on primary production (net photosynthesis, respiration) due to the actual air pollution as well as that of the recent past, or are there only indirect effects involved (e.g. soil related processes)?
- Are there any direct effects on photosynthesis performance, which result from air pollution history over many years, still ascertainable today under declining, qualitatively changing loads?
- Is there a possible compensation of the sulfur and acidic pollution-caused inhibition of photosynthesis by the input of alkaline dusts and by N eutrophication?
- Has there been a stronger impact of air pollution on stomata control or water relations than on the biochemical machinery of the photosynthesis apparatus? Is the water use efficiency changing?
- By what other conditions (e.g. temperature and seasonal adaptation, soil dryness) are the effects of air pollution superimposed?

To answer these questions the time-dependent gas exchange variation has been measured in different crown sections and needle age classes under

- regular site conditions (on-line measurement with simulation of actual temperature and humidity in their diurnal course),
- standardized light and moisture conditions (artificial illumination with PAR to assess light saturation curves immediately in the stands, optimal air humidity related to the seasonal temperature),

– an additional CO_2 supply to overcome the stomatal conductivity resistance.

The gas exchange data were condensed using models to discriminate air pollution- and site-caused differences from variation of moisture and light. In this connection, not only the photosynthetic performance under stand conditions varying over time with irradiation strength, humidity and CO_2 concentration (A_{350}, A_{2500}: e.g. Führer *et al.*, 1993), but also the total sets of parameters of complete light saturation curves under defined conditions (CO_2-assimilation vs. light intensity: A-I curves) at different seasonal conditions of adaptation were calculated ($P_{max350}(T)$ and $P_{max2500}(T)$). Furthermore, the integrated diurnal CO_2-exchange rates and the parameters on the basis of the A-I-curves related to temperature were used for the validation and sensitivity analysis of the Stand Carbon Balance Model 'FORSANA' (Grote and Bellmann, in this volume).

2. Materials and methods

2.1. *Areas of investigation*

Site parameters of the edaphic, climatic and phytosociological conditions essential to this paper are given in Table 1. Further details can be found in Hüttl and Bellmann; Schulz et al.; Lüttschwager et al., this volume.

2.2. *Experimental approaches*

Two different approaches were pursued (cf. section 3.1):

- diurnal courses of gas exchange:
  to determine in-situ gas exchange rates (net photosynthesis and transpiration) on one twig, in various sections of the crowns or trees, respectively, during a complete diurnal course each under the respective site conditions (immediate modulation with a short time lag of ambient temperature and humidity inside the PAR-permeable gas exchange cuvettes).

- capacity measurements (light-saturation-curves):
  to determine briefly a comparative host of gas exchange rates within the different crown sections and various trees of a stand under steady state conditions (setting time ca. 30 minutes) over the whole range of illumination conditions and temperature ranges that are really possible for gross photosynthesis in Central Europe (PAR: from darkness 0 up to max. 1700 $\mu mol*m^{-2}*s^{-1}$, temperature: 5 to 40°C) in the respective humidity (30 up to 80 percentage humidity). The twigs in the chamber were changed 3–4 times a day.

Table 1. Characteristics of the experimental sites

Location (Abbreviation) Load characteristics (Abbrv.)	Rösa (R) High (HL)	Taura (T) Moderate (ML)	Neuglobsow (N) Low (LL)
Deposition (kg/ha/a)[1]			
Calcium	13.31	13.88	8.05
Sulfate	14.01	16.17	8.45
Fertilization	1970–1972 1076–1979, 1983–1985 N-fertilization; 100 kg N $ha^{-1}a^{-1}$	Application of urea (quantity unknown)	1973/74 N-fertilization (quantity unknown)
Soil pH[2] (OF horizon)	4.3	3.4	3.0
Calcium mg/l (OF horizon)	5.5	1.9	1.4
Climate			
Annual precipitation (mm)	566	565	586
Mean annual temperature (°C)	8.87	8.87	8.18
Ecosystem Typ[3]	Calamagrostio Cultopinetum sylvestris	Avenello-Cultopinetum sylvestris	Myrtillo-Cultopinetum sylvestris
Stand data[3]			
Age (years)	61	45	65
Number of stems (1/ha)	935	853	1043
Basal area (m^2/ha)	34	28	36
HBD (cm)	20.7	20.6	21.0
Mean tree height (m)	16.0	18.0	20.1

[1]data from Rust et al. 1995 (measured period August 1993 to July 1994)
[2]Schaaf et al. 1995,
[3]after Hofmann, 1992

First, the capacity measurements were conducted at normal atmospheric CO_2 partial pressure ($c_a = 350$ ppm), as well as with increased CO_2 concentration ($c_a = 2500$ ppm) in the cuvette, in order to bridge the stomatal resistance. Approximately 3100 individual twig-related gas exchange measurements were made in each stand ($N_{total} = 9700$).

2.3. *Plant material and periods of investigation*

Within the scope of SANA project trial plots with 15 sample trees were selected. Crown structure and size of the single trees in the three trial stands are comparable related to light extinction (Figure 1). However, there are differences in stand density and age (Table 1). Five to seven trees were randomly selected from the parent population for recording the photosynthesis-light characteristics (A-I-curves) and complete diurnal courses. The same trees were repeatedly referred to (Table 2).

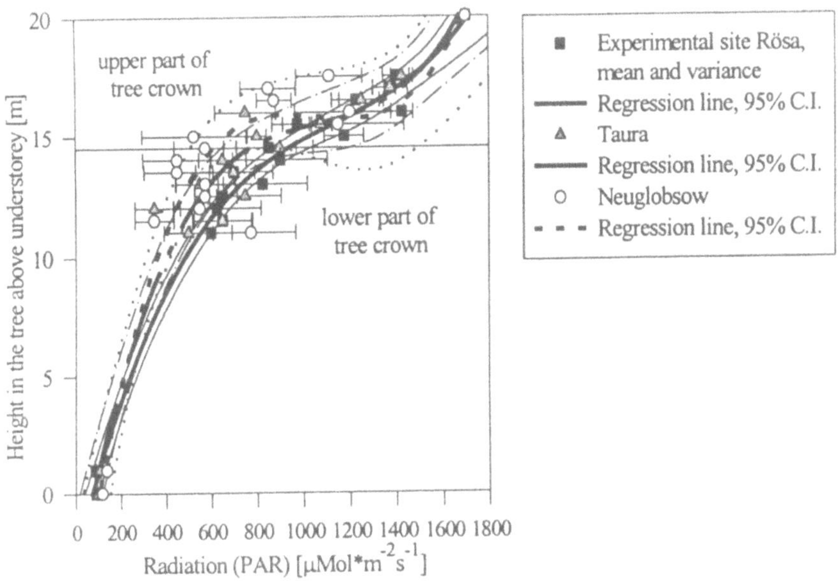

Figure 1. Light (PAR) attenuation in the Scots pine stands investigated (light extinction curves)

Table 2. Periods of investigation

Year	Jan	Feb	Mar	Apr	May	June	July	Aug	Sept	Oct	Nov	Dec
1993					***					***		
1994					***		***		***			
1995		**		***			***					

***investigations in all stands

**investigations in Rösa, Taura

The most essential phases of seasonal growth cycle, environmental load and adaptation conditions were recorded during a total of 8 selected periods of investigation from May of, 1993 to August of, 1995:

- foliation and shoot elongation (May to June)
- maximum gross photosynthesis and randomly appearing water stress (July/August)
- senescence and needle shedding (mid-August/September)
- accumulation (soluble sugar and fats) and adaptation to cold weather (October to December)

- dormant season (January/February)
- remobilization and starch accumulation (from mid-March until leafing)

These phases were defined on the basis of information on the production reallocation and mobilization of reserve materials in *Pinus sylvestris* L. (Fischer and Holl, 1991).

2.4. *Gas exchange and radiation measurements as well as weather conditions*

During each monitoring period photosynthesis-light characteristics as well as transpiration performance within a radiation intensity ranging from 0 to 1700 $\mu mol*m^{-2}*s^{-1}$ were recorded under optimal water supply from separated, intact single branches of first order, subdivided into first to third needle age classes. The differention between light and shade crown, i.e. upper and lower crown sections, took place by morphological criteria as well as by attenuation of light (criterion for distinguishing: turning point of the light attenuation curve in Figure 1). Corresponding to the climatic adaptation, a 30–80% humidity was measured in a temperature range of 5 to 40°C. Further, diurnal courses of the CO_2 and the H_2O fluxes in the upper and lower crown parts were continously recorded in one or two trees. The gas exchange was determined, using a cuvette system CMS 400/P (Fa. Walz, Effeltrich, Germany). In the course of measurements the gas fluxes, the temperature of cuvettes and needles, photon flux density as well as the concentration of CO_2, and water vapour inside and outside the cuvette were registered. The gas exchange rates were calculated from these values, using the methods according to Von Caemmerer and Farquhar (1981) (Software Diagas 2.02-CMS400c/Diagas 2.14-CMS4004P, Fa. WALZ, Effeltrich, Germany).

Water potential determination

The water potential (ψ) was recorded in five needles of every needle age class. The sampled twigs were immediately measured after they had been separated from the tree, using a pressure chamber (PMS chamber model 1000, Fa. PMS Instrument Co., Oregon, USA).

2.5. *Biomass estimation*

The biomass was estimated as follows:

- Fresh weight: The fresh weight of the single needle samples was determined immediately after the gas exchange measurement in the field using a precision balance.

- Dry weight: Following a short transportation from the trial plots, desiccation was performed in a vacuum drying oven at 30 to 40°C for a period of at least 16 hours until constant weight was reached.

- Needle surfaces: The projected surface area of the individual needles was determined after their removal from the cooling box, applying an areameter (video image analysis system DIAS II, Delta-T Devices Ltd., Cambridge, England). A magnification was chosen, through which 10 to 15 needles became visible on the screen and could be measured. The allometry constant of 2.57 determined for *Pinus sylvestris* L was used to calculate the total needle surface from the projected needle surface, according to recommendations by Küppers *et al.* (1985); Roberts *et al.* (1982), as well as by Perterer and Körner (1990).

2.6. *Modeling*

The physiological performances may be described using saturation kinetics on the basis of exponential or optimum functions, e.g. the Michaelis-Menten kinetics (Baly, 1935) . In general, has been an attempt has been made to use an empirical model that in a process-oriented view permits a causal interpretation of parameters.

Dependence of CO_2 gas exchange on light intensity
From a rather broad spectrum of models (overview see Keller, 1989), which describe the photosynthesis-light (PAR) dependence, we selected the model of Bannister (1974):

$$P = \frac{P_{max} I}{(K_I^m + I^m)^{1/m}} \qquad \text{[eq.1]}$$

The exponent m was renounced, i.e. the form parameter m=1 was adhered to further on, and the additional light compensation parameter I_o was introduced, so that the applied model can be expressed as follows:

$$P = \frac{P_{max}(I - I_o)}{K_I + (I - I_o)} \qquad \text{[eq.2]}$$

Either directly or by parameter combination it is appropriate for the assessment of performances or conditions based upon gas exchange rates measured in the field (Figure 4).

Subsequent to extension, it contains the following causally interpretable parameters:

- maximum photosynthesis capacity at light saturation P_{max}
- light affinity of photosynthesis apparatus: slope α, relating to parameter K_I
- determination of the light compensation point $I_o + K_I$
- computation of the estimated photorespiration from the model parameters

An inhibition parameter was not needed. Even at the highest radiation densities (1700 $\mu mol*m^{-2}*s^{-1}$), a slight inhibition in photosynthesis efficiency was observed in only three out of 9700 tested cases (light inhibition).

The parameters may be estimated by non-linear regression. In addition to the values of standard deviation and the variation coefficients of parameter estimation calculated by the iteration program used (SIGMAPLOT, Fa. Jandel Sci.) asymptotic ranges were determined via programming a suitable routine after Seber and Wild (1989) (data not presented).

Confidence limits (with unknown probability of covering the true value) for approximative photorespiration were computed as follows: Photorespiration rate is no immediate state variable of the model. However, it can be readily calculated, when $I=0$, provided all parameters are involved. From the parameters comparable confidence intervals for the photorespiration rates were determined by calculating the maximum and minimum of $P(I=0)$ using all further parameters within their calculated confidence limits.

Variances

Variances of the light-saturation-data (A-I curves) show that only the CO_2 assimilation data possess variance which is not negligible. Under approximately defined external conditions of light and humidity these variables show a variance level of about 3% up to max. 5%, provided that upper/lower zones of the crowns or individual trees are dealt with separately. It may be regarded as 'input' variance. The parameters of the photosynthesis-light model per A-I curve ($N=1200$), which were estimated based upon such data, attain the variance regions given in Table 3.

Table 3. Range of variance of model parameters

Variance in %	From	To
P_{max}	6.5	13.0
I_o	29.7	54.4
K_I	14.2	28.6
Respiration	31.5	44.0

Temperature dependence

Additional to its dependence on light, the gas exchange takes place according to an optimum function in regard to temperature. The photosynthesis-temperature relationship is required for a formalization to compare directly photosynthesis capacity ($P_{max350}(T)$ and $P_{max2500}(T)$) with respect to the optimal temperature of photosynthesis. A symmetrical relationship $P_{max}(T)$ was suggested by Webb (1974). More realistically, however, for the dependence of the net primary production (CO_2 assimilation) on the temperature is nonsymmetrical a function (e.g. Lawton, 1993). Here, for curve fitting, according to recommendations by Ratkowski (1990) a simple maximum model was used:

$$P_{max}(T) = \frac{T}{\alpha + \beta T + \gamma \sqrt{T}} \quad \text{[eq.3]}$$

2.7. *Test statistics*

Statistical significance was tested, using nonparametric tests (u-test: Wilcoxon, Mann and Whitney; H-test: Kruskal/Wallis, Sign test) and ANOVA methods according to Sachs (1992).

3. Results

3.1. *Problems resulting from experimental approach and their solutions*

Several control variables vary simultaneously in the multi-factorial relationships of a Scots pine ecosystem. Mainly, the effect of radiation and weather conditions, the state variables undergoing detectable changes during the diurnal course, cannot readily be distinguished from air pollution effects and/or soil-related site factors. The attempt was made to solve this problem by multiple (linear) regression models (e.g. Faensen-Thiebes *et al.*, 1991, other quotations ibid.: correlation of gas exchange data with temperature and humidity values of the ambient milieu). A solution to this problem seems to be possible, if photosynthetic capacities are investigated with respect to light and/or CO_2 saturation of photosynthesis. This needs, however, considerable experimental interference (e.g. Führer *et al.*, 1993, other quotations ibid.): The light intensity was adjusted only to one level or only bright sunny days were selected for the measurements – implying the danger of limitation of light or depression of photosynthesis at noon by stomata closure. Normally, only one temperature was proven (e.g. 20°C, Führer *et al.*, 1993). What remains unconsidered in 'photosynthesis capacity measurements' like these is the dynamic transitional range between the limitation of light and the light saturation. Moreover the seasonal change of the slope of light saturation curves was neglected.

We attempted to discriminate between air pollution effects on the gas exchange and short-term, temporary environmental fluctuations (light, humidity), and applied following method: The photosynthetic performance potential of the twigs is estimated in dependence on the adaptation condition ('light-exposed crown' vs. 'shade crown' or season). Correspondingly, in the statistical test not only original values of gas exchange measurements were compared, but also the estimated parameters for the respective temperature range after calculation (see below: $P_{max}(T)$). This requires the testing of the following two variants:

- Variant A) All CO_2 fluxes measured during the diurnal course were related to the radiation values that were surveyed at the same time (Figure 2). The photosynthetic potential can be estimated from the resultant curves (A-I curves, Figure 3 and triangles in Figure 4). This method, however, allows one to survey only some of the branches or twigs or trees, respectively, during one monitoring period. The CO_2 net uptake measured after several hours of darkness served as the calculation of dark respiration.

- Variant B) It is possible to measure during one monitoring period also a much greater number of branches or trees, respectively, in various stands using the following experimental premise: All critical variables(temperature, air humidity, light) were optimized on the twig level for various needle age classes: The radiation intensity (PAR) was varied artificially, and an air humidity, adequate to the respective state of adaptation or ambient temperature, was set in the sample cell. From the course of the curve the maximum photosynthesis performance relating to the seasonal temperature range (P_{max}) as well as the slope of curve (K_I, α) could be determined (cf. Figure 4). In addition to this, it is possible to estimate not only the light compensation point (sum of parameter values: $I_o + K_I$), but also the net CO_2 exchange under illumination. The point of intersection with the y-axis in A-I-curves has been assumed to correspond to approximate photorespiration (Caldwell, 1995).

Comparison between variants A and B

It appears that the parameters estimated by means of the two methods are comparable restricted fashion: Parameters can be estimated both from 'pooled' data of diurnal courses and from data obtained from measurements based on the given radiation intensity (compare fitted curves, triangles and circles, respectively, in Figure 4). However, depending upon the site, there was an incidental occurrence of light saturation in few diurnal courses only or not at all. In the case of non-light-saturated photosynthesis the estimation of parameter $P_{max}(T)$ is difficult. Thus, an respectively insignificant estimation of photosynthesis performance in the saturation range is obtained, if mainly data from non-saturating light intensities are involved. Furthermore, because of the huge amount of data normally obtained by equidistant diurnal gas exchange measurements during the low-light and dark diurnal phases the respiration and the slope of A-I-curves appears to be modified as compared with the characteristic lines measured with saturating light intensities (circles in Figure 4). Hence, the photosynthetic performance is, in part, considerably over-estimated, because there is a strong representation of data of diurnal or seasonal times characterized by the scarceness or absence of light (stronger weighting of data from the non-saturating light intensity range because of the frequency of data: triangles in Figure 4). This leads to the additional problem of a suitable selection of data from diurnal courses. An additional reason why variant B is more appropriate for comparing the photosynthetic capacity of the

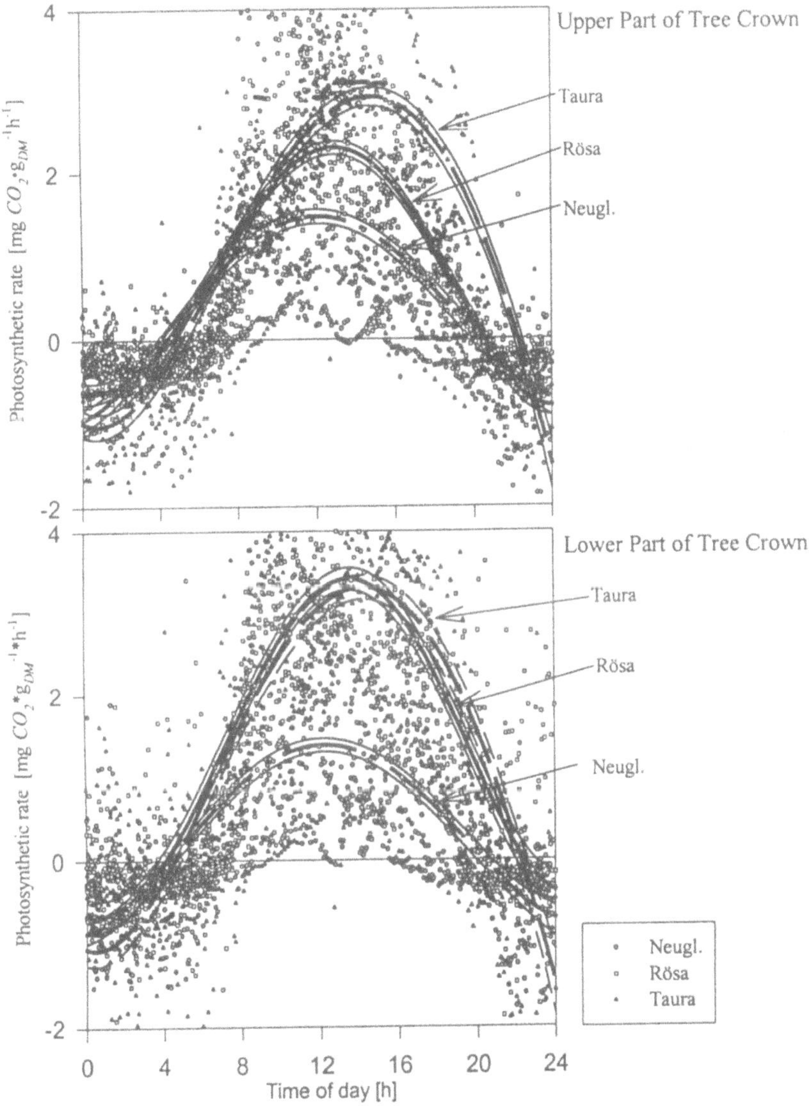

Figure 2. Diurnal course of net photosynthesis (above) and transpiration (below) of *Pinus sylvestris* L. twigs from selected seasonal periods (see Table 2) in 1993 and 1994. Interrupted bold line: sinusoidal regression line; thin full line: 95% confidence interval

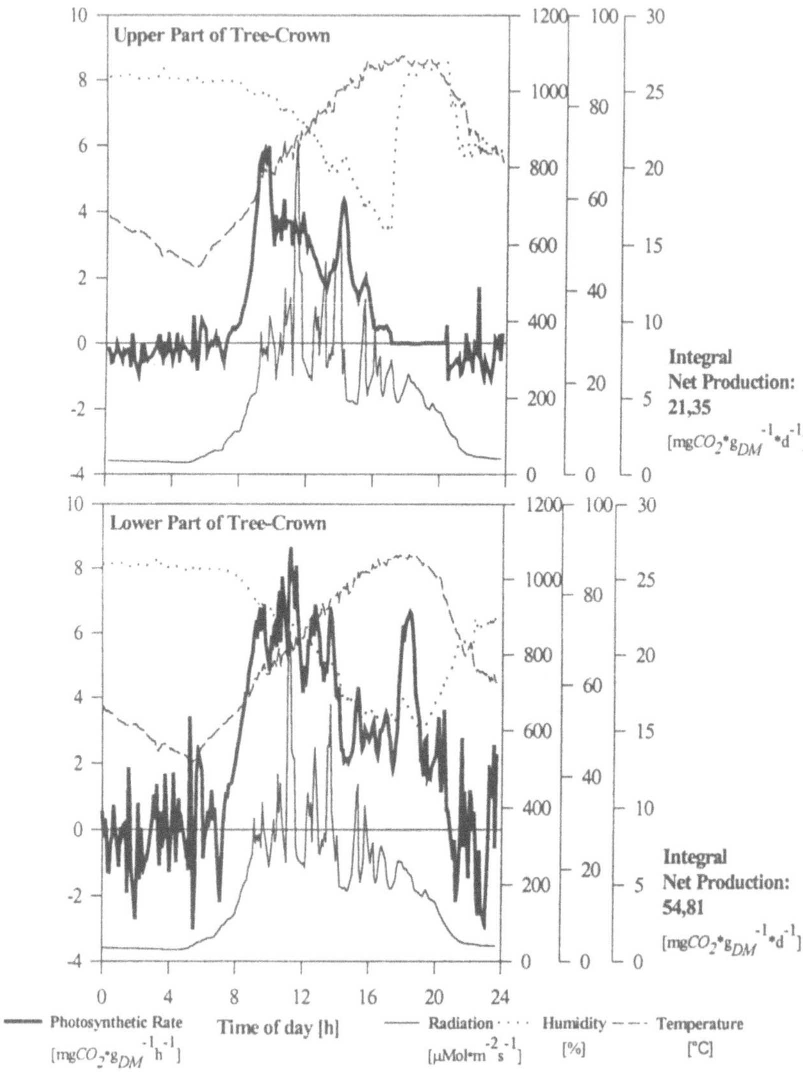

Figure 3. Selected example of net photosynthesis rates and related environmental variables of upper and lower parts of tree crowns in a formerly polluted stand (Rösa). Bold value right-hand side: calculated daily net productivity

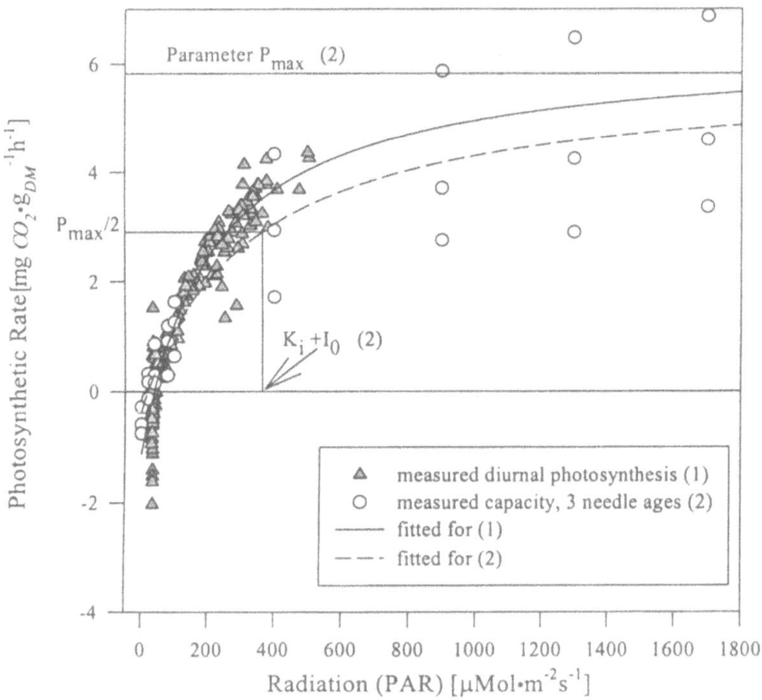

Figure 4. Light saturation curves (A-I-curves) calculated from measured diurnal photosynthesis rates and capacity measurement with simulated PAR, respectively

three stands is, that many more twigs or trees, respectively, can be measured over the monitoring period (e.g. one week) for direct comparison of the different stands. Therefore, *the assessment of photosynthetic potential determined by measurements with simulated radiation (PAR)* on the twig/needle level including their subsequent parametric estimation, *is preferable over the respective processing of diurnal course data* with natural light.

Photosynthesis capacity versus carboxylation potential

Carbon dioxide fluxes measured under light saturation and optimum water supply at 350 ppm CO_2, photosynthetic activity calculated on the basis of A-I-curves are a measure of photosynthetic capacity ($P_{max350}(T)$). During periods of drought stress (e.g. July/August of, 1994) and in several cases already occurring in May in certain years, the stomata closed at least partially due to drought of air and/or soil (compare example in Figure 3: 'depression at noon'). Comparative measurements under unnaturally high CO_2 concentrations (2500 ppm) were applied to overcome the raised stomatal resistance at any rate. A value near carboxylation capacity ($P_{max2500}(T)$) could be determined that way. The comparison of both ($P_{max350}(T)$ vs. $P_{max2500}(T)$) was meant to assess, largely independent of light and air humidity, the air pollution effects and/or site-related effects, which may immediately affect the photosynthesis apparatus (maximum rate of CO_2 assimilation-photosynthetic capacity).

In our case the CO_2 concentration by far exceeded the concentration necessary to obtain a CO_2 saturation (Lawton, 1993). The tests and continual comparisons with unsaturated CO_2 concentrations, however, led to results, that up to equilibrium the stomata openings were not significantly affected. The transpiration rates were not significantly influenced under unnaturally high CO_2 concentrations (2500 ppm) (compare Table 4: relatively small changes of the ratio between transpiration rate by 350 vs. 2500 ppm CO_2 during incubation of twigs in cuvettes).

Table 4. Ratio between transpiration rates under normal (350 ppm) and raised (2500 ppm) CO_2 concentrations (N = 2500). Column right side: mean ratio; column left side: std.dev.

	Crown part			
	Upper		Lower	
Rösa	1.13	1.1472	1.14	0.7270
Taura	1.22	1.0673	0.89	0.6969
Neuglobsow	1.39	0.9830	1.17	0.6919

3.2. *Effects of air pollution and site on photosynthetic capacity and carboxylation potential*

To differentiate between the factors that act indirectly via the soil and those with direct effects via the atmosphere a comparison between the sites was carried out involving various soil-related site conditions (cf. Table 1). A site-specific approach should partially facilitate the discrimination between the indirect soil-related effects and the direct effects on the photosynthesis apparatus. In this connection, the following sources of variances for photosynthetic performance in the field were related to the site specific differences:

- differences in the characteristic features of light and shade crown (cf. Figure 1) (see below: Comparison between upper and lower crown sections)

- seasonally changing influences as related to the corresponding temperature ($P_{max}(T)$) (see below: Seasonal adaptation)

- decrease in photosynthetic activity, depending upon the age of needles (declining pollution load since, 1990: comparison of needle age classes) (see below: Needle ageing)

Comparison by site

The Scots pine stand at Rösa, that was most heavily affected by air pollution in the past appears to have the highest photosynthetic capacity at present, especially at higher temperatures (Figure 5). However, a statistical comparison of parametric values ($P_{max350}(T)$) for each individual temperature range by means of u-test (SPSS; 1995) shows, that the differences are significant only for single temperatures or seasonal conditions of adaptation ($p = 0.05$). Also the choice of the needle surface instead of dry weight as biomass reference does not change these statements (compare Figure 6).

Both in the acidified (Taura) and in the less polluted, but also acidified control stand (Neuglobsow), specific (needle mass related) photosynthesis appeared to be more restricted by an increased stomatal resistance which resulted from water stress. The respective predawn water potentials of needles are shown in Figure 7. There are indications of water supply deficits especially for the stands in Taura and Neuglobsow in a temperature range from 25 to 40°C. Therefore, a differentiation between pollution-induced variations of site and effects caused by fluctuations in water supply and air moisture is impossible at a concentration of 350 ppm of CO_2 . However, if the stomatal resistance is overcome due to an increased CO_2 supply (2500 ppm) and thus approximately the carboxylation capacity ($P_{max2500}(T)$) is registered, the differences between the variants become negligible (Figure 5).

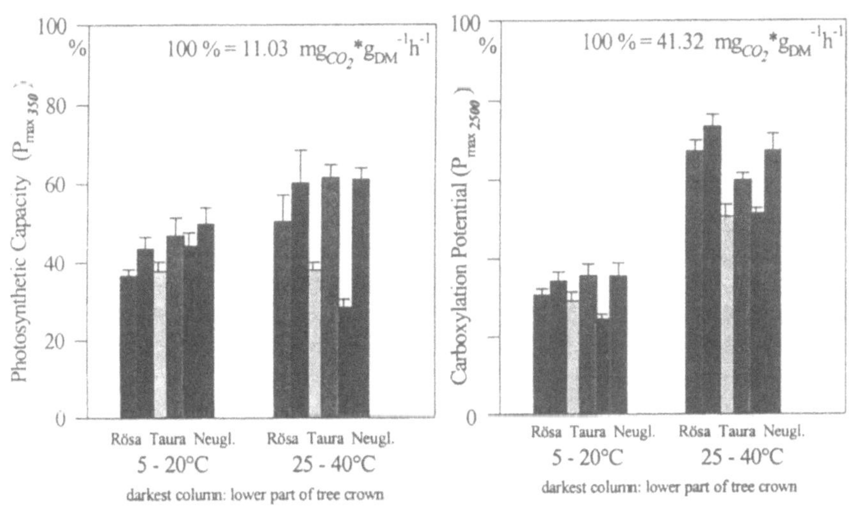

Figure 5. Percentage photosynthetic capacity in different crown parts at two different temperature levels

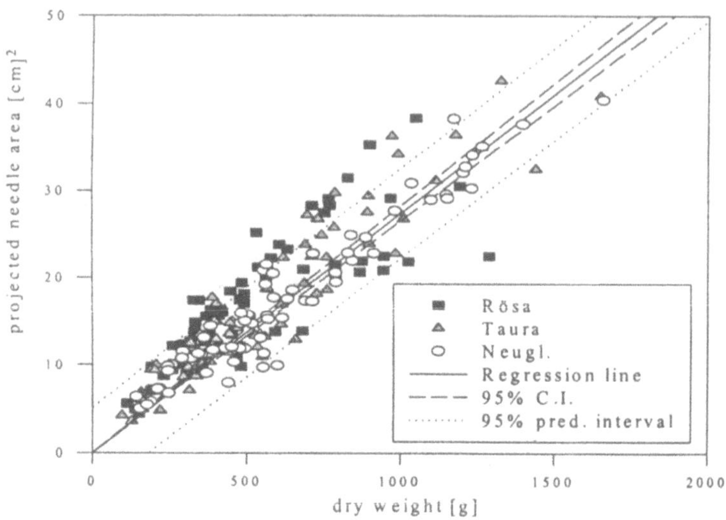

Figure 6. Relation between needle biomass and needle surface (summarized data from all investigation periods, N = 2700)

Comparison between upper and lower crown sections

In contrast to recent knowledge, on average, the estimations conducted for the upper parts of the crown (so called 'light crown') yield lower saturation values (P_{max350}(T) and $P_{max2500}$(T)) of photosynthetic performance (Figure 5). The differences occurring between crown parts and sites have been examined pairwise by a multiple comparison based upon the χ^2 procedures (H-test, Sachs, 1992; SPSS, 1996). The statistical comparisons have shown that the differences in photosynthetic capacities between upper and lower crown sections are significant ($p = 0.01$). The comparison each between the upper crown sections or each between the lower crown sections of various stands did not indicate any differences. Further details of statistical comparisons can be found in Dudel *et al.* (1997).

The photosynthetic potential of the needles or twigs, respectively, in the upper parts of the crown sections ('light needles') is not as high as in the lower more shaded inner sections. This holds true for all three sites and each period of investigation. Apparently, branches in exposed positions of the canopy are more impaired than those in the centre of crowns.

Seasonal adaptation

A differentiation of the form parameters of the light saturation curves (P_{max} vs. K_I and shifting of the light compensation parameter I_o, respectively for 'light' and 'shadow' crowns) as a measure of adaptational change of the photosyn-

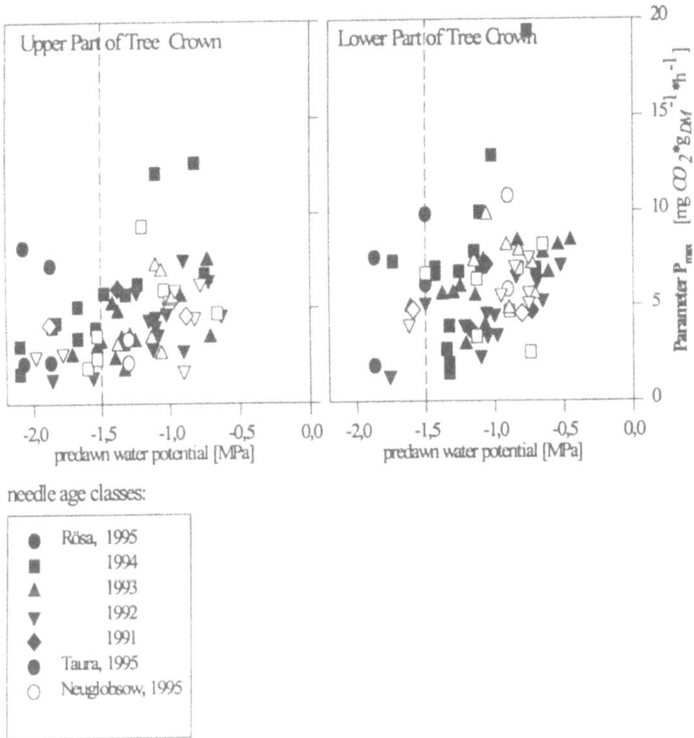

Figure 7. Predawn water potential of different needle age classes in three Scots pine stands. Left from dotted line probable water stress impacts

thetic apparatus between upper and lower crown parts is not provable at a certain point. However, the quasi-linear slope of curves (parameter value for K_I or α, respectively) increases from summer to winter season by a factor of 3–4 (Figure 8). This means, (1) a light adaptation characteristic of the whole (upper and lower) crown changes congruently from late winter to late spring, and vice versa, a shade adaptation takes place from summer to autumn, and (2) stand differences can be shown by a comparison of the complete parameter set of A-I-curves.

Needle ageing

The specific, i.e. biomass-related performance follows the seasonal sinusoidal change in total radiation, with the major part of the older needles being found at the lower limit of the 95% confidence range (Figure 9). However, the investigations did not suffice to obtain enough confidence as for the obvious trend of increasing photosynthetic capacity in general.

These comparatively slight differences between older and younger needles, also between differently polluted sites, gave rise to assume that there is a small

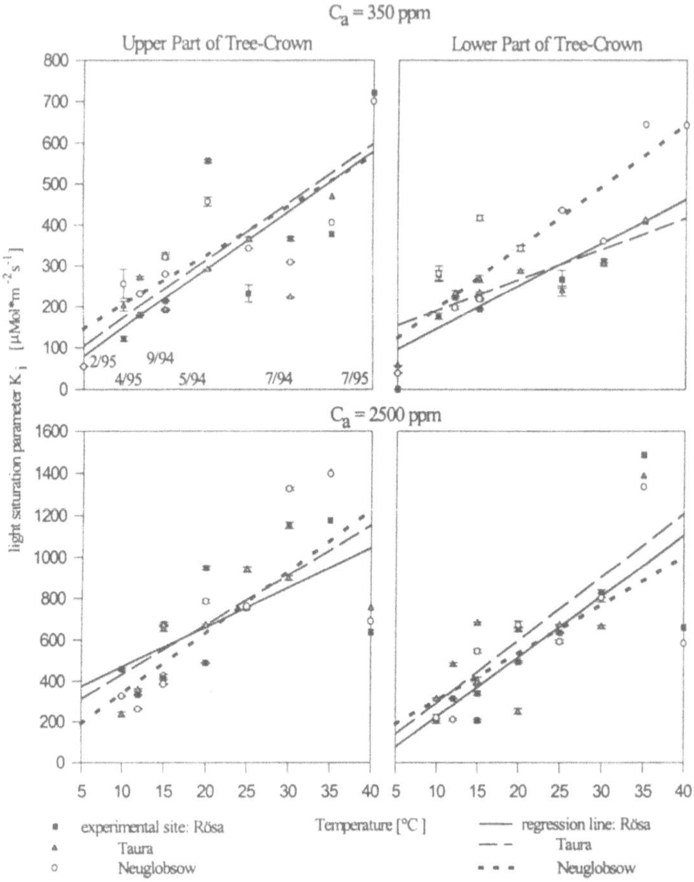

Figure 8. Temperature effects on the calculated light saturation parameter of photosynthesis (K_I, slope of A-I curve) in different crown parts (compare Figure 4)

divergence in photosynthetic capacity ($P_{max350}(T)$) and, in particular in the approximate carboxylation potential ($P_{max2500}(T)$) being largely independent of the stomatal performance. Rather, no significant differences in the photosynthetic performance of older and younger needles occurred in almost 60% of cases for $P_{max350}(T)$ and in about 40% of cases for $P_{max2500}(T)$ (Table 5). The influences of site and air pollution are relatively small. However the (maximum) transpiration rate in most of the investigated cases is highest in younger needles ($p = 0.01$). It is suggested, that needle ageing and duration of pollution contribute to a decrease of transpiration. Stand differences were not provable by comparison of single needle age classes.

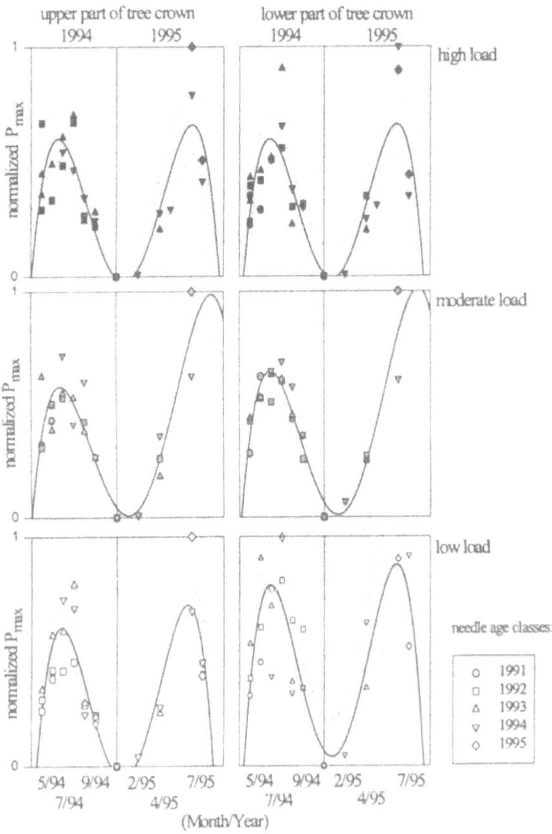

Figure 9. Seasonal-course of normalized photosynthetic capacity ($P_{max350}(T)$) in different needle age classes of three Scots pine stands. Full line: Regression line (sinusoidal model); dotted line: 95% confidence limit

3.3. Respiration losses

Contrary to expectations it has been demonstrated in the previous section, that the trees at Rösa do not photosynthesize to a lesser extent despite the higher previous and actual pollution load. In some cases photosynthetic capacity ($P_{max350}(T)$ and in general $P_{max2500}(T)$) was even higher compared with the Neuglobsow reference site.

Dark respiration

The rates of dark respiration show an exponential increase in the respective temperatures of the leafy branches at the individual sites measured in different seasons. Significant differences could not be shown to exist by statistical comparison.

Table 5. Pairwise comparison (by Sign test) of photosynthesis and transpiration of individual needle age classes of *Pinus sylvestris* L. at sites subjected to a different extent of pollution

	Rösa		Taura		Neuglobsow	
Result of test:	% of cases	var	% of cases	var	% of cases	var
Photosynthetic capacity (Ca=350ppm)						
Photosynthesis						
Younger needles > older	31.41	0.16	23.08	0.05	20.45	0.05
Younger needles = older	60.90	0.13	59.62	0.04	57.58	0.10
Younger needles < older	7.69	0.02	17.31	0.07	21.97	0.06
Carboxylation potential (Ca=2500ppm)						
Younger needles > older	22.7	0.09	37.8	0.09	40.2	0.04
Younger needles = older	43.8	0.10	39.1	0.11	40.2	0.10
Younger needles < older	33.5	0.15	23.1	0.04	19.7	0.07
Transpiration						
Ca = 350 ppm						
Younger needles > older	72.44	0.09	57.69	0.07	51.52	0.05
Younger needles = older	14.10	0.03	13.16	0.03	23.48	0.04
Younger needles < older	13.46	0.05	28.85	0.08	25.00	0.06
Ca = 2500 ppm						
Younger needles > older	42.2	0.13	62.2	0.13	68.9	0.11
Younger needles = older	29.1	0.13	20.5	0.05	16.7	0.03
Younger needles < older	28.7	0.07	17.3	0.11	14.4	0.07

Number of single values (for photosynthesis and transpiration resp.)
Total: 2942
Rösa: 983; Taura: 1024; Neuglobsow: 935

The dark respiration attains the highest values at the site Taura (compare single values plotted in Figure 10) and has a slightly higher average there, especially in the lower temperature range. It could be assumed, that due to pollution/acidification at this site, the respiration losses were increased.

In the course of one growing season respiration losses in blossoms and green cones relative to the biomass exceed the respective losses occurring in needles (Table 6). Due to the sharp increase in respiration at a temperature above 25°C, a net carbon loss can be expected for hot summer days in flowering and strongly fructifying pines.

The respiration losses of leafless twigs are low compared with the total losses of foliated branches, and can be neglected at temperatures below 20°C as regards photosynthesis (Table 7). In this context it should be known that the needles for the measurement of branch respiration were removed and that an increased share of wound respiration in total respiration losses has to be expected.

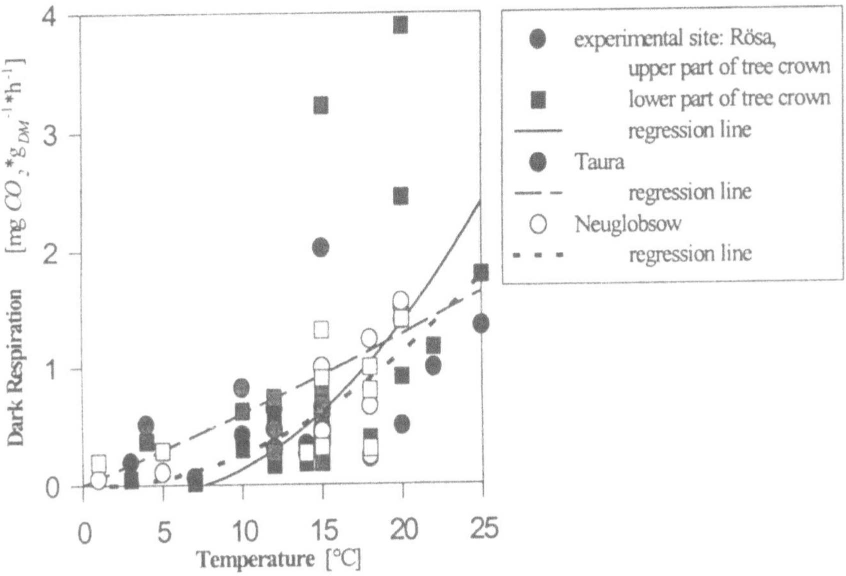

Figure 10. Temperature effect on dark respiration (predawn) in needles of three Scots pine stands

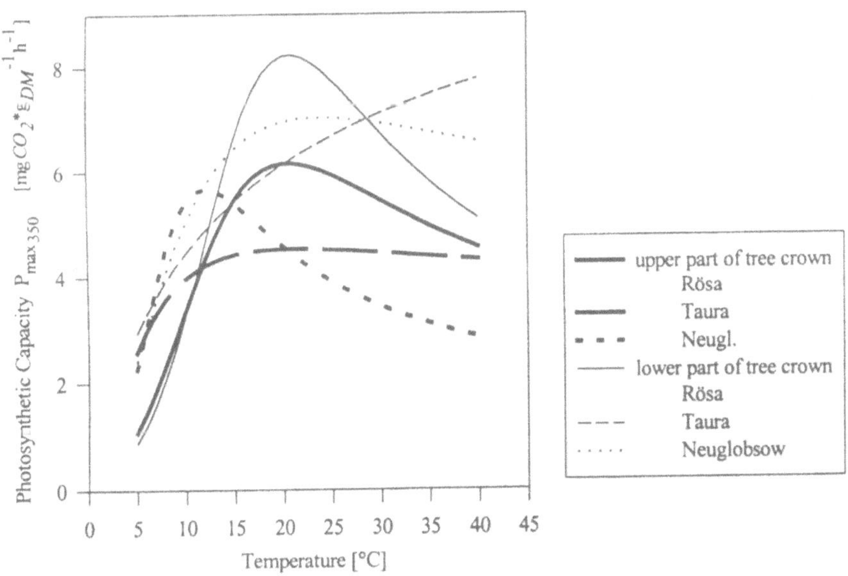

Figure 11. Temperature dependence of photosynthetic capacity ($P_{max350}(T)$) in needles of three Scots pine stands (regression lines calculated from 9700 single $P_{max350}(T)$ data based upon the model (eq. 3))

Photorespiration

The comparison of the computed gas exchange rates measured at normal (350 ppm) and increased CO_2 partial pressure shows that mainly photorespiration should be considered in connection with the calculated values (Figure 12, above and below). Apart from exceptions, photorespiration is completely suppressed by the increased CO_2 partial pressure (inhibition of the oxygenase function of the ribulose 1,5-diphosphate carboxylase, Lawton, 1993). Significant differences in photorespiration of the three sites along the pollution gradient could not be identified ($p = 0.01$).

Water potential and water use efficiency (WUE)

In general the predawn water potential of needles varied greatly. Significant differences between the sites could not be detected (Figure 13). Also during a hot period in the summer of, 1995 at noon with temperatures from 38–40°C the predawn water potential was only slightly reduced. However a clear trend of

Table 6. Dark respiration rates of Scots pine blossoms and green cones

	CO_2 (ppm)	Temp. (°C)	Rel. moisture (%)	Experimental site/ crown part	Respiration rate ($mgCO_2/gDM/h$) Mean	std.dev.	n
Blossoms	350	15	70	Taura/upper	2.90	0.2144	14
May 1994	350	15	70	Taura/lower	2.45	0.3529	14
Cones	2500	20	70	Rösa/upper	3.51	0.0010	3
July 1994	350	30	50	Neuglobsow/ upper	2.00	0.0000	3

Table 7. Dark respiration rates of Scots pine bare twigs

Twigs	CO_2 (ppm)	Temp. (°C)	Rel moisture (%)	Experimental site	Respiration rate ($mgCO_2/gDM/h$) Mean	std.dev.	n
Sept. 1994	350	15	70	Rösa	0.12	0.1450	18
July 1994	350	25	50	Rösa	0.71	0.5080	17
				Taura	0.16	0.1066	41
				Neuglobsow	0.34	0.1508	14
	350	30	40	Rösa	0.72	0.6529	35
				Taura	0.27	0.1750	20
				Neuglobsow	0.37	0.1650	21
	2500	30	40	Taura	0.16	0.0214	10

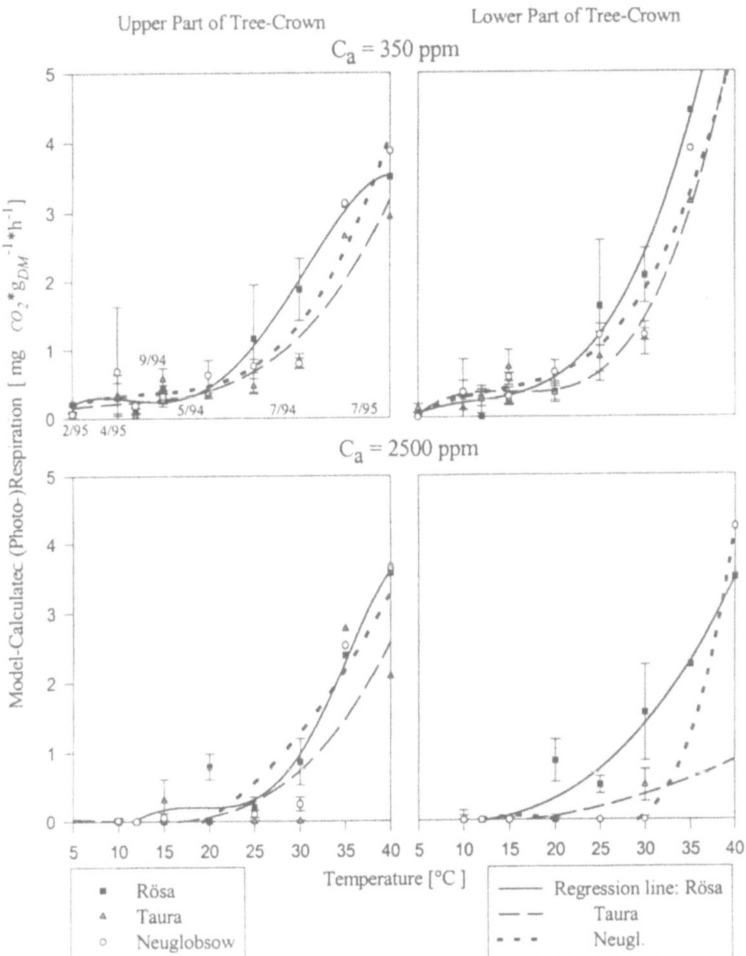

Figure 12. Temperature effect on approximate light respiration (compare intersection of A-I-curves with y-axis in Figure 4). Error bars: 95% confidence interval

decreasing P_{max} can be detected.Only in single cases in the summer of, 1994 (end of July and beginning of August) there was an unambiguous indication of water stress that followed a prolonged period of drought (Figure 7, values on the left-hand side of the dotted vertical line).

The normalized data of photosynthetic performances measured at normal atmospheric (P_{max350}(T)) and at increased CO_2 partial pressure ($P_{max2500}$(T)) – at least up to 25°C – do not differ (Figure 5, see also Dudel *et al.*, 1997) and suggest the conclusion, that the water supply restricted the C assimilation on a

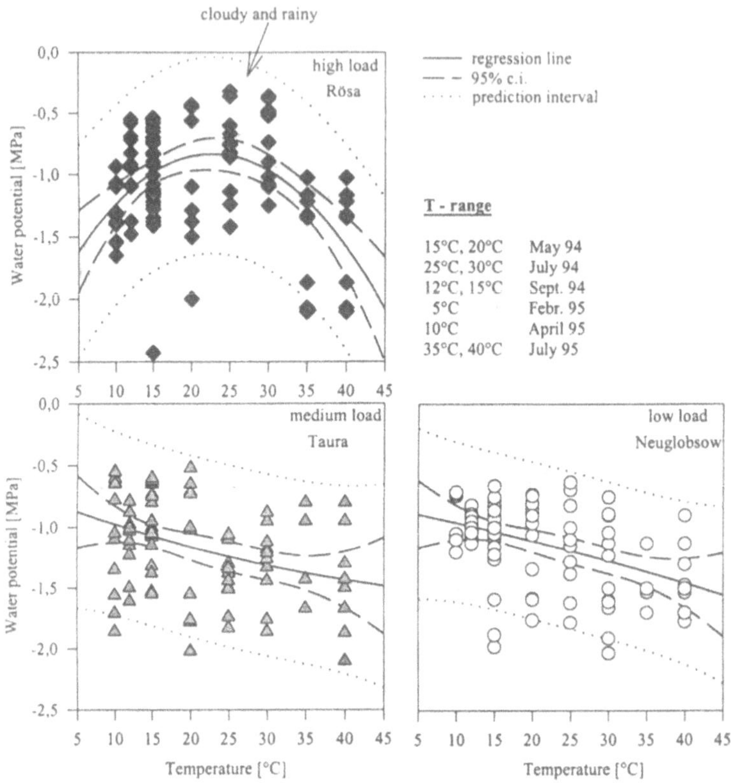

Figure 13. Seasonal differences (arranged on the x-axis according to ambient temperature) of water potential in different crown parts of three Scots pine stands

few days of some weeks only (July and August of, 1994). This statement is likewise corroborated by the analyses of the diurnal courses of photosynthesis and transpiration (compare huge of dottes in Figures 2 and 3): An afternoon depression has to be regarded as a rare event in this connection also if there was a high deficit of water vapour pressure (Figure 2: average sinusoidal diurnal course of photosynthesis and transpiration rate). This finding holds true for all site conditions and air pollution impacts investigated.

Contrary to the non-verifiable site differences regarding water potential in some cases, CO_2 assimilation-related transpiration performance (WUE) appears to be significantly higher at Rösa as compared with the more acidified stands at Taura and Neuglobsow (Figure 14).

In addition, it is apparent, that the WUE is higher in the upper, i.e. more exposed parts of the crown, than in the lower ones. This finding is not related to and independent of air pollution and site conditions, respectively.

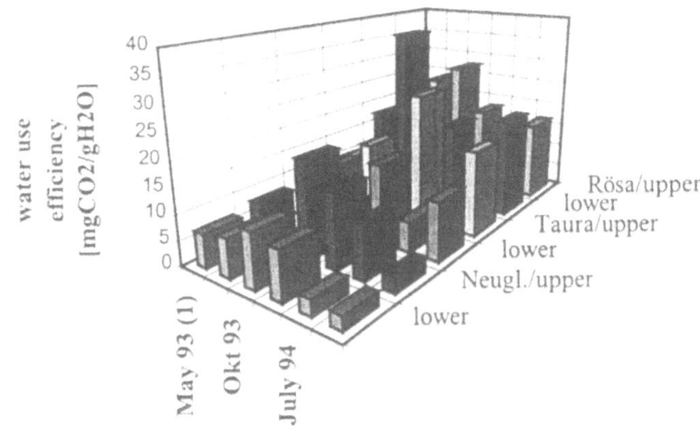

Figure 14. Water use efficiency in different crown parts of three Scots pine stands each subject to various pollution extents, calculated from 36 diurnal courses of CO_2 assimilation rates and transpiration rates

4. Discussion

The trees at the stand in Rösa do not photosynthesize to a lower extent, despite higher previous and recent pollution load. The former damage is compensated. This result, with even partly raised photosynthetic performances at the polluted sites may be interpreted in the following way: At those sites that were subject to a high SO_2 load, air pollution was accompanied with alkaline dusts and nitrogen. The latter caused an improved nutrient supply and a well balanced supply of the soil with bases (Table 1). This suggestion is corroborated by findings of Schaaf *et al.* (1995) and Ende *et al.* (this volume), that the nitrogen nutrition is on the upper level. The fact is underlined that mineral N input in a nutrient-limited system may immediately increase the photosynthesis performance (Ingestad, 1979; Küppers *et al.*, 1985; Lawton, 1993; Dudel *et al.*, 1995). Hence, it is possible, that a pollution-caused increased activity of the antioxidative system (superoxide dismutase activity) as well as an altered N and S balance in needles, as they were verified for that site by Schulz *et al.* (in this volume), can be supported by the surplus photosynthesis. This is related to the comparatively high increment performances at the polluted site of Rösa (Wenk *et al.*, this volume). Thus, it could be proven both via the photosynthetic performance (P_{max350}(T) and $P_{max2500}$(T)) and the magnitude of radial increment measurement, that the former heavily polluted but better nutrient-supplied and actually less acidified site Rösa has a primary production superior to the less loaded and nutrient supplied comparison sites, respectively. Recently, the plot at Taura which was subject to intermediate pollution loads over the past years did not markedly differ in its pollution load or depositions from the former highly loaded area of Rösa. However, its photosynthetic performance is at present comparable to the reference site at Neuglobsow. In parts – as it is statistically evident – it falls even short of it. The two sites have one thing in common: Their soils are more acidified. In contrast to this, the former input and recent supply with bases of Rösa plot is still distinctly higher (Schaaf *et al.*, 1997; Schulz *et al.*, this volume). However, continued decreasing acid inputs and extremely low inputs of alkaline dusts are supposed to cause conditions at the presently most efficient site Rösa similar to Taura.

At stand level the lower biomass specific photosynthetic performance of the less fertile stand could be compensated by a larger amount of needles per unit sap wood area (Vanninen *et al.*, 1996). The lower primary production per unit needle surface and biomass, respectively, of the reference site Neuglobsow achieved in branches or individual trees could be largely compensated at stand level, resulting from a higher stand density here (compare Table 1, Lüttschwager *et al.* in this volume). These findings correspond to the tree-related transpiration rates based on xylem sap flux measurement (Rust *et al.*, this volume). However, the latter results are not comparable with the needle weight and needle surface related (e.g. specific) transpiration rates of this study. Furthermore, it cannot be excluded, that the respective individual trees at Rösa were supplied with minerals and water more because of the lower stand density and differences in understorey vegetation (cf. Ende *et al.*, this volume; cf. Table 1 and Figure 1 as well as Lüttschwager *et al.* this volume).

It has to be admitted, that pollution-caused differences may be superimposed by a different genetic potential and adaptation capacity, respectively (Scholz, 1984; Kriebitzsch and Scholz, 1996). The provenances of the pines examined, however, are unknown.

That air pollution leads to morphological modifications of the needles and, being dependent on the extent of the pollution load, also to early needle shedding, is a well known fact (e.g. Gluch, 1988). Normally, Scots pine needle age reaches three or four years (Fraude, 1987; Gussone, 1986). In general, the pine needles at the highly pollution-loaded plots of Rösa and Taura do not attain high needle age classes. Likewise, however, at the reference site of Neuglobsow the average needle age is about three years. By long-term investigations pronounced necrosis could be shown to exist due to the pollution loads in the area around Leipzig/Bitterfeld (Gluch, 1988). The extent of needle necrosis was used as an indicator of air pollution damage. Supposedly, the oldest needles (age class 3, dating from, 1991 and, 1992, respectively) at Rösa and Taura, which meanwhile are subject to reduced pollution levels, differ from the younger ones by their photosynthesis performance: Thus, a respective gradation from site to site and year to year should be detectable. Occasionally, the Rösa and Taura trees are characterized by needle necrosis, even now (Schulz *et al.*, this volume). Contrary to expectations, significant differences in the photosynthetic performance (P_{max350}(T)) between needles of various age classes could be identified in single cases only. The former pollution effect were completely compensated at the level of the photosynthetic apparatus (twig/needle level). This finding corresponds to the critical conclusions drawn by

Kandler (1992, 1993), regarding forest damage assessment. He has ascertained that visible damage to (single) branches does not necessarily restrict the performance in the overall tree or stand.

It would seem feasible to assume that these comparatively high net-CO_2-assimilation rates were accompanied by an enhanced maintenance respiration due to the raised metabolic activity of the detoxification systems (e.g. SOD activity, Schulz *et al.*, this volume). To verify this statement (1), dark respiration was determined during the late night phase and (2) the respiration losses during the daytime light phase (photorespiration) with the aim of modeling (eq. 2) based upon A-I curves. Especially at higher temperatures respiration achieves relevant amounts as compared with the net CO_2 assimilation (cf. y-axis in Figure 10 and Figure 11). The amounts as well as their dependence on temperature are comparable to the findings of Künstle and Mitscherlich (1975, 1976), Troeng and Lindner (1982), Faensen-Thiebes *et al* (1991). The site differences of respiration are small and not significant. A higher maintenance respiration at the Taura acidified site could be an effect of disease and/or pollution-caused premature ageing due to insufficiently supplied bases (Dizengremel and Citerne, 1988). The raised respiration may be attributable to individual needle ageing which having commensed too early and/or to an increased mitochondrial activity of all photosynthetically active parts as a result of relatively premature senescence (Weidmann *et al.*, 1990). The raised maintenance respiration may also be ascribed to the additional energy requirement for detoxification processes (SO_2 in terms of SO_4, Schulz *et al.*, this volume) discussed above. However, more experimental studies are required to prove this hypothesis.

Due to a relatively high respiration rate at a temperature above 25°C and the unfavourable surface-volume ratio of green cones a low net-C-income can be expected for hot summer days in flowering and strongly fructifying pines. The present sample is very limited to allow a straight forward conclusion to be drawn and further measurements are required to examine the universality of these results. Further insights into the actual proportion of these respiration losses in the C-balance and related stand differences may be facilitated only by quantifying of the generative biomass proportion at tree and stand level. In analogy with this, comparable conclusions are suggested to hold true for the share of respiration of the living parts of branches and wood (bark) in the total C balance (cf. Künstle and Mitscherlich, 1976).

The small site differences regarding water potential and transpiration rate are not significant. Contrary to this non-verifiable effect, the water use efficiency (WUE) is significantly higher at the well nutrified and less acidified stand of Rösa. This finding coincides with a great number of others, which prove that air pollutants and climatic stress (belong to those environmental factors that) considerably affect (1) the internal CO_2 concentration (ci) via the net photosynthesis, and (2) the stomatal conductivity via the stomata regulation (Boutton and Flagler, 1990; Greitner and Winner, 1988; Martin *et al.*, 1988; Martin and Southerland, 1990; Lösch and Schulze, 1994). In summary, it has been shown repeatedly, also in various pine species – at least in fumigation experiments – that the WUE and the $^{12}C/^{13}C$ ratio ($\sigma^{13}C$), respectively, are enhanced by air pollutants (Elsik *et al.*, 1993). From this, one may conclude, that especially at the site of Taura, however, also at Neuglobsow, i.e. at the strongly acidified (loss of earth alkali elements, increase of solouble Al) trial plots, the stomata regulation ability could have changed to a lower efficiency. The osmoregulation of the stomata could have been influenced indirect via the soil system, which would be attributable to an altered base supply or modifications in the electrolyte balance of the needles due to a increased sulfate concentration and loss of Mg (Schulz *et al.*, this volume; Ende *et al.*, this volume). In comparison with this, the conditions at Rösa, subject to the highest air pollution, appear to be different. Here, the specific photosynthetic capacity is higher as well as WUE. Additionally the optimal N and bases supply which is given at that site characterized by favourable values of soil pH. These hypotheses of stomatal and non-stomatal limitation of photosynthesis related to macro and micro element supply, sulfate and nitrate storage, respectively, need to undergo further verification.

In summary the effects of air pollution on primary production and related processes, which in some areas still occur in large amounts in east Germany are hardly detectable at the leaf/twig level any more. Contrary to this, noxious substances are supposed to have an indirectly significant influence via soil on the efficiency of photosynthesis by the control of stomatal conductivity and water balance, respectively. The eutrophic site at Rösa near the center of chemical industry Halle/Bitterfeld which used to be heavily polluted earlier appears to have improved because of the actual growth increment comparable due to higher photosynthetic capacity. The following trends can be assumed at present: The input of alkaline dusts and the storage of bases in the soil are declining. Acidification, however, is not yet decreasing because of low-level large-distance sulfur and (ammonia-) nitrogen input will continue (Ende *et al.*, Schaaf *et al*,. in this volume). A similar decrease in primary production must be predicted: this has already appeared in some cases for the intermediary polluted, strongly acidified Taura site where the bases accumulated in the soil have presently reached a very low level. The indirect effects occurring via the soil are supposed to enhance or may be compensated by better air-transported (nitrogen?) nutrition. The rate of change and extent of compensation is unknown.

In the course of the strongly changing input of bases (dusts) in eastern Germany which resulted from the implementation of TA Luft (Technical Instruction about Air), the lasting SO_2 pollution (though at a reduced level, mainly from Poland and the Bohemian Basin) is supposed to be the source of the continued damage, having a greatly photosynthetic and growth efficiency-reducing, perhaps even destabilizing effect. The verification of these statements require secondary investigations to be conducted in the poorly base-supplied and nitrogen supersaturated pine and spruce stands of the eastern parts of Saxony.

Acknowledgements

This research was supported by the Bundesministerium für Bildung und Wissenschaft Bonn, Germany.

References

Baly ECC. 1935. The kinetics of photosynthesis. Proc R Soc L Ser B. 117, 218–239.

Bannister TT. 1974. A general theory of steady state phytoplankton growth in a nutrient-saturated mixed layer. Limnol Oceanogr. 19, 457–473.

Boutton TW, Flagler RB. 1990. Growth and water-use efficiency of shortleaf pine as affected by ozone and acid rain. Proceedings of the 83rd Annual Meeting and Exhibition of the Air and Waste Management Association 90 – 187.7. Air and Waste Management Association, Pittsburgh, PA.

Caldwell MM. 1995. personal communication.

Dizengremel P, Citerne A. 1988. Air pollutant effects on mitochondria and respiration. In: Air Pollution and Plant Metabolism. Eds. Schulte-Hostede S, Darall NM, Blank LW, Wellburn AR. Elsevier, London, 169–188.

Dudel EG, Pietsch M, Solger A, Zentsch W. 1995. Photosynthese, Atmung und Transpiration in Kiefernbeständen an der Schnittstelle Atmosphäre-Zweig unter abnehmender Immissionsbelastung. Eds. Hüttl RF, B Bellmann, and W Seiler. Atmosphärensanierung und Waldökosysteme Blottner-Verlag, Taunusstein, 87–126.

Dudel EG, Federbusch U, Solger A. 1997. Variability and photosynthesis performance in pine stands: limits of the differentiation of site and air pollution-caused influences. Ecosys Suppl. 20, 37–46.

Elsik CG, Flagler RB, Boutton TW. 1993. Carbon isotope composition and gas exchange of loblolly and shortleaf pine as affected by ozone and water stress. In: Stable Isotopes and Plant Carbon – Water Relations. Eds. JR Ehleringer, AE Hall, GD Farquhar. Academic Press, Inc.

Faensen-Thiebes A, Cornelius R, Meyer G, Bornkamm R. 1989. Ecosystem study in a Central European pine forest. In: Coniferous Forest Ecology From An International Perspective. Eds. Nakagoshi N, Golley FB. SPB Academic Publishing by The Hague, The Netherlands, 137–150.

Faensen-Thiebes A, Cornelius R. 1991. Modellierung des Gaswechsels der Kiefer in seiner Abhängigkeit von klimatischen Faktoren. Verhandlungen der Gesellschaft für Ökologie (Osnabrück, 1989), 19/III: 681–691.

Fischer CH, Holl W. 1991. Food reserves of Scots pine (*Pinus sylvestris* L.) Trees. 5, 1871–1895.

Fraude H-J. 1987. Zur Anzahl der Nadeljahrgänge der Waldkiefer. Untersuchungen im Rahmen von Waldschadenserhebungen in Rheinland-Pfalz. Der Forst- und Holzwirt. 15, 415–417.

Führer G, Payer H-D, Pfanz H. 1993. Effects of air pollutants on the photosynthetic capacity of young Norway spruce trees. Response of single needle age classes during and after different treatments with O_3, SO_2 or NO_2. Trees. 8, 85 92.

Gluch W. 1988. Zur Benadelung von Kiefern (*Pinus sylvestris* L.) in Abhängigkeit vom Immissionsdruck. Flora. 181, 395–407.

Greitner CS, Winner WE. 1988. Increases in ^{13}C values of radish and oybean plants caused by ozone. New Phytol. 108, 489–494.

Gussone HA. 1986. Wieviel Nadeljahrgänge sind normal? Der Forst- und Holzwirt. 15, 415–417.

Heber U, Kaiser W, Luwe M *et al.* 1994. Air pollution, photosynthesis and forest decline: interactions and consequences. In: Ecophysiology of Photosynthesis. Eds. E-D Schulze, MM Caldwell. Ecol Stud. 100, 279–296.

Heinsdorf D. 1993. The role of nitrogen in decling Scots pine forests (*Pinus sylvestris* L.) in the Lowland of East Germany. Water Air Soil Pollut. 69, 21–35.

Ingestad T. 1979. Mineral nutrient requirement of Pinus sylvestris and Pices abies seedlings. Physiol Plantarum. 45, 373–380.

Jandel Scientific. 1993. Bd. 1: Sigma Plot – Users Manual, Bd. 2: Sigma Plot - Transforms and Curve Fitting.

Jassby AD, Platt T. 1976. Mathematical formulation of the relationship between photosynthesis and light for phytoplankton. Limnol Oceanogr. 21, 540–547.

Kandler O. 1992. The german forest decline situation: a complex desease or a complex of deseases. In: Forest Decline Concepts. Eds. Manion PD, Laechence D. ASP Press, St Paul, Minnesota, 59–84.

Kandler O. 1993. The air pollution/forest decline connection: the Waldsterben theory refuted. Unasilva. 44, 39–48.

Keller T, Häßler R. 1986. The influence of prolonged SO_2 fumigation on the stomatal reaction of spruce. Eur J For Path. 16, 110–115.

Kriebitzsch W-U, Scholz F. 1996. Zur Wirkung erhöhter SO_2 Konzentration auf den Gaswechsel von Fichtenklonen (*Picea abies* (L.) Karst.) aus dem Fichtelgebirge. Verhandlungen der Gesellschaft für Ökologie. 26, 121–126.

Künstle E, Mitscherlich G. 1975. Photosynthese, Transpiration und Atmung in einem Mischbestand im Schwarzwald. III. Teil: Atmung. Allg Forst-u J-Ztg. 147(9), 169–177.

Künstle E, Mitscherlich G. 1976. Photosynthese, Transpiration und Atmung in einem Mischbestand im Schwarzwald. IV. Teil: Bilanz. Allg Forst-u J-Ztg. 148(12), 227–239.

Küppers M, Zech W, Schulze E-D, Beck E. 1985. CO_2-Assimilation, Transpiration und Wachstum von *Pinus sylvestris* L. bei unterschiedlicher Magnesium-Versorgung. Forstw Cbl. 104, 23–36.

Lange OL, Gebel J, Walz H, Schulze E-D. 1985. Eine Methode zur raschen Charakterisierung der photosynthetischen Leistungsfähigkeit von Bäumen unter Freilandbedingungen – Anwendung zur Analyse neuartiger Waldschäden bei der Fichte. Forstw Cbl. 104, 186–198.

Lawton DL. 1993. Photosynthesis. Longman Sci. and Techn. Essex, UK.

Lerch G. 1991. Pflanzenökologie. Teile I und II. Akademie-Verlag. Berlin.

Linder S, Troeng E. 1981. The seasonal variation in stem and coarse root respiration of a 20-year-old Scots pine (*Pinus sylvestris* L.). In: Radial Growth in Trees. Ed. W Tranquillini. Mitteil der Forst Bundesversuchsanstalt, Wien. 142(1), 125–140.

Lösch R, Schulze E-D. 1994. Internal coordination of plant responses to drought and evaporational demand. In: Ecophysiology of Photosynthesis. Eds. E-D Schulze, MM Caldwell. Ecol Stud. 100, 185–204.

Lux H. 1965. Die großräumige Abgrenzung von Rauchschadenszonen im Einflußbereich des Industriegebietes um Bitterfeld. Wiss Zeitschr Techn Univ Dresden. 14(2), 433–442.

Lux H, Stein G. 1977. Die forstlichen Immissionsschadgebiete im Lee des Ballungsraumes Halle und Leipzig. Hercynia. 14, 413–421.

Marquardt W, Brüggemann E. 1995. Long-term trends in acidity of precipitation after longscale transport-effects of atmospheric rehabilitation in East-Germany. Water Air Soil Pollut. 85, 665–670.

Martin B, Bytnerowicz A, Thorstenson YR. 1988. Effects of air pollution on the composition of stable carbon isotopes, ^{13}C, of leaves and on leaf injury. Plant Physiol. 88, 213–217.

Martin B, Sutherland EK. 1990. Air pollution in the past recorded in width and stable carbon isotope composition of annual growth rings of Douglas-fir. Plant Cell Environ. 13, 839–844.

Matyssek R. 1986. Carbon, water and nitrogen relations in evergreen and deciduous conifers. Tree Physiol. 2, 177–187.

Matyssek R, Schulze E-D. 1988. Carbon uptake and respiration in aboveground of a Larix decidua × leptolepis tree. Trees. 2, 233–241.

Norby RJ, Weerasuriya Y, Hanson PJ. 1989. Induction of nitrate reductase activity in red spruce needles by NO_2 and HNO_3 vapor. Can J For Res. 19, 889–896.

Perterer J, Körner Ch. 1990. Das Problem der Bezugsgröße bei physiologisch-ökologischen Untersuchungen an Koniferennadeln. Forstw Cbl. 109, 220–241.

Pfanz H, Heber U. 1986. Buffer capacities of leaves, leaf cells, and leaf cell organelles in relation to fluxes of potentially acidic air pollutants. Plant Physiol. 81, 597–602.

Pfanz H, Beyschlag W. 1993. Photosynthetic performance and nutrient status of Norway spruce [*Picea abies* (L.) Karst.] at forest sites in the Ore Mountains (Erzgebirge). Trees. 7, 115–122.

Platt T, Gallegos CL, Harrison WG. 1980. Photoinhibition of photosynthesis in natural assemblages of marine phytoplankton. J Mar Res. 38, 687–701.

Ratkowski DA. 1990. Handbook of nonlinear regression models. Marcel Dekker Inc., New York.

Roberts J, Pitman RM, Wallace JS. 1982. A comparison of evaporation from stands of Scots pine and Corsican pine in Thetford Chase, East Anglia. J Appl Ecol. 19, 859–872.

Rust S, Lüttschwager D, Hüttl RF. 1995. Transpiration and hydraulic conductivity in three Scots pine (*Pinus sylvestris* L.) stands with different air pollution histories. Water Air Soil Pollut. 85, 1677–1682.

Sachs L. 1992. Angewandte Statistik. 7. völlig überarb. Aufl., Springer, Berlin.

Schaaf W, Weisdorfer M, Hüttl RF. 1995. Soil solution chemistry and element budgets of three Scots pine ecosystems along a deposition gradient in North-Eastern Germany. Water Air Soil Pollut. 85, 1197–1202.

Schaaf W, Weisdorfer M, Httl RF. 1997. Recovery of Scots pine ecosystems in NE-Germany affected by long-term pollution with SO_2, N, and alkaline dust. Geochim Cosmochim Acta. in press.

Scholz F. 1984. Wirken Luftverunreinigungen auf die genetische Struktur von Waldbaumpopulationen? Forstarchiv. 55(2), 43–45.

Schulz H, Huhn G, Jung K, Härtling S, Schürmann G. 1995. Biochemical responses in needles of Scots pine (*Pinus sylvestris*) from air polluted field sites in Eastern Germany Air Pollution III. Volume 4: Observation and Simulation of Air Pollution: Results from SANA and EUMAC. Eds. A Ebel, N Moussiopoulos. 33–42. Computational Mechanics Publications, Southhampton, Boston.

Schulze E-D, Oren R, Zimmermann R. 1987. Die Wirkung von Immissionen auf 30jährige Fichten in mittleren Höhenlagen des Fichtelgebirges auf Phyllit. Allg Forstzeitg. 27/28/29, 725–729.

Schulze E-D, LangeOL, Oren R. 1989. Forest decline and air pollution. Ecol Stud. 77, Springer, Berlin-Heidelberg-New York.

Seber and Wild. 1989. Nonlinear Regression. Wiley and Sons, New York.

Smith WH. 1990. Air Pollution and Forests. Interactions between Air Contaminants and Forest Ecosystems. 2nd ed. Springer, New York-Berlin-Heidelberg.

SPSS 6.1 Guide to Data Analysis. 1995. Prentice Hall, Englewood Cliffs, NJ.

Troeng E, Lindner S. 1982. Gas exchange in a 20-year-old stand of Scots pine. II Variation in net photosynthesis and transpiration within and between trees. Physiol Plant. 54, 15–23.

Ulrich B, Matzner E. 1983. Ökosystemare Wirkungsketten beim Wald- und Baumsterben. Der Forst- und Holzwirt. 18, 468–474.

Umweltbundesamt and Senatsverwaltung für Stadtentwicklung und Umweltschutz (Hrsg.), 1990 Ballungsraumnahe Waldökosysteme. Abschlußbericht. 258 S.

Ulrich B, Pirouzpanah D, Murach D. 1984. Beziehungen zwischen Bodenversauerung und Wurzelentwicklung von Fichten mit unterschiedlich starken Schadsymptomen. Forstarchiv. 55, 127–134.

Vanninen P, Ylitalo H, Sievänen R, Mäkelä A. 1996. Effects of age and site quality on the distribution of biomass in Scots pine (*Pinus sylvestris* L.). Trees. 10, 231–238.

Von Caemmerer, Farquhar GD. 1981. Some relationships between the biochemistry of photosynthesis and the gas exchange of leaves. Planta. 153, 376–387.

Wallin G, Skärby L. 1992. The influence of ozone on the stomatal and non-stomatal limitation of photosynthesis in Norway spruce, *Picea abies* (L.) Karst, exposed to soil moisture defizit. Trees. 6, 128–136.

Webb WL, Newton N, Starr D. 1974. Carbon dioxide exchange of Alnus rubra: A mathematical model. Oecologica. 17, 281–291.

Weidmann P, Einig W, Egger B, Hampp E. 1990. Contents of ATP and ADP in needles of Norway spruce in relation to their development, age, and to symptoms of forest decline. Trees. 4, 68–74.

6
Tree canopy and field layer transpiration in Scots pine stands

D. LÜTTSCHWAGER, M. WULF, S. RUST, J. FORKERT and R.F. HÜTTL

1. Introduction

Stand transpiration is a major component of the hydrological and biogeochemical cycles. It is controlled by species composition, LAIs and stomatal regulation. Stand structure is influenced by, for example, site, management, air pollution, and interspecific interactions like competition and herbivory.

In the context of SANA, the impact of air pollution and N-fertilisation were of special importance. Although the three sites Rösa, Taura, and Neuglobsow are similar with respect to their mineralogical substrate, soil physical properties, climate, and stand structure, the ecosystem types are very different (cf. Table 1, Hüttl and Bellmann, this volume). This is the result of many years of differences in atmospheric deposition rates. Moreover, the sites were, especially those close to the industrial region of Halle and Leipzig, fertilised with high amounts of nitrogen in an effort to stabilize the forests (Krauß, 1975). From 1970 through 1985, almost a ton of nitrogen per hectare was applied in Rösa (Wenk *et al.*, 1994). Even ten years after the last application, the level of N in the pine needles of Rösa was higher than in Neuglobsow by 30 to 50% (Ende and Hüttl, this volume).

N-fertilisation can result in higher LAI and higher drought susceptibility (Linder and Rook, 1984). The effect of nitrogen fertilisation on water use efficiency (WUE) is not clear. Brix and Mitchell (1986) report an increase in WUE in fertilised *Pseudotsuga menziesii* and Squire *et al.* (1987) in *Pinus radiata*. Sheriff *et al.* (1986) and Hillerdahl-Hagströmer *et al.* (1982) found a higher sensitivity to drought stress caused by tighter stomatal control. Experiments by Mitchell and Hinckley (1993) show that N increased transpiration and photosynthesis, resulting in constant WUE.

The large pool of available nitrogen allows lush growth of a nitrophilous field layer vegetation that competes with the trees for water. For the last 20 years, the grass *Calamagrostis epigeios* had spread over increasing areas (Bergmann, 1993; Hofmann, 1994; Kopp, 1986). Some authors assume, that *Calamagrostis epigeios* is a better competitor for water than pine, or that stands with a dense cover of this grass consume more water than those dominated by *Avenella flexuosa* (Hofmann, 1994).

In this study, species composition and cover, LAI, and biomass, of the three

R.F. Hüttl and K. Bellmann, Changes of Atmospheric Chemistry and Effects on Forest Ecosystems, 97–110.

sites Rösa, Taura and Neuglobsow were analyzed. Moreover, transpiration of the tree canopy and field layer vegetation were estimated and tree water relations monitored. The aim of our investigation was to identify the impact of N-fertilisation and air pollution on the transpiration of pine ecosystems. Our special focus was the competition for water between *Calamagrostis epigeios* and mature Scots pine.

2. Materials and methods

2.1. Tree canopy transpiration

Tree transpiration was estimated by sap flow measurements in single trees, and scaled up to stand level. From the summer of 1993 through the autumn of 1995, sap flow was recorded for 15 representative trees per stand, selected in proportion to the frequencies in the diameter size class distribution.

Sap flow density in the trunks of the trees was measured using a constant heating method (Granier, 1985). Since in pine the entire sap wood conducts water, two gauges were installed in each tree ranging from 0 to 2.2 cm and 2.2 to 4.4 cm from the cambium, respectively. Automatic readings are taken every 30 s and averaged over 30 min periods. Sap flow was calculated as the product of sap flow density and conductive sap wood area as measured by computer-tomography (Edwards and Jarvis, 1983; Habermehl *et al.*, 1986; Rust *et al.*, 1995; Rust *et al.*, this volume) in collaboration with the Center for Radiology of the Phillips-University Marburg.

2.2. Water status of the trees

Tree drought stress was assessed by periodical measurements of pre-dawn water potentials. Shoots of ten trees per stand were collected with a shotgun from the upper crown and measured in a pressure chamber (Scholander *et al.*, 1965). A drought stress index was calculated as a time weight mean.

2.3. Investigation of the field layer vegetation

In each site, three permanent plots of 9 m^2 were installed in the summer of 1994. The plots were divided into four quadrants to estimate cover degree of all plant species to one percent. Moss cover was estimated without differentiating for species. Because the exact estimation of all plant species within a minimum area of at least 200 m^2 was impossible without disturbance, the plots were selected so that all relevant plant species of that forest type were represented. Exact estimation of cover degrees was secured by randomly selecting additional plots of 9 m^2 outside of the permanent plots. All plots were pooled to calculate monthly averages of cover. To characterise the sites by the indicator values of their species, all species with more than two individuals within a 2500 m^2 plot were recorded.

2.4. *Biomass and leaf area index*

Five trees per stand were sampled as a stratified random sample for needle mass in September 1995. All branch diameters and the needle mass of one branch per whorl were measured. Using the close correlation of branch diameter and needle mass (Kaibiyainen *et al.*, 1986; Mencuccini and Grace, 1995), data were scaled to tree level. Specific needle area (projected) was estimated with an image analysis system on samples stratified for crown location, age, and length. A regression of needle area on sapwood area was used to scale to stand level (Albrektson, 1984; Whitehead, 1978).

In three plots (0.25 m) per site all living herbaceous plants were collected in height strata of 10 cm, dried at 104°C and weighed. Means were scaled to a hectar basis. Specific needle area for each species was estimated with an image analysis system and used to calculate the leaf area index of the species (LAI_{part}). The LAI of the herb layer is the sum of the LAI_{part}'s.

2.5. *Transpiration of the field layer vegetation*

Transpiration was measured for species with at least 10% cover within an minimum area of 200 m. In Rösa, these were *Brachypodium sylvaticum*, *Calamagrostis epigeios* and *Rubus idaeus*, in Taura *Avenella flexuosa* and in Neuglobsow *Avenella flexuosa* and *Vaccinium myrtillus*. Measurements were taken with a porometer (Walz, Effeltrich). In the late summer of 1994, 20 min cycles, and in the growing season of 1995 entire days were measured. Five minute averages of exposed leaves of one species were recorded from dawn till dusk. Transpiration was scaled up to the stand level with the LAI of the species. The ratio of field layer transpiration to tree transpiration was interpolated through periods without measurements and used to estimate herb layer transpiration from tree transpiration during these times.

2.6. *Radiation, temperature, and humidity*

Starting in spring 1994, temperature, humidity, and photosynthetic radiation were recorded as 30 min averages in three levels:

a. 1 m below maximum tree height,

b. variably installed between 2 m and 15 m, and

c. variably installed between 0.3 m and 1.8 m above ground.

3. Results

3.1. *Needle mass and leaf area index of the trees*

Needle mass was highest in Neuglobsow (6.54 t/ha) and lowest in Taura (5.41

Table 1. Specific needle area (cm^2/g)

	Upper crown			Lower crown		
Needle age (yr)	1	2	3	1	2	3
Rösa	54.40	46.95	38.34	54.07	51.66	46.48
Taura	44.69	43.30	35.39	50.27	45.00	33.11
Neuglobsow	42.95	43.41	35.15	52.46	43.95	43.75

t/ha, Rösa 6.15 t/ha). The specific leaf area (Table 1) resulted in a higher LAI of the canopy in Rösa (3.16) than in the other stands (Neuglobsow 2.87, Taura 2.43).

3.2. *Sap flow densities*

In addition to climatic variables such as vapor pressure deficit and radiation, sap flow densities varied with predawn water potential and available soil water. Below 35% available soil water, daily maxima of flux density fell strongly, above this threshold trees in all stands reached values of around 80 kg m^2 h^{-1}.

The variation within the 15 sample trees differed between stands and position of the sensors. The average coefficient of variation of cumulated daily values ranged from 7% (outer sensors in Neuglobsow) to 33% (inner sensors in Taura).

Tree size had an impact on sap flux densities at the outer sensors in Taura and Neuglobsow. On days with higher than average flow rates, the daily sums of sap flux density increased significantly with sapwood area at 1.3 m.

3.3. *Water status of the trees*

Predawn water potentials differed substancially between 1994 and 1995 (Figure 1). During a long period of drought in 1994 predawn water potential fell from above –0.5 MPa in spring to below –1 MPa at the end of July. In Neuglobsow trees reached the lowest needle water potentials with single trees as low as –2.6 MPa, on average -1.65 ± 0.24 MPa as compared to Rösa with -1.16 ± 0.21 MPa. The drought stress indices were –1.1, –0.9 and –0.6 MPa in Neuglobsow, Taura and Rösa, respectively. In 1995, predawn water potentials never fell below –1 MPa and drought stress indices were between –0.5 and –0.6 MPa.

3.4. *Tree canopy transpiration*

Daily transpiration for 1994 and 1995 is shown in Figure 2. On fine days, transpiration reached approximately 1 mm/d, in Neuglobsow up to 1.5 mm/d. A period with almost no precipitation and high radiation and temperature in the summer of 1994 is indicated by the lack of days with low transpiration. Because of declining soil water availability, transpiration in Neuglobsow fell to

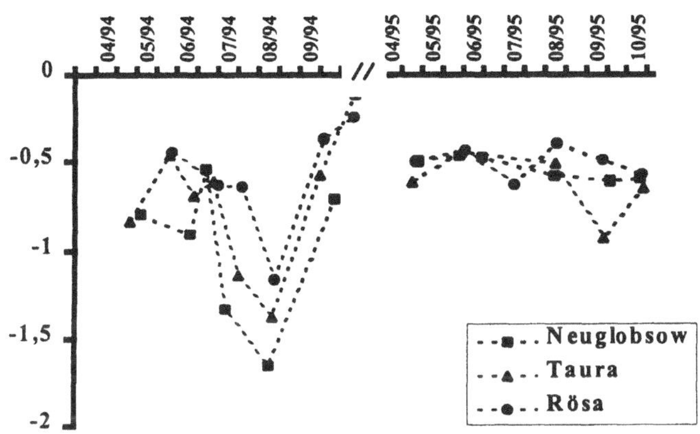

Figure 1. Pre-dawn water potential for 1994 and 1995

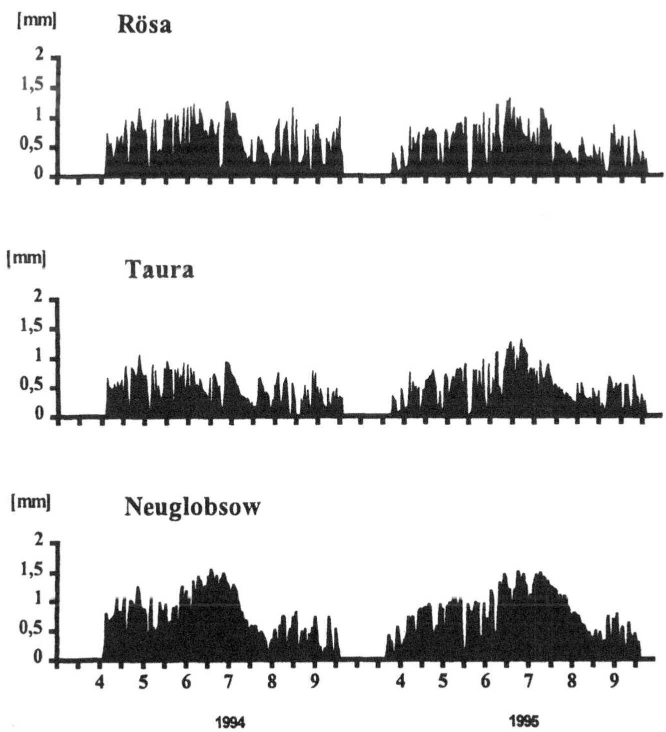

Figure 2. Daily transpiration of the pines during the growing seasons 1994 and 1995

less than one third from mid July to mid August 1994, in spite of fairly constant climatic conditions. Tree canopy transpiration during the growing season of 1994 (April to September) was 106 mm in Rösa, 82 mm in Taura, and 113 mm in Neuglobsow. In 1995, the values were Rösa 100, Taura 94, and Neuglobsow 124 mm. Transpiration per needle area was lowest for the nitrogen fertilised stand Rösa in all three years. This could mean that stomatal conductance in Neuglobsow was higher than in Rösa. Further, it was found that sapflow densities started to decline at higher soil water availability in Rösa than in Neuglobsow.

3.5. Field layer vegetation

In Rösa the number of N-indicator species and the number of species was highest (57 species in 2.500 m^2). Species diversity in Taura (22) and Neuglobsow (24) was very similar (Table 2). The field layer vegetation of Neuglobsow, which was dominated by *Avenella flexuosa* and *Vaccinium myrtillus* indicates a site without major deposition impacts (Table 3). Rösa, however, was dominated by *Calamagrostis epigeios* and *Brachypodium sylvaticum*, showing the influence of N-fertilisation and Ca-deposition. The species in Taura were a mix of N-indicators like *Calamagrostis epigeios* and *Rubus idaeus* and acid-tolerant species like *Avenella flexuosa*. Of the indicator values, only the N-value showed

Table 2. Indicator values, plant life forms and anatomy of the leaves of the plant species with high cover

Number of species (per 2.500 m^2)	Rösa 57	Taura 22	Neuglobsow 24
Average of the indicator values (Ellenberg *et al.*, 1991)			
Indicator of light	5.9	6.4	6.0
Indicator of moisture	5.1	5.5	4.9
Indicator of soil reaction	5.1	3.2	3.1
Indicator of nitrogen	4.9	4.3	3.3
Plant life forms (%) (Ellenberg *et al.*, 1991)			
Geophytes	14.1	9.1	8.4
Hemikryptophytes	43.8	40.9	29.1
Nanophanerophytes	10.6	4.5	4.2
Phanerophytes	12.4	22.8	25.0
Dwarf-shrubs	1.7	9.1	12.3
Other	17.4	13.6	21.0
Anatomy of the leaves (%) (Ellenberg, 1979)			
Hygromorphic	7.0	0	0
Mesomorphic	80.8	81.9	75.0
Scleromorphic	8.8	13.6	25.0
Other	3.4	4.5	0

Table 3. Cover degree of plant species of the herb layer during the growing season 1995

Stand	Herb layer (species with high cover)	Cover degree (%) End of April	End of May	End of June	End of July
Rösa	*Calamagrostis epigeios*	12.2	24.9	21.8	17.5
	Brachypodium sylvaticum	3.5	16.1	18.4	16.4
	Rubus idaeus	1.2	10.3	12.4	11.5
	Rubus fabrimontanus	0.3	6.7	8.3	10.9
	Avenella flexuosa	1.1	1.4	1.3	1.7
	Oxalis acetosella	3.8	10.2	9.3	10.6
	Senecio fuchsii	1.3	3.9	5.3	5.7
	Viola reichenbachiana	0.3	1.6	1.6	1.6
	Vaccinium myrtillus	<0.1	1.2	1.5	1.2
	Dryopteris carthusiana	<0.1	1.1	1.1	1.2
Taura	*Avenella flexuosa*	44.3	57.0	56.1	35.9
	Betula pendula (juv.)	<0.1	1.0	1.0	1.9
Neuglobsow	*Avenella flexuosa*	16.4	15.9	22.3	23.1
	Vaccinium myrtillus	7.9	9.0	12.2	12.5
	Dryopteris carthusiana	0.1	0.6	1.0	1.4

a clear gradient. Plant life forms, however, indicated clear gradients. There was a trend for scleromorphic plants to dominate in Neuglobsow and, but less so, in Taura. In Rösa, however, hygromorphic plants had a high number of species (Tablc 2).

Table 4 indicates that in Taura most of the species have a high demand for light, but only a small part of these requires high nitrogen supply. In Rösa many species grow, which need high irradiation, but only 15% of these need higher amounts of nitrogen. Neuglobsow is characterized by species, which occur at a low level of light and nitrogen supply.

Table 4. Distribution of the indicators of light and nitrogen for all species

Rösa (R) Taura (T) Neuglobsow (N)	Indicator of nitrogen 1 to 4 = low nitrogen level		Indicator of nitrogen 5 to 9 = moderate to high nitrogen level	
Indicator of light 1 to 4 = low light level	R:	7.0%	R:	16.3%
	T:	0.0%	T:	6.3%
	N:	6.3%	N:	6.3%
Indicator of light 5 to 9 = moderate to high light level	R:	46.5%	R:	30.2%
	T:	62.5%	T:	31.2%
	N:	75.0%	N:	12.4%

At the end of April, biomass in Taura and Neuglobsow, which where *Avenella flexuosa* dominated, was twice that of Rösa. Maxima of circa one ton per hectar were found in June/July (Table 5). Moss species were irrelevant in Taura, and reached low levels of biomass in Rösa. In Neuglobsow, however, moss species with 1.5 t/ha surpassed the herbaceous species (Figure 3).

Table 5. Biomass of herbaceous species during the growing season 1995

	Herbaceous species	End of April		End of May		End of June		End of July	
	(Species with high	Biomass	Portion	Biomass	Portion	Biomass	Portion	Biomass	Portion
	cover at least in	(t/ha)	n	(t/ha)	n	s	(%)	s	n
	one month)		(%)		(%)	(t/ha)		(t/ha)	(%)
R	*Calamagrostis epigeios*	0.093	59	0.458	55	0.488	54	0.292	30
	Brachypodium sylvaticum	0.038	24	0.304	36	0.246	27	0.429	44
	Rubus idaeus	0.000	0	0.030	4	0.098	11	0.095	10
	Rubus fabrimontanus	0.027	17	0.033	4	0.068	7	0.078	8
	Avenella flexuosa	0.000	0	0.010	1	0.010	1	0.092	9
	Total	0.158		0.835		0.910		0.986	
T	*Avenella flexuosa*	0.437	100	0.840	100	1.177	100	0.838	100
	Total	0.437		0.840		1.177		0.838	
N	*Avenella flexuosa*	0.287	65	0.351	87	0.492	74	0.568	65
	Vaccinium myrtillus	0.153	35	0.051	13	0.175	26	0.308	35
	Total	0.440		0.402		0.667		0.876	

R = Rösa; T = Taura; N = Neuglobsow

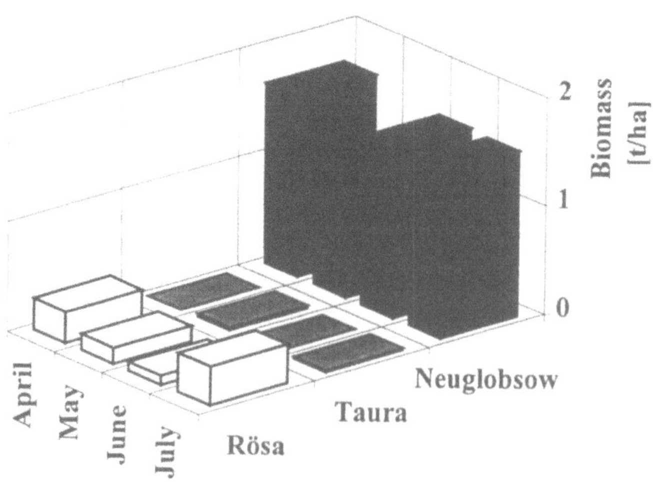

Figure 3. Biomass of mosses during the growing season 1995

Table 6. Proportional leaf area index (LAI part) of herbaceous species during the growing season 1995

Herbaceous species (Species with high cover at least in one month)		End of April LAI part (m^2/m^2)	End of April Portion (%)	End of May LAI part (m^2/m^2)	End of May Portion (%)	End of June LAI part (m^2/m^2)	End of June Portion (%)	End of July LAI part (m^2/m^2)	End of July Portion (%)
R	*Calamagrostis epigeios*	0.26	54	0.96	45	1.02	49	0.62	25
	Brachypodium sylvaticum	0.16	33	1.03	48	0.81	39	1.42	58
	Rubus idaeus	0.00	0	0.07	3	0.15	7	0.14	6
	Rubus fabrimontanus	0.06	13	0.07	3	0.10	5	0.11	4
	Avenella flexuosa	0.00	0	0.02	1	0.01	0	0.16	7
	Total	0.48		2.15		2.09		2.45	
T	*Avenella flexuosa*	0.71	100	1.40	100	1.65	100	1.42	100
	Total	0.71		1.40		1.65		1.42	
N	*Avenella flexuosa*	0.51	88	0.46	91	0.60	79	0.78	77
	Vaccinium myrtillus	0.07	12	0.05	9	0.16	21	0.23	23
	Total	0.58		0.51		0.76		1.01	

R = Rösa; T = Taura, N = Neuglobsow

Large differences between sites were found for the LAI (Table 6). Rösa, because of the prevalence of big-leafed species, had two to three times the LAI of Neuglobsow.

Since Rösa, due to its species diversity, was highly structured, some information on phenology, biomass dynamics and distribution of selected species is given: Biomass development of *Calamagrostis epigeios* started early and ended in May 1995. Plants reach more than 20 g per 10 cm height in 50 × 50 cm Plots between 0 and 40 cm. Thereafter, only shifts in vertical distribution were recorded. Biomass decreased in the first 20 cm due to death, and was constant up to 50 cm. The ratio of leaf mass to total biomass was constant between May and July.

Biomass of *Brachypodium sylvaticum* and *Rubus fruticosus* increased from April to July. Biomass of *Brachypodium sylvaticum* steadily decreased from the soil to the tip, concentrating in the first 30 cm. Here, too, the ratio of leaf mass to total biomass was constant between May and July.

After a concentration of biomass of between 20 and 30 cm in May, *Rubus idaeus* showed an even distribution between 0 and 40 cm until July. In July, most of the biomass was found between 30 and 50 cm in height. Apparently, this light demanding species avoid the competition with *Calamagrostis epigeios*. The shade tolerant species *Rubus fabrimontanus* had its biomass in July concentrated between 10 and 30 cm.

3.6. *Contribution of the field layer vegetation to stand transpiration*

The field layer transpiration increases from April to July corresponding to increasing LAI. Comparing the results of Tables 6 and 7, the relative contribution of a species to stand transpiration is mainly controlled by leaf area index and specific transpiration rates. While in July the LAI of *Rubus idaeus* did not exceed 6% of the total herb layer in Rösa, this species contributed 12% to herb transpiration. *Vaccinium myrtillus*, however, transpired less than 18% of the herb layer, although its partial LAI was 23%. During bright summer days, field layer transpiration exceeded tree transpiration. In Neuglobsow, where tree transpiration rates were highest, the transpiration of the field layer vegetation (excluding mosses) reached half the tree transpiration (Table 7).

Table 7. Contribution of the species to stand transpiration

Stand	Species*	Contribution to stand transpiration (%)			
		End of April	End of May	End of June	End of July
Rösa	*Pinus sylvestris*	72	44	52	50
	Brachypodium sylvaticum	7	11	11	18
	Calamagrostis epigeios	20	41	31	18
	Avenella flexuosa	0	0	1	3
	Rubus idaeus	0	1	4	6
	Rubus fabrimontanus	1	2	3	5
Taura	*Pinus sylvestris*	67	57	49	45
	Avenella flexuosa	33	43	51	55
Neuglobsow	*Pinus sylvestris*	74	76	58	66
	Avenella flexuosa	17	21	29	28
	Vaccinium myrtillus	9	3	13	6

Since transpiration data of the field layer vegetation were only available for some days, stand transpiration estimates have to be rather rough. For the growing season of 1995, these are 188 mm in Rösa, 181 mm in Taura and 185 mm in Neuglobsow. The estimated contribution of the field layer vegetation is 88 mm in Rösa, 87 mm in Taura and 61 mm in Neuglobsow.

4. Discussion

The number of species per 2500 m^2 in Rösa (57) is significantly higher than in Taura (22) or Neuglobsow (24). This can be explained with the deposition of alcaline ashes, a fact that is emphasized by the significantly higher average soil reaction indicator value. A high cover of *Calamagrostis epigeios* is usually explained by better nutrition caused by deposition, fertilisation and/or liming (e.g. Lux, 1964; Heinsdorf, 1984; Kopp, 1986; Hofmann, 1994). This is supported by the high N-content of *Calamagrostis epigeios* (Bergmann, 1993). Höpfner (1966) and Werner (1983), however, have shown that even on N-deficient sites this grass can reach high dominance. As early as 1976 Schmidt reported, that it took *Calamagrostis epigeios* 6 years to increase in cover after N-fertilisation, while N-mineralization was decreasing. This can also be seen in the data of Hofmann (1972 and 1987). In both, fertilised and unfertilised Scots pine plots *Calamagrostis epigeios* increased rapidly 20 years after the experiment had been started. No data on light regime in these stands were given by Hofmann, however. It is plausible, though, that light availability had increased because of enhanced tree mortality. Seidling (1993) and Brünn and Schmidt (1995) could show that N-availability is secondary to *Calamagrostis epigeios*, whereas light is of major importance. According to Eisenhauer and Strathhausen (1995) this reaction is restricted to moderately fertile sites.

All of these studies have in common, that they try to concentrate the attention on the impact of abiotic factors on this species, but neclect an essential biotic factor. As early as 1987 Hofmann observed, that in a fenced plot *Rubus idaeus* and *Epilobium angustifolium* gained dominance over *Calamagrostis epigeios*. Ellenberg (1989) reported that *Calamagrostis epigeios* was less grazed than other nitrophiluous plants. Two years after the experimental plot in Rösa had been fenced, *Calamagrostis epigeios* had disappeared and the site was dominated by *Brachypodium sylvaticum*, *Rubus fruticosus*, *Rubus idaeus* and *Senecio fuchsii*.

The high cover of *Avenella flexuosa* in Taura may result from the low light extinction in the canopy (middle L-value 6.4, Table 1), which could be a consequence of less LAI caused by air pollution. Similarly, the higher (by 5% as compared to Taura and Neuglobsow) share of geophytes in Rösa can be explained by atmospheric nutrient inputs, and is comparable to results of Paucke and Lux (1982). The highly structured herbaceous vegetation of Rösa allows species that require constantly high relative humidity, so-called hygromorphics, to thrive. In Neuglobsow the much higher contribution of scleromorphic species could be a consequence of relatively low water availability near the ground. This effect could be increased by the thick moss layer.

In Taura, the LAI of the herb layer in summer is nearly equivalent to the thirty-year-old Hartheim Scots pine plantation located in the Southern Rhine Valley (Wedler *et al.* 1996). The LAI for the field layer vegetation was comparatively high in Rösa and relatively low in Neuglobsow, although the mosses were not taken into account.

Similarly, the significantly higher specific needle area of the tree canopy in Rösa indicated the higher nitrogen supply as a result of repeated fertilisation (cf. Ende and Hüttl, this volume). In spite of a higher LAI, higher soil water availability and similar potential evapotranspiration, in 1994 transpiration of the pine in Rösa was lower than in Neuglobsow. The largest difference in transpiration occurred in the very dry month of July and August. Again, in

1995 the differences in transpiration were largest in the dry periods. In times of low water availability, the trees in Rösa consumed less water than those of Neuglobsow, without any losses in productivity (see Neumann and Wenk, this volume). This might result from differences in hydraulic architecture (Rust *et al.*, 1995; Rust *et al.*, this volume).

Data on transpiration rates of herb species of pine forests are rare. Müller (1967) reports 0.41 mm/d (1961) and 0.44 mm/d (1962) for a closed cover of *Vaccinium myrtillus.* Considerably higher are his values for *Calamagrostis epigeios* on bright summer days: 8 mm/d for moist soil and up to 3 mm/d when soil water availability was low.

In mid-summer, the forest floor contributed 50 to 60% to the stand transpiration in Rösa and Taura. The small data base and the lacking knowledge of root distribution and the effects of soil water availability on herb layer transpiration make these estimates rather uncertain. At least during droughts *Calamagrostis epigeios*, with its potentially 2 m deep roots (Rebele, 1995) could have a competitive advantage. The tap roots of some pine might serve the same purpose.

Black (1980), Roberts *et al.* (1980), and Tan and Black (1976) report a similarly high contribution of the field layer vegetation. Whole in times of high water availability *Pteridium aquilinum* transpired only 20% of the total in a pine stand in east England, its share rose to 57% in July (Roberts *et al.*, 1980). In periods of low soil water potential, the field layer vegetation contributed as much as two thirds to stand transpiration. This has been explained by tighter stomatal control in trees (Roberts *et al.*, 1980; Tan and Black, 1976).

The results show, that the field layer vegetation in the sites affected by deposition has a greater share in biomass, LAI, as well as stand transpiration compared with the background site in Neuglobsow. There were, however, no indications of increased drought stress in the Scots pine attributable to competition (Lüttschwager *et al.*, 1995). Stand transpiration reached similar levels in all three stands, merely the relative contribution of trees and herbs varied, as has been reported by Roberts (1983) and Wedler (personal communication). Hence, the stand transpiration was counter-balanced by competition effects of field layer vegetation.

Acknowledgements

The authors thank Bodo Grossmann for the construction of the stem flow gauges and Lothar Löwe for the measurements of gas exchange and biomass in the field. This study has been funded by the Ministry for Education and Science, Bonn, Germany.

References

Albrektson A, 1984. Sapwood basal area and needle mass of Scots pine (*Pinus sylvestris* L.) trees in Central Sweden. Forestry. 57, 36–43.

Bergmann JH. 1993. Das Sandrohr (*Calamagrostis epigeios* (L.) ROTH.). Forschungsbericht Zeneca Agro Frankfurt/M. 69 p.

Black T A, Tan CS, Nnyamah JU. 1980. Transpiration rate of Douglas fir trees in thinned and unthinned stands. Can J Soil Sci. 60, 625–631.

Brix H, Mitchell AK. 1986. Thinning and nitrogen fertilization effects on soil and tree water stress in a Douglas-fir stand. Can J For Res. 16, 1334–1338.

Brünn S and Schmidt W 1995 Reaktion von *Calamagrostis epigeios* (L.) ROTH. auf Unterschiede im Nährstoff- und Lichtangebot. GfÖ-Abstraktband 1995, p. 52.

Edwards WRN, Jarvis PG. 1983. A method for measuring radial differences in water content of intact tree stems by attenuation of gamma radiation. Plant Cell Environ. 6, 255–260.

Eisenhauer D-R, Strathhausen R. 1995. Sandrohr und Kiefernwirtschaft. I. Einfluß von Standortsfaktoren und Bestockungsstruktur auf die Konkurrenzsituation in der Krautschicht. AFZ/ Der Wald. 23, 1255–1261.

Ellenberg H. 1979. Zeigerwerte der Gefäßpflanzen Mitteleuropas. Scripta Geobotanica. 9, 122 p.

Ellenberg H. 1989. Eutrophierung – das gravierendste Problem im Naturschutz? Ein Dutzend illustrierte Informationen. NNA-Ber. 2. Jg. H. 1, 8–13.

Ellenberg H, Weber HE, Düll R, Wirth V, Werner W, Paulißen D. 1991. Zeigerwerte von Pflanzen in Mitteleuropa. Scripta Geobotanica, 18, 248 p.

Granier A. 1985. Une novelle méthode pour la mesure du flux de sève brute dans le tronc des arbres. Ann Sci For. 42(2), 193–200.

Habermehl A, Ridder H-W, Schmidt S. 1986. Mobiles Computer-Tomographie Gerät zur Untersuchung ortsfester Objekte. Atomenergie Kerntechnik. 48(2), 94–99.

Heinsdorf D. 1984. Wirkung von Mineraldüngung auf Ernährung und Wachstum von Winterlinden (*Tilia cordata* MILL.) auf Kippenbodenformen der Niederlausitz. Beitr Forstw. 18, 28–36.

Hillerdal-Hagströmer K, Mattson-Djöse E, Hellkvist J. 1982. Field studies of water relations and photosynthesis in Scots pine. II. Influence of irrigation and fertilization on needle water potential of young pine trees. Physiol Plant. 54, 295–301.

Höpfner B. 1966. Ökophysiologische Untersuchungen in einem extrem sauren Calamagrostidetum epigeios. Dissertation, Gießen, 243 p.

Hofmann G. 1972. Vegetationsveränderungen in Kiefernbeständen durch Mineraldüngung und Möglichkeiten zur Nutzanwendung derErgebnisse für biologische Leistungsprüfungen. Beitr Forstw. 4, 29–36.

Hofmann G. 1987. Vegetationsveränderungen in Kiefernbeständen durch Mineraldüngung. Hercynia N. F. 243, 271–278.

Hofmann G. 1994. Der Wald. Sonderheft Waldökosystem-Katalog. Deutscher Landwirtschaftsverlag, Berlin GmbH, 52 p.

Kaibyainen LK, Khari P, Sazonova T, Myakel YA. 1986. Balance of water transport in *Pinus sylvestris* L. III. Conducted xylem area and needles amount. Lesowedenje. 1, 31–37.

Kopp D. 1986. Vegetationsveränderungen auf Waldstandorten des Tieflandes durch Immission basischer Flugaschen und Zementstäube. Arch Naturschutz und Landschaftsforschung (Berlin). 26, 105–115.

Krauß HH. 1975. Stickstoffdüngung von Kiefernbeständen im Flachland der DDR - entscheidende Intensivierungsmaßnahme zur Stabilisierung und Erhöhung der Produktivität der Kiefernwälder. Landwirtschaftsausstellung der DDR, Markkleeberg, 51p.

Linder S, Rook DA. 1984. Effects of mineral nutrition on carbon dioxide exchange and partitioning of carbon in trees. In: Nutrition of Plantation Forests. Eds. Bowen GD, Nambiar EKS. Academic Press. pp 211–236.

Lüttschwager D, Rust S, Hüttl RF. 1995. Transpiration und Wasserleitfähigkeit von drei Kiefernbeständen bei unterschiedlicher sich ändernder Belastung mit Luftschadstoffen In

Atmosphärensanierung und Waldökosysteme Ed. Hüttl R F, Bellmann K and Seiler W, pp 75–86. UmweltWissenschaften UW Band 4, Eberhard- Blottner-Verlag, Taunusstein.

Lux H. 1964. Beitrag zur Kenntnis des Einflusses der Industrie-Exhalationen auf die Bodenvegetation in Kiefernforsten (Dübener Heide). Arch Forstwes. 13, 1215–1223.

Mencuccini M, Grace J. 1995. Climate influences the leaf area/ sapwood area ratio in Scots pine. Tree Physiol. 15, 1–10.

Mitchell AK, Hinckley TM. 1993. Effects of foliar nitrogen concentration on photosynthesis and water use efficiency in Douglas-fir. Tree Physiol. 12, 403–410.

Müller H. 1967. Standortsökologische Wasserhaushaltsuntersuchungen an *Vaccinium myrtillus* L. Arch Forstwes. 16(6/9), 587–590.

Paucke H, Lux E. 1982. Physiologische und ökologische Betrachtungen zur Wirkung von Immissionen auf Wälder. Hercynia N. F. (Leipzig). 19(2), 249–272.

Rebele F. 1995. *Calamagrostis epigejos* (L.) ROTH auf anthropogenen Standorten – ein Überblick. Vortrag auf der 25. Jahrestagung der GfÖ 11.09. bis 16.09.1995 in Dresden.

Roberts J. 1983. Forest transpiration: a conservative hydrological process? J Hydrol. 66, 133–141.

Roberts J, Pymar CF, Wallace JS, Pitman RM. 1980. Seasonal changes in leaf area stomatal and canopy conductance and transpiration from bracken below a forest canopy. J Appl Ecol. 17, 409–422.

Rust S, Lüttschwager D, Hüttl RF. 1995. Transpiration and hydraulic conductivity in three Scots pine (*Pinus sylvestris* L.) stands with different air pollution histories. Water Air Soil Pollut. 85, 1677–1682.

Schmidt W. 1976. Änderungen in der Stickstoffversorgung auf Dauerflächen im Brachland. Vegetatio. 36, 105–113.

Scholander PF, Hammel HT, Badsteet ED, Hemmingsen EA. 1965. Sap pressure in vascular plants. Science. 148, 339–346.

Seidling W. 1993. Zum Vorkommen von *Calamagrostis epigeios* und *Prunus erotina* in den Berliner Forsten. Verh Bot Ver Berlin Brandenburg. 126, 113–148.

Sheriff DW, Nambiar EKS, Fife DN. 1986. Relationships between nutrient status carbon assimilation and water use efficiency in *Pinus radiata* (D. Don.) needles. Tree Physiol. 2, 131–142.

Squire RO, Attiwill PM, Neales TF. 1987. Effects of changes of available water and nutrients on growth root development and water use in *Pinus radiata* seedlings. Aust For Res. 17, 99–111.

Tan CS, Black TA. 1976. Factors affecting the canopy resistance of a Douglas Fir forest. Boundary Layer Meteorology. 10, 475–488.

Wedeler M, Heindl B, Hahn S, Köstner B, Bernhofer C, Tenhunen JD. 1996. Model-based estimates of water loss from 'patches' of the understorey mosaic of the Hartheim Scots pine plantation. Theor Appl Climatol. 53(1–3), 135–144.

Wenk G, Wätzig H, Neumann U. 1994. Zuwachsuntersuchungen zur veränderten Immissionssituation im Lee des Gebietes Halle/Dessau. SANA-report 1993, E1.7.

Werner W. 1983. Untersuchungen zum Stickstoffhaushalt einiger Pflanzenbestände. Scripta Geobotanica. XVI, 95 p.

Whitehead D. 1978. The estimation of foliage area from sapwood basal area in Scots pine. Forestry. 51, 137–149.

7
Hydraulic architecture of Scots pine

S. RUST, D. LÜTTSCHWAGER and R.F. HÜTTL

1. Introduction

The hydraulic architecture of plants is controlled by internal and external variables like, *inter alia*, species (Huber, 1928; Zimmermann, 1978; Yang and Tyree, 1993), genotype (Neufeld *et al.*, 1992), competition (Sellin, 1993), site fertility (Espinosa-Banclari *et al.*, 1987; Long and Smith, 1989), stand management (Pothier and Margolis, 1988), climate (Whitehead *et al.*, 1984; Mencuccini and Grace, 1995), and air pollution (Happla *et al.*, 1986a,b; 1987a,b; Gruber, 1995). In the 1980s, concern about the decline of forests due to air pollution was widespread in the industrialised countries. Only rarely, though, have parameters of hydraulic architecture been included in studies on the causes of forest damage related to air pollution. Some investigations have focused on the relationship between crown transparency, then understood to be an indicator of tree vigour, and heartwood formation. In Scots pine, a lower sapwood/heartwood ratio was sometimes associated with high crown transparency (Happla and co-workers (1986a,b; 1987a,b), but see Bauch *et al.* (1986) and Bues and Schulz (1988)).

The SANA-stands, which had been chosen to represent two major deposition types and the background, not only significantly differed with regard to their air pollution histories, but have been fertilised with nitrogen to various degrees (see Ende and Hüttl; Lüttschwager *et al.*; Schaaf *et al.*, this volume). Therefore, the formulation of hypotheses on the likely direction and magnitude of differences in the hydraulic architecture was difficult.

We studied those parameters of hydraulic architecture, that were expected to be most sensitve to air pollution and to have the highest explanatory value for the stand water relations. We measured the proportion of sapwood, leaf area to sapwood area ratio, leaf specific conductivity, xylem hydraulic conductivity, and vulnerability to embolism. In addition, we analysed the sapflow density profiles of the three stands.

2. Materials and methods

In 1993 and 1994, conductive sapwood area was measured by computer-tomography (Edwards *et al.*, 1983; Habermehl *et al.* 1978, 1986) for all the 45 trees used for sap flow measurements in collaboration with the Center for

R.F. Hüttl and K. Bellmann, Changes of Atmospheric Chemistry and Effects on Forest Ecosystems, 111–118.

Radiology of the Phillips-University Marburg. Additionally, the method allowed to recognize hidden fungal infections in the trunks. In 1995, five trees per stand were felled, and conducting sapwood area at breast height measured by staining the heartwood with dye.

Sap flow density in the trunks of the trees was measured using a constant heating method (Granier, 1985). Preliminary experiments showed that in Scots pine, the entire sapwood contributes to sap flow. Therefore two gauges were installed in each tree ranging from 0 to 2.1 cm and 2.2 to 4.4 cm from the cambium, respectively. The gauges were installed facing north and insulated with Styrofoam and aluminum foil to avoid errors by direct sunlight. Automatic readings were taken every 30 s and averaged over 30 min periods.

The constant length of the probes and their installation at fixed distances of the cambium in connection with the varying diameter of the trees places them at a range of relative positions within the sapwood, therefore allowing the calculation of a sapflow profile (Rust *et al.*, 1995).

In 1995, 10 small (basal diameter 0.5 cm) and two larger (basal diameter 2.5 cm) branches per tree were collected from the top of the crown of 5 trees per stand and immediately re-cut under water. On the small branches hydraulic conductivity K_h (kg s^{-1} m MPa^{-1}) and vulnerability to embolism were measured in 2 years old segments 5 mm in diameter (including bark) and 40 mm length using a conductivity apparatus as described by Sperry *et al.* (1988). Branches were bench-top dried in the laboratory to reach a specified xylem water potential. Hydraulic conductance (K_T (kg s^{-1} MPa^{-1}) and K_h of the larger branches were measured in the field with a high-pressure flowmeter (Yang and Tyree, 1993; Tyree *et al.*, 1995). All needles were removed, dried and weighed. The shoots were infiltrated with de-ionised, degassed and filtered (0.2 μm) 0.01 N HCl for 24 hours to remove some embolism. Whole-shoot conductance (K_T, kg s^{-1} MPa^{-1}) was determined by dividing the rate of water flow through the shoot measured with the high-pressure flowmeter (F, kg s^{-1}) by the pressure drop (P, MPa) from the base of the shoot to the apices with needles removed (Yang and Tyree, 1993). Then, twigs were removed in steps of 2 mm of diameter, i.e. first all twigs smaller than 3 mm, then 5, 7 and so forth. The length of each excised segment and the age were recorded to calculate hydraulic conductivity (K_h, kg s^{-1} m MPa^{-1}) (Yang and Tyree, 1993) and mean age of a size class. Needle area was measured with a computerised image analysis system for a subsample and leaf specific conductance (LSC, kg s^{-1} MPa^{-1} m^{-2}) calculated as the ratio of hydraulic conductance and needle area as calculated from dry weights.

From other batches of branches, collected in May, September and October 1995, segments 40 mm in length and 1 to 10 years of age were excised under water and their hydraulic conductivity measured with a conductivity apparatus described by Sperry *et al.* (1988). We used de-ionized, degassed, filtered (0.2 μm) 0.01 N HCl and a pressure of 6 kPa. These data were used to estimate correlations of K_h and LSC with diameter and age.

In September 1995, five trees per stand were felled and discs cut at breast height. Out of each disc, cores 2.54 cm in diameter were cut in various positions of the sapwood. Relative position of the core was calculated as the ratio of its distance to the heartwood and total sapwood width. Hydraulic conductivity (K_h, kg s^{-1} m MPa^{-1}) was determined as above.

3. Results

In the pictures yielded by computer tomography (CT), there was a clear delineation between sapwood and heartwood. Staining gave similarly clear results. However, the results of these methods differed significantly. The heartwood area estimated by CT was more than twice that given by staining (22% vs. 9% of basal area).

As a consequence of the differences in stand density, stand basal area at breast height (calculated from the CT data) was highest in Neuglobsow with 22.5 ± 0.43 m^2, followed by Rösa with 17.9 ± 0.55 and Taura with 16.1 ± 0.64.

The correlations between needle and sapwood areas were close but significantly different for the three stands. Rösa showed the highest needle area per unit sapwood area ratio (0.207 ± 0.008), whereas the ratio was lowest for Neuglobsow (0.106 ± 0.005; Taura: 0.164 ± 0.01).

The diameter size class of the sample trees had a significant impact on K_h of segments of twigs from the top of the crown: in Neuglobsow and Rösa, K_h declined with size class; in Taura it increased. For all data combined, there was a clear decline from the largest to the third size class. The K_h of the third to fifth class were similar. Differences between size classes were only significant at high water potentials, at low water potentials, K_h was low for all trees.

In 2-year-old segments with an outer diameter of ca. 5 mm and water potentials close to 0 MPa, the hydraulic conductivity K_h was significantly higher in the least polluted stand ($p < 0.013$). The values (10^{-6} kg m s^{-1} MPa^{-1}) were in Neuglobsow 8.4 (± 1.2), in Taura 5.31 (± 0.64) and in Rösa 5.29 (± 0.51). The LSC in Neuglobsow was 52% higher than in the other stands (significance of difference $p < 0.005$).

There was a site-specific relation between K_h and xylem diameter from 2 to 15 mm. In xylem of low diameter, K_h was highest in Neuglobsow, whereas Rösa had the highest K_h at higher diameters. In Taura, there was a weak correlation.

Neuglobsow:	$Kh = 10^{1.39}\, D^{2.68}\quad R^2 = 0.91$
Rösa:	$Kh = 10^{2.40}\, D^{3.15}\quad R^2 = 0.90$

The Huber value (sapwood area/needle area) of segments from Rösa $2.41*10^{-4} \pm 9.44*10^{-6}$) was significantly lower than in Neuglobsow ($3.35*10^{-4}$ $\pm 1.62*10^{-5}$). Since the specific conductivity (K_s, kg s^{-1} m^{-1} MPa^{-1}) was similar, this resulted in higher LSC in Neuglobsow over the range to 15 mm.

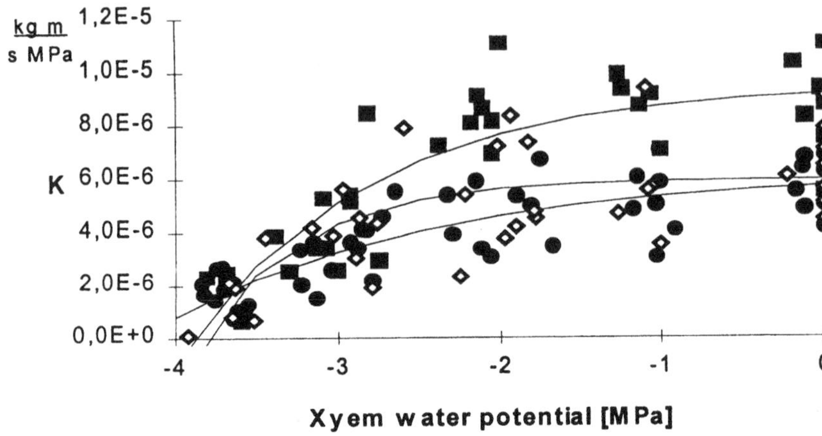

Figure 1. Xylem hydraulic conductivity K_h in 2-yr-old, ca. 5 mm outer diameter twigs as a function of xylem water potential. Closed squares, Neuglobsow; open diamonds, Taura; closed circles, Rösa

The R^2 of the fitted model was between 0.66 (Taura) and 0.86 (Neuglobsow). There was a significant difference between the sites, which was mainly due to the differences in maximum K_h (Figure 1).

Total conductance of whole shoots increased with basal diameter in a way that significantly differed between Rösa and Neuglobsow. At low diameters, K_T of Neuglobsow was higher than Rösa, at higher diameter there was no significant difference. Since branches from Rösa had higher needle mass and

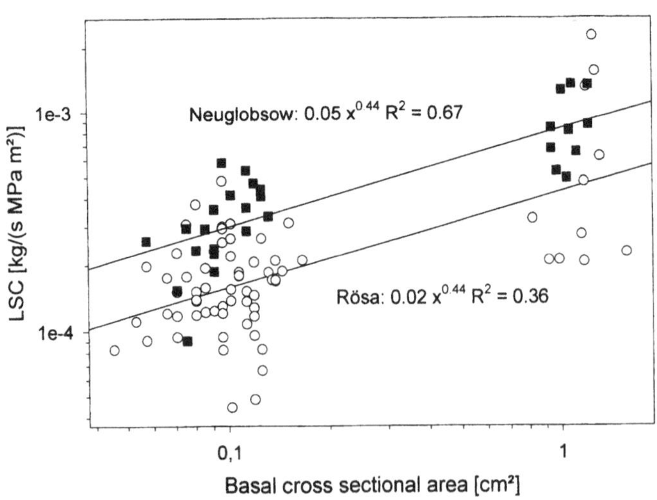

Figure 2. Leaf specific conductance LSC of branches from Rösa and Neuglobsow as a function of basal diameter. Closed squares: Neuglobsow, open circles: Rösa

specific needle areas, LSC was higher in Neuglobsow over the entire range measured (Figure 2).

Over the range 1 to 13 mm diameter, and 1 to 11 years in mean age, K_h significantly increased with diameter, but decreased with age for both, Rösa and Neuglobsow. One year old twigs had a much larger conductivity than older ones. Age accounted for 24 to 36% of the residual variation of the regression of K_h against diameter.

K_S, K_T and needle area decreased significantly with age, whereas HV increased, so that LSC remained fairly constant over the range 1 to 11 years of shoot age.

In Rösa and Taura, K_h of sapwood at breast height was significantly, although not strongly, correlated with relative position (Figure 3). For all data combined, K_h declined from the cambium to the heartwood. However, the regressions for the sites Neuglobsow and Rösa were significantly different.

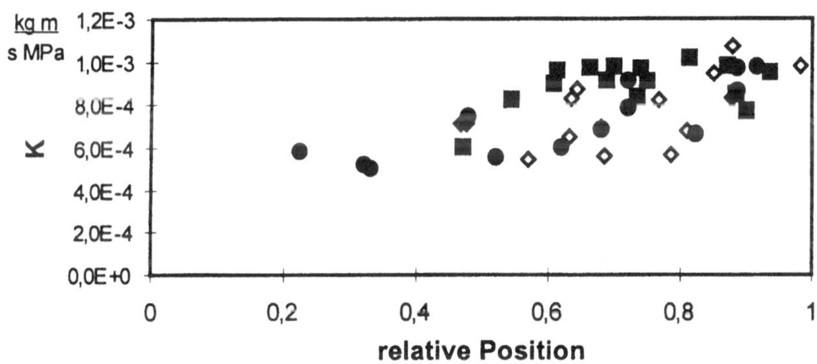

Figure 3. Xylem hydraulic conductivity K_h in cores cut out of stem discs at chest height $d_{1.3}$

The ratio of sap flow densities of inner and outer sapwood differed significantly between the stands. In the stand receiving the highest deposition rate, the mean flow density in the outer sapwood was higher than at the other sites, but decreased much more steeply towards the heartwood than at the low pollution site. During four weeks of comparable climatic conditions, the ratio of sap flow densities of inner and outer sapwood was 0.88 in the stand receiving low deposition rates, but 0.40 in the highly polluted stand. At the site with intermediate deposition rates, we found a ratio of 0.63 (all differences significant at $p < 0.001$, Table 1). For the entire growing season of 1994, sapflow densities at the outer sensors in Rösa were significantly higher than in Neuglobsow, but significantly lower at the inner sensors. On average, sap flow per tree in Rösa was 90% of that in Neuglobsow.

These ratios were, however, not constant, but changed from year to year and increased to approach unity in periods with low flow rates, e.g. at the beginning

Table 1. Mean sap flow density (kg/dm^2/d) in June 1994

	Rösa	Taura	Neuglobsow
Outer sensor	4.91 ± 0.04	3.68 ± 0.08	4.68 ± 0.08
Inner sensor	1.96 ± 0.09	2.32 ± 0.06	4.14 ± 0.08

and the end of the growing season. Because of the fixed length of the sensors and their constant distance to the cambium, the sap flow gauges were placed at different relative positions in the sap wood. Therefore, the data of all 30 sensors per stand were combined to estimate a sap flow profile, relating the average flux density to the relative sensor position.

For all sites and at high (75% quartile of flux densities) and low (25% quartile) flux densities, there was a significant ($p < 0.001$) correlation between the relative position of the sensor and flux density. In Taura and Rösa, there were significant differences between the flow profiles at high and low flux densities ($p < 0.001$). At high flux densities, 65% of the variation in Rösa can be attributed to sensor position, in Taura it is 40%. In Neuglobsow the correlation was weak with coefficients of determination of 0.37 at low flux densities and 0.15 for higher densities.

4. Discussion

Several authors have studied conducting sapwood area in Scots pine in the context of air pollution and forest decline. While Bauch *et al.* (1986) and Bues and Schulz (1988) found no differences between vigorous and declining trees with respect to their sapwood/heartwood ratio and moisture content, Hapla and co-workers (1986a,b; 1987a,b) report significant effects on moisture, wood density, and sapwood area. In contrast to the studies cited above, trees in Rösa showed no obvious signs of decline. Although we found no significant differences between the stands using computer tomography, the sapflow profile and the conductivity data indicate, that water transport was shifted from the inner, older tree rings to the outer sapwood close to the cambium. Despite the fact that flow density in the outer xylem was highest in Rösa, this could not make up for the loss in the inner xylem. From 1993 to 1995, mean flow per tree in Rösa was lower than in Neuglobsow, in spite of higher soil water availability in 1994.

The difference between CT and staining results can be explained by a relatively dry, but chemically not yet altered transition zone between sapwood and heartwood, which cannot be detected by staining. The use of staining techniques to determine an active xylem area could result in a significant inflation of stand transpiration because the extent of the sapwood is overestimated.

The larger conductivity in one year old xylem is similar to the findings of Sellin (1987). He also found lower permeability in various compartments of a suppressed Norway spruce (Sellin, 1993), similar to our results for tree size class. Although in Rösa branches of very small diameter (<7 mm) had a lower conductance than those of Neuglobsow, this difference disappeared at diameters above 10 mm. This probably results from the heigher conductivity at larger diameters and the lower average age in Rösa. LSC, however, was lower in Rösa over the entire range measured.

In small, distal twigs the polluted and N-fertilized sites Taura and Rösa had only half the LSC of Neuglobsow and, especially in Rösa, a much lower Huber value of the stem. The low sapwood area/needle area ratio may have been caused by the lower stand density and the nitrogen fertilisation (Pearson *et al.*, 1984; Keane and Weetman, 1986; Shelburne *et al.*, 1993; Valinger, 1993). The stand suffered from 45% loss of needles in 1989 (Hüttl *et al.*, 1995). Its high LAI in 1995 indicates a recovery since then. Since sapwood area is much more conservative than needle area, the increase in needle area caused by N-input and the decreasing deposition rates might have been out of proportion with the increase in sapwood area. A long-term acclimation to differences in evaporative demand as found by Mencuccini and Grace (1995) is another possible cause, since the climate in Rösa has been cooler and not as dry as in Neuglobsow (see data on climate in Lüttschwater *et al.*, this volume). The largest differences between the tree canopy transpiraton rates of the sites occurred during periods of low precipitation (Lüttschwager *et al.*, this volume). As a consequence of their less favorable hydraulic architecture, especially their lower LSC in distal twigs, lower Huber values and lower conductivity in older sections of the stem, the stands in Rösa and Taura may have reduced stomatal conductance to avoid the development of steeper water potential gradients, resulting in higher rates of embolism (Jones and Sutherland, 1991).

Acknowledgements

We thank Mel Tyree for giving Steffen Rust the chance to study his methods at the Proctor Maple Research Centre.

References

Bauch J, Götsche-Kühn H, Rademacher P. 1986. Anatomische Untersuchungen am Holz von gesunden und kranken Bäumen aus Waldschadensgebieten. Holzforschung. 40, 281–288.

Bues CT, Schulz H. 1988. Festigkeit und Feuchtegehalt von Kiefernholz aus Waldschadensgebieten. Holz als Roh- und Werkstoff. 46, 41–45.

Edwards WRN, Jarvis PG. 1983. A method for measuring radial differences in water content of intact tree stems by attenuation of gamma radiation. Plant Cell Environ. 6, 255–260.

Espinosa Bancalari MA, Perry DA, Marshall JD. 1987. Leaf area-sapwood area relationships in adjacent young Douglas-fir stands with different early growth rates. Can J For Res. 17, 174–180.

Granier A. 1985. Une nouvelle methode pour la mesure du flux de seve brute dans le tronc des arbres. Ann Sci For. 42(2), 193–200.

Gruber F. 1995. Splintermittlung und Splint-Nadelmasse-Beziehungen. AFZ. 15, 807–808.
Habermehl A, Ridder H-W. 1978. Ein neues Verfahren zum Nachweis der Rotfäule. Proc 5th Intern Conf on Root and But Rot in Conifers, 340–347. Hess Forst Versuchsanstalt, Hann.-Münden.
Habermehl A, Ridder H-W, Schmidt S. 1986. Ein mobiles Computertomographiegerät zur Untersuchung ortsfester Objekte. Kerntechnik. 48, 94–99.
Hapla F. 1986. Holzfeuchtigkeit von Kiefern unterschiedlicher Immissionsschadstufen. Holz als Roh- und Werkstoff. 44, 432.
Hapla F. 1986. Splint- und Kernanteile an Kiefern unterschiedlicher Immissionsschadstufen. Holz als Roh- und Werkstoff. 361.
Hapla F, Kottwitz K. 1987. Holzqualität von Kiefern aus einem Waldschadensgebiet. Holz-Zentralblatt. 95/96, 1333–1335.
Hapla F, Knigge W, Rommerskirchen A. 1987. Physikalische Holzeigenschaften und Zuwachs von schadsymptomfreien und immissionsgeschädigten Kiefern. Forstarchiv. 58, 211–216.
Huber B. 1928. Weitere quantitative Untersuchungen ueber das Wasserleitungssystem der Pflanzen. Jahrbuch Wiss Bot. 67, 877–959.
Jones HG, Sutherland RA. 1991. Stomatal control of embolism. Plant Cell Environ. 14, 607–612.
Keane MG, Weetman GF. 1987 Leaf area- sapwood cross-sectional area relationship in repressed stands of lodgepole pine. Can J For Res. 17, 205–209.
Long JN, Smith FW. 1989. Estimating leaf area of *Abies lasiocarpa* across ranges of stand density and site quality. Can J For Res. 19, 930–932.
Mencuccini M, Grace J. 1995. Climatic influences the leaf area/sapwood area ratio in Scots pine. Tree Physiol. 15, 1–10.
Neufeld HS, Grantz DA, Meinzer FC, Goldstein G, Crisosto GM, Crisosto C. 1992. Genotypic variability in vulnerability of leaf xylem to cavitation in water-stressed and well-irrigated sugarcane. Plant Physiol. 100, 1020–1028.
Pearson JA, Fahey TJ, Knight DH. 1984. Biomass and leaf area in contrasting lodgepole pine forests. Can J For Res. 14, 259–265.
Pothier D, Margolis HA, Waring RH. 1988. Patterns of change of saturated sapwood permeability and sapwood conductance with stand development. Can J For Res. 19, 432–439.
Rust S, Lüttschwager D, Hüttl RF. 1995. Transpiration and hydraulic conductivity in three Scots pine (*Pinus sylvestris* L.) stands with different air pollution histories. Water Air Soil Pollut. 85, 1677–1682.
Sellin AA. 1987. Hydraulic conductivity of the water transport system in Norway spruce. Fiziologiya Rastenii. 34(3), 545–553.
Sellin AA. 1993. Resistance to water flow in xylem of *Picea abies* (L.) Karst. trees grown under contrasting light conditions. Trees. 7, 220–226.
Shelburne VB, Hedden RL, Allen RM. 1993. The effects of site, stand density, and sapwood permeability on the relationship between leaf area and sapwood area in loblolly pine (*Pinus taeda* L.). For Ecol Management. 58, 193–209.
Sperry JS, Donnelly JR, Tyree MT. 1988. A method for measuring hydraulic conductivity and embolism in xylem. Plant Cell Environ. 11, 35–40.
Tyree MT, Sinclair B, Lu P, Granier A. 1993. Whole shoot hydraulic resistance in *Quercus* species measured with a new high-pressure flowmeter. Annales des Sciences Forestieres. 50, 417–23.
Tyree MT, Patino S, Bennink J, Alexander J. 1995. Dynamic measurements of root hydraulic conductance using a high-pressure flowmeter in the laboratory and field. J Exp Bot. 46, 83–94.
Valinger E. 1993. Effects of thinning and nitrogen fertilization on growth of Scots pine trees: total annual biomass increment, needle efficiency, and aboveground allocation of biomass increment. Can J For Res. 23, 1639–1644.
Whitehead D, Edwards WRN, Jarvis PG. 1984. Conducting sapwood area, foliage area, and permeability in mature trees of *Picea sitchensis* and *Pinus contorta*. Can J For Res. 14, 940–947.
Yang S, Tyree MT. 1993. Hydraulic resistance in *Acer saccharum* shoots and its influence on leaf water potential and transpiration. Tree Physiol. 12, 231–42.
Zimmermann MH. 1978. Hydraulic architecture of some diffuse-porous trees. Can J Bot. 56, 2286–2295.

8
Estimating fine root production of Scots pine stands

F. STRUBELT, B. MÜNZENBERGER and R.F. HÜTTL

1. Introduction

Photosynthesis, allocation and consumption of photosynthate are the main processes in many recently developed models for trees and stands (Zhang, 1994). A substantial part of the tree's overall carbon economy is represented by the root system. A thorough investigation of root requirements must include tissue synthesis, maintenance costs of the roots and costs of the symbionts over the lifetime of the roots (Eissenstat, 1992). A considerable proportion of the total costs of the root system can be quantified by the annual tissue synthesis in fine root production (Agren *et al.*, 1980). Additional maintenance costs are relevant in coarse roots, but in short living very fine and fine roots most of the required photosynthates for life are stored with the root formation as initial starch concentration (Persson, 1992). Marshall and Waring (1985) have shown that fine roots of seedlings die when all their stored carbohydrates are respired. Schneider *et al.* (1989) found nonstructural carbohydrates in dead Spruce roots and concluded that fine roots died from causes other than shortage of carbohydrates. That means a proper estimation of fine root production (FRP) contains both, the tissue synthesis and the later maintenance costs. For the carbon allocation to the mycorrhizal fungi only rough estimates exists. The lack of adequate methods for quantifying mycorrhizal mycelium impedes an accurate estimation (Finlay and Söderström, 1992).

Because of technical difficulties it is presently impossible to directly measure the annual FRP in situ of forest ecosystems (Santantonio and Grace, 1987). Therefore the FRP has to be estimated from other measurable root parameters or in specially designed experiments. The methods for these estimations vary widely and there are still considerable conceptual problems concerning what assumptions are appropriate regarding the growth and death of fine roots and the mechanisms that control these reactions (Orlov, 1968; Persson, 1978, 1983; McClaugherty *et al.*, 1982; Fogel, 1983; Alexander and Fairly, 1983; Nadelhoffer *et al.*, 1985; Vogt *et al.*, 1986; Santantonio and Grace, 1987). Moreover growth and death can occur simultaneously in different microsites (cells) of the soil, depending on favourable or unfavourable local conditions (Reynolds, 1970).

Numerous studies are based on the live and dead root amounts of sequential core samples. In this case the standing crop is known, but the mean lifetime of

R.F. Hüttl and K. Bellmann, Changes of Atmospheric Chemistry and Effects on Forest Ecosystems, 119–136.

the roots has to be estimated. The calculation of FRP from samples can be done in several ways which are described in Kurz and Kimmins (1986), Nadelhoffer and Raich (1992), Neill (1992), Persson (1978) and Publicover and Vogt (1993). Publicover and Vogt distinguish the max-min method, which is described by the subtraction of the annual minimum from the annual maximum of live root amount, from the decision-matrix method which considers the direction and relative magnitudes of changes in both live and dead root mass. The latter generally gives higher estimates than the max-min method, but both methods have an inherent tendency to underestimate the production (Neill, 1992; Publicover and Vogt, 1993).

Several authors have used the mesh bag (ingrowth core) method with the intention of directly measuring the FRP in a root free soil volume. After a time of incubation the mesh bags can be withdrawn and the amount of ingrown roots determined. To avoid critical changes of soil conditions, fresh soil from local origin can be carefully sieved. Nevertheless the resulting mesh bag FRP reflects an artificial environment in which a very low root mortality occurs (Persson, 1979; Kalhoff and Bornkamm, 1992).

If the standing crop of live and dead roots is known from sequential root sampling, the lifetime of roots can be estimated by measuring the root decomposition time. The mean ratio of live root mass to dead root mass should be the same for a long standing as the ratio of lifetime to decomposition time. This assumption of course only concerns the equilibrium case of a forest ecosystem where the root senescence is balanced by the root production. For a comparison of different sites, the decomposition rates need not to be determined exactly in any case. Considering the problems to measure the velocity of root decomposition, which is dependent on the preparation of the incubated roots, the humidity, the temperature and the depth in the soil, a proportional comparison of different sites can be accomplished without determination of the decomposition rates as values. Even a rough determination of relative decomposition in combination with dead root amounts can give information about relative FRP of different plots.

Three differently polluted and fertilized pine stands (*Pinus sylvestris*) were investigated in the research project SANA over the years 1993 to 1995. Rapidly decreasing input of alkaline fly ashes, together with a less rapidly decreasing input of SO_2 and increasing input of traffic NO_x since 1989, initiated a change in the adapted ecosystem status especially at higher polluted areas. Because of the high annual variation in root systems most of these long-term changes cannot be measured in two or three years of investigation (Persson, 1980; Vogt *et al.*, 1980). Therefore, the results of the reported research on roots have to be seen as a structural description of the current status of the investigated sites with special regard to FRP estimation.

2. Material and methods

Studies were carried out in three Scots pine ecosystems, at Rösa (61 years old) and Taura (45 years old) and Neuglobsow (65 year old), located next to the industrial centre of Bitterfeld/Halle, and at Neuglobsow in the remote area 75 km north of Berlin. Soils are sandy and the humus form is classified as moder or mor-moder. The throughfall precipitation at the Rösa site was 620 mm in 1994 (11/93–10/94) and 430 mm in 1995 (11/94–10/95). At the Taura site 690 and 510 mm and at the Neuglobsow site 520 and 490 mm were measured in the same periods.

For the Rösa site highest loads of SO_2 and fly ashes were reported for the pollution history (cf. Table 2, Hüttl and Bellmann, this volume). The forest floor is a mor-moder with a C:N ratio of 22 in the Of and 21 in the Oh and a total depth of 10 cm (Bergmann *et al.*, this volume). The pH(H_2O) is about 5 in the forest floor and down to a depth of more than 60 cm in the mineral soil. The substrate is sand with 13% clay and silt down to 50 cm depth. The mean element concentrations in the soil solution at the mineral soil surface are Ca: 32.17, Mg: 2.37, K: 1.94, Na: 3.24, Fe: 0.26, Mn: 0.09, Al: 1.01, NH_4: 0.38, SO_4: 49.84, NO_3: 28.69, Cl: 3.80, DOC: 35.42 (mg l^{-1}). The Ca/Al molar ratio in the Aeh is 2.66 (cf. Weisdorfer, this volume). In 1995 a leaf area index (LAI) of of 3.16 was measured, which was the highest of all stands under investigation (Lüttschwager *et al.*, this volume). The basal area of the trees is 34 m^2/ha. The bottom and the field layer is dominated by *Calamagrostis epigeios*, *Brachypodium sylvaticum*, *Rubus idaeus* and *Rubus fabrimontanu*.

The Taura site was polluted by lower amounts of SO_2 and almost no fly ashes. The forest floor is a moder with a C:N ratio of 25 in the Of and 30 in Oh and a total thickness of 7 cm (cf. Table 1, Hüttl and Bellmann, this volume). Highest contents of clay and silt of about 18% of the total mass to 50 cm depth were measured at this stand. The mean element concentrations in the soil solution at the mineral soil surface are Ca: 10.07, Mg: 1.32, K: 1.67, Na: 2.52, Fe: 0.29, Mn: 0.10, Al: 3.24, NH_4: 0.44, SO_4: 33.49, NO_3: 4.69, Cl: 2.13, DOC: 38.82 (mg l^{-1}). The Ca/Al molar ratio in the Aeh is 0.46 (Weisdorfer, this volume). The basal area is only 28 m^2 ha^{-1} depending on the lower number of trees. The field layer is dominated by *Avenella flexuosa* and patches of *Calamagrostis epigeios*.

The Neuglobsow site received the lowest amounts of SO_2 and no fly ashes, and, therefore, served as a reference site. The forest floor is a moder with a thickness of about 5 cm and a C:N ratio of 28 (Bergmann *et al.*, this volume). The pH(H_2O) varies between 3.8 in the humus layer and 4.6 at a depth of 50 cm. Different from the other plots soil organic matter distribution indicates a fAp horizon from former tillage. The content of clay and silt is only 10% in the upper 50 cm. The mean element concentrations in the soil solution at the mineral soil surface are Ca: 3.88, Mg: 0.94, K: 1.25, Na: 3.45, Fe: 0.31, Mn: 0.44, Al: 2.15, NH_4: 0.19, SO_4: 9.02, NO_3: 0.78, Cl: 4.25, DOC: 48.49 (mg l^{-1}). The Ca/Al molar ratio in the Aeh is 0.40 (Weisdorfer *et al.*, this volume). The basal area is 36 m^2/ha. The bottom and field layer is dominated by *Avenella flexuosa*, *Vaccinium myrtillus* and several mosses. X-ray fluorescense measurement of P in the forest floor results in about 400 mg·kg^{-1} at Rösa, 720 mg·kg^{-1} at Taura and 670 mg·kg^{-1} at Neuglobsow (Bergmann *et al.*, this volume).

The plots Rösa, Taura and Neuglobsow were investigated by core sampling for the root parameters live root mass, dead root mass and number of root tips. Sampling points were located at randomly selected coordinates within a permanently marked grid. Every 4 to 6 weeks during the growing season ten to twelve 80 mm diameter cores were taken from the top 55 cm (1993: 40 cm) of the mineral soil and of the humus layer. Cores were divided into samples of forest floor and mineral soil of 0–5 cm, 5–20 cm, 20–40 cm and 40–55 cm. To compare the amounts of roots in different soil depths the first two samples of the mineral soil were counted together as 0–20 cm. Samples were cool stored or deep frozen until root extraction. All live and dead root material longer than 15 mm was extracted manually from the soil. The material was divided into (i) live roots <1 mm in diameter (finest roots), (ii) live roots 1–2 mm (fine roots), (iii) live roots 2–5 mm (coarse roots), (iv) live roots 5–10 mm and dead roots in the same diameter classification. Definition of dead roots was done by judging the colour of the stele and the tensile strength of the root. Subdivision into diameter classes has been controlled by digital image analysis which was carried out for root surface determination (not reported in this paper). Roots were oven-dried at 65°C and weighed. Live roots up to 2 mm in diameter were used for counts of live root tips. Live root tips were assessed by judging turgescence and colour of the tip. Only tips with a white or yellow top were counted except in the case of single black or brown mycorrhizas or in case of mycorrhizal aggregates ('ball mycorrhiza') which were judged by turgescence.

FRP was calculated for finest and fine roots at every root sampling date in 1994 and 1995 (i) as the difference between annual maximum and minimum of finest and fine root biomass and (ii) by using the decision-matrix of McClaugherty and Aber (1982):

1. $FRP = \Delta L + \Delta D$ in case of increased live and dead root amounts
2. $FRP = \Delta L$ in case of increased live and decreased dead root amounts
3. $FRP = \Delta L + \Delta D$ or 0 in case of decreased live and increased dead root amounts (negative results must be replaced by 0)
4. $FRP = 0$ in case of decreased live and dead root amounts

where ΔL is the change in live root mass and ΔD is the change in dead root mass.

In July 1993 for installation of the mesh bag experiment, cylindrical nylon net bags of 80 mm in diameter and a mesh size of 10 mm were filled with coarsely sieved soil (7 mm sieve) of local origin and incubated at the sites down to a mineral soil depth of 40 cm. After each stage the filling was stamped down. Roots from the humus layer were sorted out manually before refilling. It was assumed that short root pieces left in the soil would have decomposed when the first mesh bags were taken from the plot. On six dates in 1994 and 1995 ten mesh bags were taken from each site and handled in the same way as described for the cores.

For testing the velocity of decomposition, a basic in situ experiment with litter bags was carried out in cooperation with Bergmann (Bergmann *et al.*, this volume). Live roots from the forest floor of each plot were sampled. Fresh root pieces, 0–1 mm and 1–2 mm in diameter and 50–70 mm in length were incubated in traditional litter bags without soil. The bags were made of fine mesh (2 mm) nylon cloth. The fresh weight of the roots of each litter bag was 0.5 g. Thus, in the case of fine roots (1–2 mm) only three to five roots were enclosed in each bag. The small mass of the sample was necessary to avoid matting of the roots within the bags resulting in artificial conditions for decomposition. The litter bags were incubated in the Of horizons of the sites in August and recollected monthly until November, 1995.

Most of the data could be transformed to normal distribution and equal variances by $x = \lg (x + 3/8)$ (Sachs, 1992). Results of ANOVA are based on the Scheffé-test with $p = 0.05$.

3. Results

The standing crop of live very fine roots (0–1 mm) was found to be significantly higher at the sites Rösa and Neuglobsow (46/53 $g{\cdot}m^{-2}$) compared to the Taura site (39 $g{\cdot}m^{-2}$). Referring to 20% smaller tree basal area at the younger stand at Taura, the amount of very fine roots can be calculated as 47 $g{\cdot}m^{-2}$ and is the same as at Rösa. Also the amount of live fine roots (1–2 mm) was significantly lower at Taura (Figure 1). Calculated fine root amount at Taura is 62 $g{\cdot}m^{-2}$ and still lower than at Rösa or Neuglobsow. About 1 t/ha live coarse roots (2–5 mm) were found at all sites. The amount of dead roots differed in all diameter classes. At Neuglobsow, two to three times more dead very fine roots and three to five times more dead fine and coarse roots were found.

Significantly more live finest roots were found in 1995, the least in 1994. On the other hand, significantly more live fine roots were found in 1994. Coarse root amounts differed not significantly between the years. Coarse roots from 5 to 10 mm rarely appeared, so data should be judged with caution (Table 1).

The live root mass distribution in several soil depths showed a similar pattern at Rösa and Neuglobsow (Figure 2). About 75% of live very fine roots were distributed in equal parts at the humus layer and the upper 20 cm of mineral soil. The humus layers in Rösa and Neuglobsow were three times better rooted with finest roots than at Taura. The highest finest root concentration ($g*100\ ml^{-1}$) was found in the smaller humus layer of Neuglobsow.

At all plots highest fine root amounts (1–2 mm) were found in the upper 20 cm of the mineral soil. The same distribution was found for coarse roots. The deepest examined horizon (40 to 55 cm) contained very few roots at Neuglobsow.

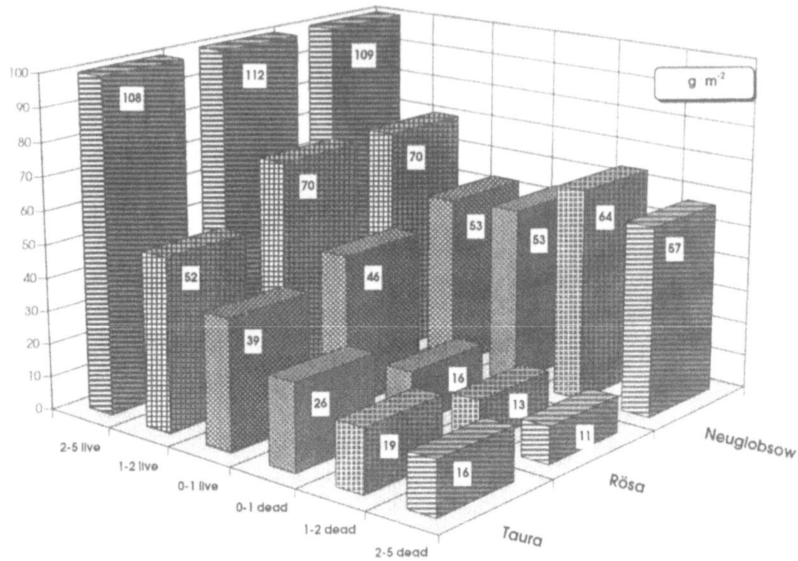

Figure 1. Mean amounts of live and dead root dry weight in three years of investigation

Table 1. Mean annual amounts of live root dry weight. Standard errors given in parentheses. Data of 1993 were complemented by mean 1994/1995 values from a depth of 40–55 cm

Biomass (g·m^2)		Taura	Rösa	Neuglobsow
0–1 mm	1993	40.0 (3.1)	47.2 (2.9)	50.8 (3.8)
	1994	34.7 (2.7)	41.5 (3.0)	47.3 (3.3)
	1995	47.7 (2.6)	52.4 (3.1)	62.0 (5.4)
1–2 mm	1993	52.5 (4.5)	69.6 (4.7)	70.7 (5.1)
	1994	62.8 (4.1)	86.3 (5.0)	79.1 (4.6)
	1995	47.9 (2.9)	64.1 (3.5)	65.7 (5.0)
2–5 mm	1993	104.8 (15.7)	83.9 (8.7)	93.4 (9.1)
	1994	104.9 (13.0)	122.0 (14.7)	119.0 (11.5)
	1995	126.6 (19.9)	144.9 (20.5)	121.1 (17.2)
5–10 mm	1994	45.6 (10.5)	43.5 (11.3)	10.8 (8.5)
	1995	72.9 (24.3)	65.5 (20.5)	84.9 (29.8)

Root settling in mesh bags is illustrated in Figure 3. After a phase of initialization in the summer and autumn of 1993 with very few ingrown roots (data not shown), the process of root settling started in early spring of 1994. After the growing season, about the same amounts of live finest roots as in soil cores were found at Rösa and Neuglobsow, while at Taura the expected

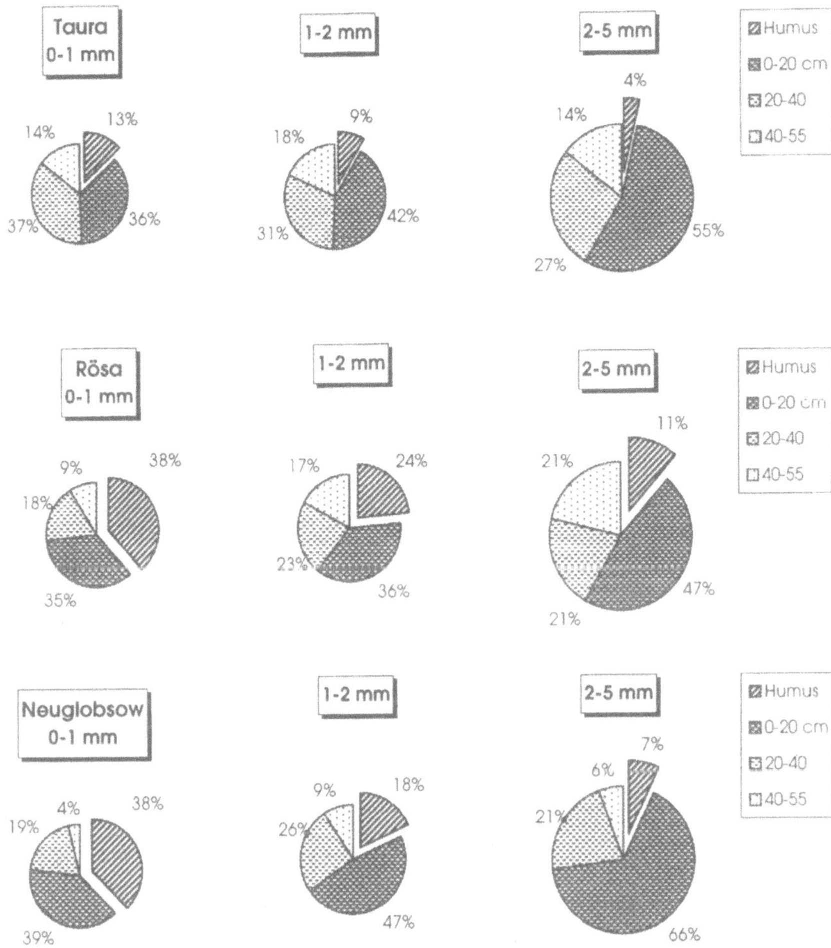

Figure 2. Distribution of live root dry weight at several soil depths

biomass did not appear until 1995. The live fine roots (1–2 mm) grew in at much slower rates at Rösa and Taura and even in 1995 at Taura the expected amount was reached only by the half. Dead root amounts in the mesh bags were generally low.

Estimates of FRP differed substantially depending on the method of calculation employed (Table 2). The turnover was calculated by the mean amounts of live very fine or fine roots from core samples (data of Figure 1).

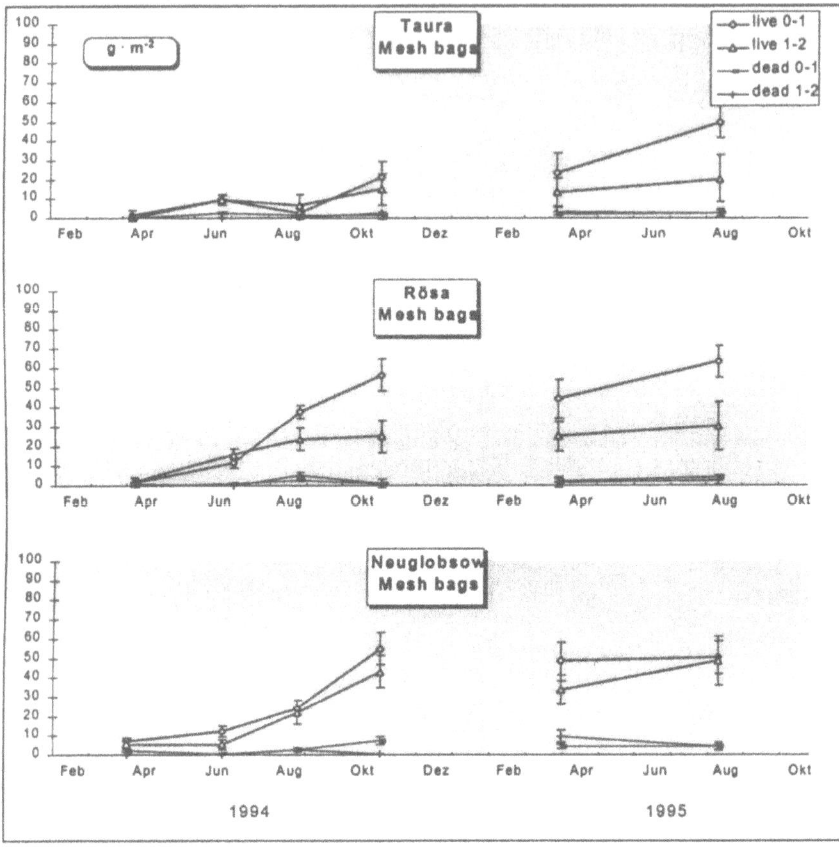

Figure 3. Ingrown finest and fine root dry weights of the mesh bag experiment. Standard errors

The fewest differences between the plots are indicated by the calculated turnover of the max-min results. At the sites Rösa and Neuglobsow, the mesh bag results of finest roots were closer to the decision-matrix calculation while the ingrown fine root amounts (1–2 mm) were at all sites of the same magnitude as the max-min values. At the Taura site, even less very fine roots emerged in the mesh bags than calculated using the conservative max-min method.

Decision-matrix values of FRP (0–1 mm) at Rösa and Neuglobsow of about 75 $g{\cdot}m^{-2}$ were higher than standing crops of 46/53 $g{\cdot}m^{-2}$, indicating a turnover of 1.6/1.5. Calculated turnover by mesh bag results almost agreed (1.3/1.2). At the Taura site the decision-matrix value of 39 $g{\cdot}m^{-2}$ for very fine roots was the same as the standing crop, which means a turnover of 1.0, while very few roots grew in the mesh bags. Calculated turnover by mesh bag result was only 0.6.

Table 2. Finest and fine root production (FRP) as max-min and decision-matrix calculations (mean of 1994 and 1995) in comparison with mesh bag results

Calculated FRP ($g \cdot m^2 \cdot year^{-1}$) and turnover		Taura	Rösa	Neuglobsow
0–1 mm	Max-min	29.0 (0.7)	34.9 (0.8)	43.1 (0.8)
	Mesh bag	24.1 (0.6)	61.0 (1.3)	65.3 (1.2)
	Decision matrix	38.9 (1.0)	75.1 (1.6)	77.2 (1.5)
1–2 mm	Max-min	21.5 (0.4)	29.3 (0.4)	35.1 (0.5)
	Mesh bag	18.6 (0.4)	30.9 (0.4)	46.0 (0.7)
	Decision matrix	29.9 (0.6)	54.1 (0.8)	100.6 (1.4)

Decision-matrix values of FRP (1–2 mm) differed more strongly between the plots than the FRP (0–1 mm). Calculated turnover from decision-matrix values was highest at Neuglobsow (1.4) and the same as for the finest roots. At Rösa and Taura the turnover of 0.8/0.6 of fine roots was only half as fast as for the finest roots.

The velocity of fine root decomposition in litter bags was measured to be about the same at all sites. Data are shown in Bergmann *et al.*, this volume.

Seasonal fluctuations of live very fine roots were highest in the humus layer at Rösa and Neuglobsow and in 0–20 cm in Taura (data not shown).

Since these horizons also included the highest root amounts of this diameter class (Figure 2), comparisons of live roots 0–1 mm, dead roots 0–1 mm and number of root tips in relationship to soil water content were made at these soil depths (Figures 4–6). The time series of the measured live finest root standing crop showed a significant maximum in September 1995 during a period of high precipitation after a long drought. A general decrease was found in autumn. The course of development of the dead finest root mass was in agreement rather than giving contrasting results. No reactions to biomass changes were found.

The number of root tips showed more sensitive reactions of the same pattern as the biomass of finest roots except in the late summer and autumn 1993 when, despite good water conditions, the number of root tips decreased. At Taura, the root system showed the highest annual number of root tips in 1994 with the lowest water content at the investigated depth of 0–20 cm.

The mean number of root tips was significantly higher at Rösa ($46 \cdot 10^3 \cdot m^{-2}$) than at Neuglobsow ($37 \cdot 10^3 \cdot m^{-2}$) or Taura ($33 \cdot 10^3 \cdot m^{-2}$).

No significant differences were found between the years with the exception of Taura, where the maximum occurred in 1994. The vertical distribution of the root tips in the soil varied strongly (Figure 7). At Neuglobsow, two-thirds of all measured root tips were concentrated in the humus layer. At the other sites, the distribution of root tips was about the same as for finest root biomass.

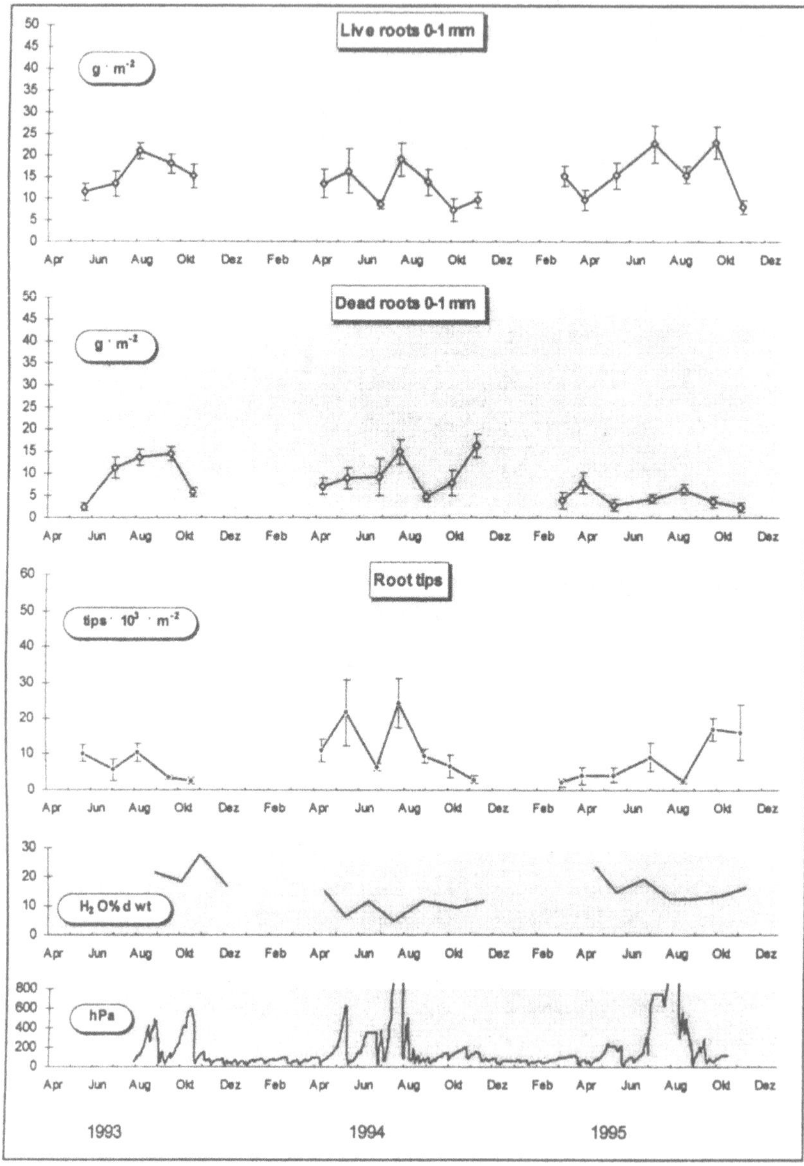

Figure 4. Comparison of live finest root dry weight, dead finest root dry weight and number of root tips in the mineral soil at a depth of 0–20 cm at the Taura site with H_2O% of soil dry weight at the same depth (Bergmann, personal communication) and tensions in the depth of 20 cm (Weisdorfer, personal communication). Standard errors

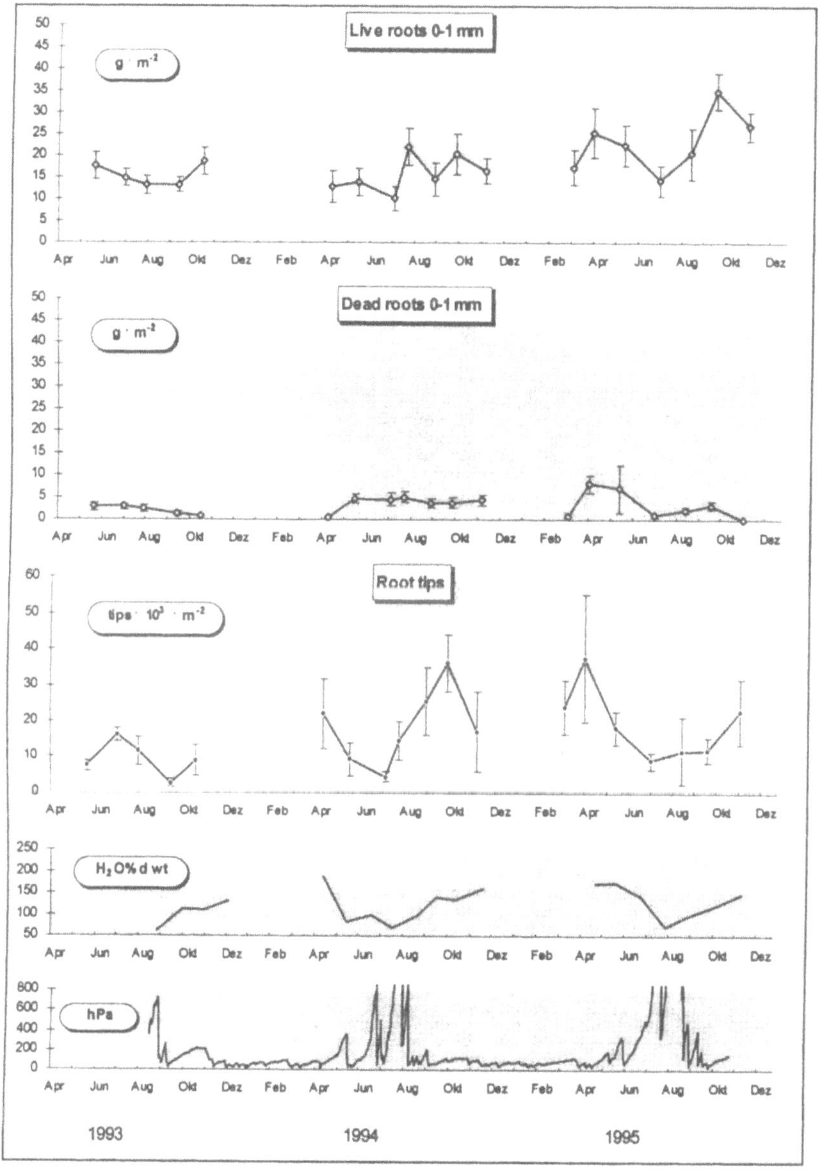

Figure 5. Comparison of live finest root dry weight, dead finest root dry weight and number of live root tips in the humus layer at the Rösa site with H_2O% of humus layer dry weight (Bergmann, personal communication) and tensions at the depth of 20 cm (Weisdorfer, personal communication). Standard errors

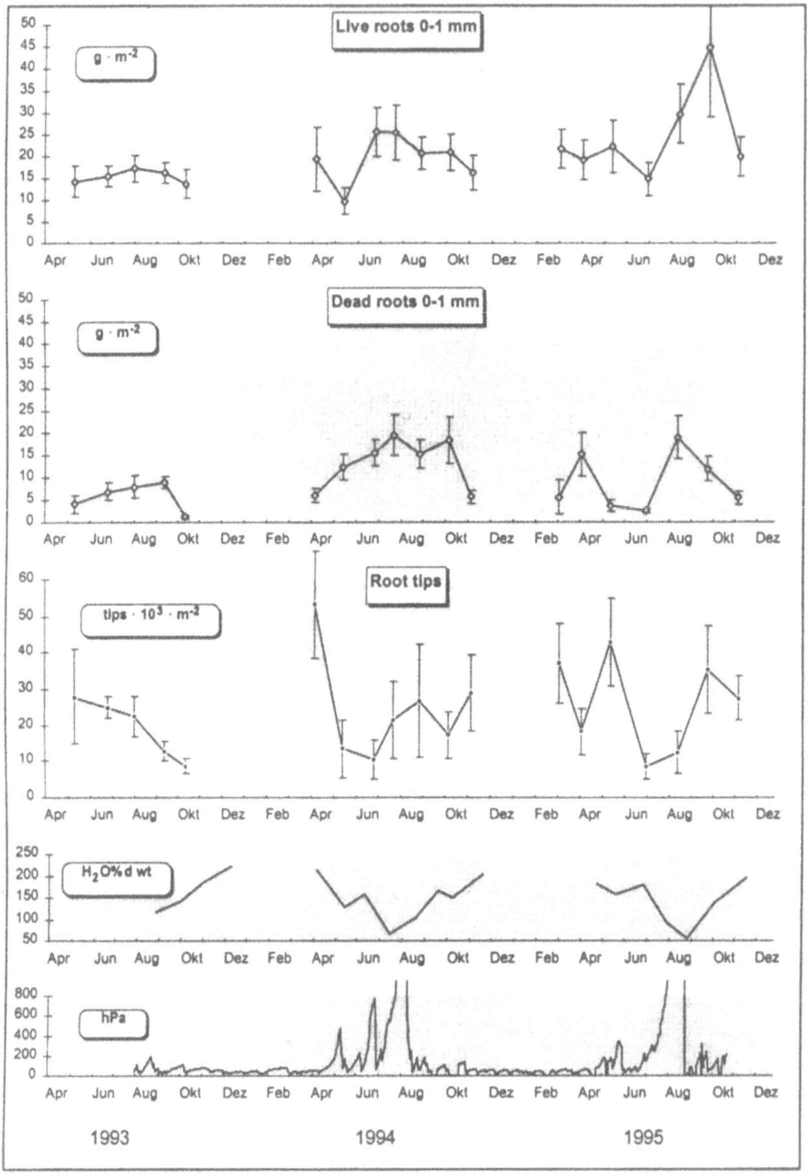

Figure 6. Comparison of live finest root dry weight, dead finest root dry weight and number of root tips in the humus layer at the Neuglobsow site with $H_2O\%$ of humus layer dry weight (Bergmann, personal communication) and tensions in at depth of 20 cm (Weisdorfer, personal communication). Standard errors

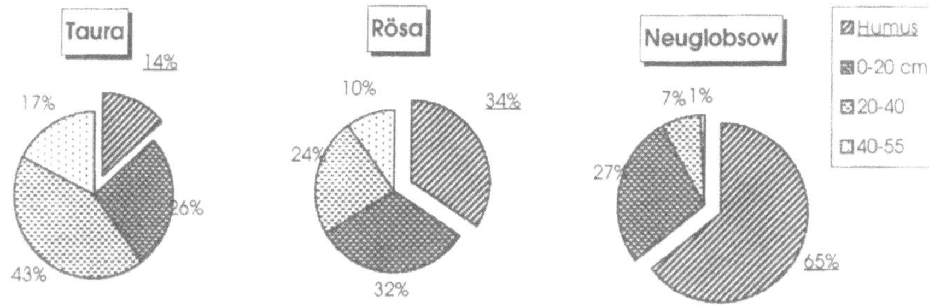

Figure 7. Distribution of root tips at several soil depths

4. Discussion

The fertilization and pollution at the investigated sites Rösa and Taura did not affect the total amount of live roots in the fractions finest, fine and coarse. For the time of investigation the biomass was measured to be about the same as at the background site Neuglobsow. Also the total number of root tips was little affected. The main difference between the sites are the changed vertical distribution of finest roots and root tips at Taura and the lower amounts of dead roots at Taura and Rösa.

McClaugherty *et al.* (1982) discussed the max-min method of FRP estimation as assuming a single annual pulse of finest or fine root production, while the decision matrix method assumes rapid root turnover throughout the year. If fluctuations are annual, then the max-min approach would yield more realistic results. If the fluctuations are short term, the FRP could be calculated only by considering monthly or shorter term changes as in the decision matrix approach. Following McClaugherty *et al.* in all three investigated sites three lines of evidence support short-term rather than annual patterns: (1) observation of new root tips at several dates throughout the year, (2) growth of roots into the mesh bags not only in spring, and (3) increases of the standing crop in 1994 and 1995 at least two times a year. According to this, the calculated FRP from the decision matrix method should represent the most accurate estimation.

It is questionable why the mesh bag results are generally lower than the probably underestimating decision-matrix results. A possible reason for the lower mesh bags results is seen in the disturbed structure of the sieved soil in the mesh bags resulting in a delayed or reduced root ingrowth. Even the careful sieving procedure and the thorough refilling in several steps with soil of the original horizon could not prevent the disintegration of soil conglomerations. After the colonization of the mesh bags a better nutrient availability probably resulted in preventing extensive short root formation. As mentioned before, consequently the roots in the mesh bags showed a very low mortality as found in other mesh bag experiments. Mesh bag results have to be seen only as an approximate estimation of root settlement behaviour under special conditions.

The velocity of decomposition in relation to the dead root amounts can be used for an estimation of the relative turnover. The decomposition rates of root material in litter bags was measured to be about the same at all stands. This corresponds with Bergmann *et al.* (thisvolume) who found different decomposition rates at the sites Rösa, Taura and Neuglobsow with respect to the different quality of the field layer necromasses, but not with respect to the soil conditions. This indicates that the decomposition rates of roots should have the same order of magnitude at all sites because pine roots represent the same quality of litter. However, the measured decomposition rate counts only for the forest floor. Decomposition in deeper horizons could be slower. Because of the diverging root distribution at Taura with more finest and fine roots in deeper horizons, the higher total amounts of dead roots compared to Rösa could be attributed to the distribution and may not indicate a different turnover.

The threefold higher dead finest root amounts and the fivefold higher dead fine root amounts (1–2 mm) at the background site Neuglobsow in comparison to the polluted plot Rösa, have to be discussed at the same decomposition rates, same live standing crops and same vertical root distributions. This would indicate a three respectively five times higher turnover and FRP for finest and fine roots at the site Neuglobsow. In contrast to this, no difference was found in the decision matrix FRP (0–1 mm) between Rösa and Neuglobsow. The FRP (1–2 mm) at Neuglobsow was calculated to be only double the size of Rösa.

It is assumed that the decision matrix values for the Neuglobsow site are too low. Max-min and decision-matrix calculations depend only on the changes in live or live and dead root amount. These changes result partly from the high variation in the samplings, which depends on the heterogenous root distribution and is the same at all sites. The other part is dependent on the sensibility of the root system to climatic changes and other influences. In years with higher variation between dry and wet periods a higher FRP will be calculated. The permanent growing and dying processes are not reflected in the calculation.

On the other hand the estimation from dead root amounts and relative decomposition rate can fail, mainly due to two reasons: (1) The measured decomposition rate from the first four months reflects only the changes of weight. Structural changes of the tensile strength of the roots which determine if a dead root falls to pieces or not are not considered. (2) Outgrowing processes were not taken into account. If most fine roots at the Rösa site are former finest roots while at Neuglobsow finest roots die and fine roots grow directly from thicker long root tips, consequently less dead finest roots must be found at Rösa. The FRP (0–1 mm) of Rösa has to be estimated higher and the FRP (1–2 mm) would be further reduced. On the other hand the turnover of fine roots cannot be higher than the turnover of finest roots, because almost all finest roots emerge from fine roots.

Short lifetime of the roots is probably the result of permanent root growing in search for unexploited microsites of the soil while roots in exploited compartments have to die or pass into a quiescent period (Persson, 1992; Reynolds, 1974, 1975). High mycorrhiza turnover in the forest floor at Neuglobsow as described in Schmincke (1995) is the direct consequence of this behaviour. Nutrient conditions at Rösa are better than at Taura which are better than at Neuglobsow. At Rösa, more nutrients can be absorbed with less water, and thus rooted soil compartments get less dry and can be exploited longer. This is in agreement with measured pine transpiration which is lowest at Rösa in periods of drought (Lüttschwager *et al.*, this volume). The gradient of soil nutrient conditions is reflected by a gradient of root longevity, and converse to the dead root amount.

Numerous studies have shown that most roots are found in the upper 50 cm of soil, and most activity and mycorrhizas in the top 20 cm depending on soil aeration and fertility (Fogel, 1983; Kalela, 1950; Persson, 1980; Vogt *et al.*, 1993), especially on sandy soils without claypans or attainable ground water table. Ehrenfeld *et al.* (1992) reviewed several papers and summarized an allocation of 30–60% of finest and fine roots to the forest floor as a general feature of conifer forests, regardless of tree species. At the Taura site only 13% of finest roots and 9% of fine roots were found in the forest floor. Low amounts of fine roots in the usually best rooted forest floor either indicate bad conditions in this horizon or much better conditions in other soil parts. Apart from little higher contents of clay and silt, there are no reasons to suppose much better conditions in the mineral soil at Taura than at Rösa. On the other hand for a Ca/Al molar ratio below one, an injuring effect to the root system is discussed (Ulrich *et al.*, 1984), but the ratio of the backgound site Neuglobsow was the same while the vertical root distribution completely differed.

Similar seasonal and annual changes in the amount of live roots at all investigation sites point to responses to climatic conditions rather than to a general ecosystem change which is expected at least at the Rösa site according to the decreasing air pollution. As mentioned before, such a long-term change cannot be measured within three years of investigation. In general, increasing moisture availability affects the increment of biomass and the root longevity (Vogt *et al.*, 1993). Corresponding to this, lowest amounts of live finest roots were found in dry periods and at the beginning and the end of the growing season depending on low soil temperatures. On the other hand, mean annual amounts of live finest roots were higher in 1995 with lower annual throughfall. Since in the same year lower fine root concentrations (1–2 mm) were found, it is assumed that a part of the thicker fine roots were replaced by more absorbing finest roots in the drier year. The deficit of fine roots (1–2 mm) at the Taura site was probably caused by the younger age of that stand. Kalela (1950) found a steady increase of this diameter class in 16 *Pinus sylvestris* stands aged between 10 and 100 years.

The number of root tips is strongly dependent on the method of root sorting on soil (Fogel, 1983) and from the distinction between vital and dead tips.

Kalela (1955) found values of vital root tips per square meter in a 75 years old *Pinus sylvestris* stand, similar to ours, between about 10 000 and 100 000 tips with a white or yellow top. After summer drought an irrigation plot led to a 50 times higher number of root tips in only three weeks. In this study the reactions of the number of root tips on soil drought and rewetting can be proved too. Especially in spring root tips responded sensibly to low water contents. Less than 150% H_2O of humus dry weight in spring resulted in a strong decrease at the plots Rösa and Neuglobsow, whereas in summer remoistening to lower contents could stimulate root tip growth again. On the other hand, unfavourable soil water conditions might stimulate root tip growth. The latter was found in 1994 at Taura when the number of root tips was unproportionally high although the soil water content was low.

In contrast to other authors who found decreasing numbers of root tips as the consequence of a better nitrogen supply (Ritter and Tölle, 1978; Uebel, 1982; Ritter, 1990; Majdi and Persson, 1995) or forest decline (Meyer, 1987) the highest total number of root tips was found at the polluted site Rösa. Because of the larger turnover a corresponding higher root tip production has to be assumed for the background site Neuglobsow. It is assumed that comparable numbers of root tips and with it mycorrhiza at all sites are affected by generally low precipitation and very low concentrations of P of between 400 and 700 $mg \cdot kg^{-1}$. This agrees with the opinion of Eissenstat (1992) that mycorrhizal infection is only beneficial for plant growth when P supply is more limiting than carbon supply.

Acknowledgements

This study was supported by the Federal Ministry of Education, Science, Research and Technology (BMBF), Bonn, Germany, as part of the SANA-project.

Root material from three dates in 1995 was placed at authors disposal by courtesy of Dr. sc. J. Lehfeldt.

References

Agren GI, Axelsson B, Flower-Ellis JGK *et al.* 1980. Annual carbon budget for a young [14-yr-old] Scots pine. In: Structure and Function of Northern Coniferous Forests – An Ecosystem Study. Ed. T Persson. pp. 307–313. Ecol Bull. 32.

Alexander IJ, Fairly RI. 1983. Effects of N fertilisation on populations of fine roots and mycorrhizas in spruce humus. Plant Soil. 71, 49–53.

Bergmann C, Fischer T, Hüttl RF., this volume.

Ehrenfeld JG, Kaldor E, Parmelee RW. 1992. Vertical distribution of roots along a soil toposequence in the New Jersey Pinelands. Can J For Res. 22, 1929–1936.

Eissenstat DM. 1992 Costs and benefits of constructing roots of small diameter. J Plant Nutr. 15, 763–782.

Finlay R, Söderström B. 1992. Mycorrhiza and carbon flow to the soil. In: Mycorrhizal Functioning - An Integrative Plant-Fungal Process. Ed. F Allen. pp 134–160. Chapman and Hall, New York.

Fogel R. 1983. Root turnover and productivity of coniferous forests. Plant Soil. 71, 75–85.

Kalela EK. 1950. The horizontal roots of pine and spruce stands. Acta Fenn. 57, 68.

Kalela EK. 1955. Changes in the root system of Pine stands during the growing season. Acta Fenn. 65, 42.

Kalhoff M, Bornkamm R. 1992. Distribution and development of fine roots in a pine oak forest ecosystem close to conurbation. In: Root Ecology and its Practical Application, 3. ISSR Symp. Wien, 1991. Eds. L Kutschera, E Hübl, E Lichtenegger, H Persson, M Sobotik. pp 501–504. Verein für Wurzelforschung, Klagenfurt.

Konopatzky A. 1996., this volume.

Kurz WA, Kimmins JP. 1987. Analysis of some sources of error in methods used to determine fine root production in forest ecosystems: a simulation approach. Can J For Res. 17, 909–912.

Lüttschwager D, Rust S, Wulf M, Hüttl RF., this volume.

Majdi H, Persson H. 1995. Effects of ammonium sulphate application on the chemistry of bulk soil, rhizosphere, fine roots and fine-root distribution in a *Picea abies* (L.) Karst. stand. Plant Soil. 168–169, 151–160.

Marshall JD, Waring RH. 1985. Predicting fine root production and turnover by monitoring root starch and soil temperature. Can J For Res. 15, 791–800.

McClaugherty CA, Aber JD, Mellillo JM. 1982. The role of fine roots in the organic matter and nitrogen budgets of two forest ecosystems. Ecology. 63, 1481–1490.

Meyer FH. 1987. Der Verzweigungsindex, ein Indikator fur Schäden am Feinwurzelsystem. Forstw Cbl. 106, 84–92.

Nadelhoffer KJ, Aber JD, Melillo JM. 1985. Fine roots, net primary production, and soil availability: a new hypothesis. Ecology. 66, 1377–1390.

Nadelhoffer KJ, Raich JW. 1992. Fine root production estimates and belowground carbon allocation in forest ecosystems. Ecology. 73, 1139–1147.

Neill C. 1992. Comparison of soil coring and ingrowth methods for measuring belowground production. Ecology. 73, 1918–1921.

Persson H. 1978. Root dynamics in a young Scots pine stand in Central Sweden. Oikos. 30, 508–519.

Persson H. 1979. Fine root production, mortality and decomposition in forest ecosystems. Vegetatio. 41, 101–109.

Persson H. 1980. Spatial distribution of fine root growth, mortality and decomposition in a young Scots pine stand in Central Sweden. Oikos. 34, 77–87.

Persson H. 1983. Root dynamics in a young Scots pine stand in central Sweden. Oikos. 30, 508–519.

Persson H. 1992. Factors affecting fine root dynamics of trees. Suo. 43, 163–172.

Publicover DA, Vogt KA. 1993. A comparison of methods for estimating forest fine root production with respect to sources of error. Can J For Res. 23, 1179–1186.

Orlov AJ. 1968. Development and life duration of the pine feeding roots. In: Methods of productivity studies in root systems and rhizosphere organisms. International Symposium USSR 28/8–12/10/1968, 139–145.

Reynolds ERC. 1970. Root distribution and the cause of ist spatial variability in *Pseudotsuga taxifolia* (Poir.) Br Plant Soil. 32, 501–517.

Reynolds ERC. 1974. The distribution pattern of fine roots of trees. In: International Symp. Ecology and Physiology of root growth, Potsdam. Ed. G Hoffmann. pp 101–112. Akademie-Verlag, Berlin.

Reynolds ERC. 1975. Tree rootlets and their distribution. In: The Development and Function of Roots. Eds. JG Torrey, DT Clarkson. pp 163–177. Academic Press, London.

Ritter G. 1990. Zur Wirkung von Stickstoffeinträgen auf Feinwurzelsystem und Mykorrhizabildung in Kiefernbeständen. Beitr Forstwirtschaft. 24, 100–104.

Ritter G, Tölle H. 1978. Stickstoffdüngung in Kiefernbeständen und ihre Wirkung auf Mykorrhizabildung und Fruktifikation der Symbiosepilze. Beitr Forstwirtschaft. 12, 162–166.

Sachs L. 1992. Angewandte Statistik. 7. Auflage. Springer Berlin Heidelberg New York.
Santantonio D, Grace JC. 1987. Estimating fine-root production and turnover from biomass and decomposition data: a compartement-flow model. Can J For Res. 17, 900–908.
Schmincke B, Strubelt F, Münzenberger B, Hüttl RF. 1995. Auswirkungen unterschiedlicher Schadstoffkonzentrationen in Kiefernökosystemen auf Mykorrhizaformen, Mykorrhizavitalität und Mykorrhizahäufigkeit. In: Mikroökologische Prozesse im System Pflanze – Boden. 5. Borkheider Seminar zur Ökophysiologie des Wurzelraumes. Ed. W Merbach. Teubner-Verlag, Stuttgart.
Schneider BU, Meyer J, Schulze E-D, Zech, W. 1989. Root and mycorrhizal development in healthy and declining Norway Spruce stands. In: Ecological Studies Vol. 77. Eds. E-D Schulze, OL Lange, R Oren. pp 370–391. Springer-Verlag, Berlin Heidelberg.
Uebel E. 1982. Einfluß einer Mineraldüngung auf höhere Pilze und die Mykorrhizabildung auf einer aufgeforsteten Ackerfläche. Arch Naturschutz u. Landschaftsforsch. 22, 169–175
Ulrich B, Pirouzpanah D, Murach D. 1984. Beziehungen zwischen Bodenversauerung und Wurzelentwicklung von Fichten mit unterschiedlich starken Schadsymptomen. Forstarchiv. 55, 127–134.
Vogt KA, Edmonds RL, Grier CC, Piper SR. 1980. Seasonal changes in mycorrhizal and fibrous-textured root biomass in 23- and 180-year-old Pacific silver fir stands in western Washington. Can J For Res. 10, 523–529.
Vogt KA, Grier CC, Gower ST, Sprugal DG, Vogt DJ. 1986. Overestimation of net root production: a real or imaginary problem? Ecology. 67, 577–579.
Vogt KA, Publicover DA, Bloomfield J, Perez JM, Vogt DJ, Silver WL. 1993. Belowground responses as indicators of environmental change. Environ Exp Bot. 33, 189–205.
Weisdorfer M, Schaaf W, Hüttl RF., this volume.
Zhang Y, Reed DD, Cattelino PJ *et al.* 1994. A process-based growth model for young red pine. For Ecol Management. 69, 21–40.

9
Mycorrhizal morphotypes of Scots pine

B. MÜNZENBERGER and R.F. HÜTTL

1. Introduction

The significance of the ectomycorrhizal symbiosis for temperate forest stability is stressed by several authors (Read, 1991; Vogt *et al.*, 1991; Dighton, 1995). In the presence of mycorrhizae litter decomposition is faster and release of nutrients from the litter is higher than in the absence of mycorrhizae (Zhu and Ehrenfeld, 1996). Mycorrhizal fungi seem to play a role in the breakdown of humus, where N and P are largely present in organic forms (Bending and Read, 1995a,b). Some mycorrhizal fungi excrete enzymes like proteases (Griffiths and Caldwell, 1992; Bending and Read, 1995b), phosphatases (Dighton, 1983; Antibus *et al.*, 1986; Griffiths and Caldwell, 1992; Dighton and Coleman, 1992) and phenoloxidase (Giltrap, 1982) to mobilize nutrients from organic compounds. 'Protein-fungi' are able to use peptides and proteins as a nitrogen source (Abuzinadah *et al.*, 1986; Abuzinadah and Read, 1986a,b; Abuzinadah and Read, 1989). Dähne *et al.* (1995) found in mycorrhizae growing in the organic layer about twice the amount of protein compared to those taken from the upper mineral soil. Thus, mycorrhizal fungi play an important role in the nutrient cycling of the organic layer.

The decline of mycorrhizal fungi is described to be connected with effects of air pollution by many authors (Termorshuizen and Schaffers, 1991; Rühling and Tyler, 1990; Jansen, 1991; Arnolds, 1991; Gulden *et al.*, 1992; Shaw *et al.*, 1992; Sastad and Jenssen, 1993; Fellner and Pešková, 1995). High nitrogen deposition causes a decrease in the number of carpophores of mycorrhizal fungi as well as the number of fruiting species (Kuyper, 1989; Termorshuizen and Schaffers, 1991; Arnolds, 1991). It seems likely that most ectomycorrhizal fungi are adapted to low inorganic nitrogen levels in the soil (Arnebrant, 1994). Ectomycorrhizal fungi and their mycorrhizae are highly sensitive to impact on forest ecosystems such as air pollution and fertilization and are thus suitable bioindicators of forest stability (Fellner and Pešková, 1995).

The negative effect of high nitrogen levels on mycorrhiza formation is well documented (Termorshuizen and Ket, 1989; Haug, 1990; Haug and Feger, 1990; Ritter, 1990; van Dijk *et al.*, 1990; Wallander and Nylund, 1992; Arnebrant, 1994). High nitrogen concentrations lead to a reduction of short roots (Haug, 1990) and of mycorrhizal frequency (Menge and Grand, 1978; Ritter and Tölle, 1978; Haug and Feger, 1990; van Dijk *et al.*, 1990; Arnebrant

R.F. Hüttl and K. Bellmann, Changes of Atmospheric Chemistry and Effects on Forest Ecosystems, 137–150.

and Söderström, 1992). Further, they inhibit formation of extramatrical mycelium of mycorrhizae in the soil that can also cause a reduction in mycorrhiza formation (Arnebrant, 1994; Wallander, 1995).

Recently, comprehensive fertilization experiments or transect studies have been conducted to investigate effects of high nitrogen levels on fruitbody formation of mycorrhizal fungi and on the infection patterns of fine roots (Jansen, 1991; Wiklund *et al.*, 1995; Brandrud, 1995; Taylor and Read, 1996). In the north of The Netherlands, where atmospheric concentrations of SO_2 and NH_3 were relatively low, numbers of mycorrhizal species and fruitbodies as well as numbers of mycorrhizae were higher than in the south where concentrations of atmospheric pollutants were high (Jansen, 1991). Fertilization, especially with nitrogen, resulted in a strong reduction in basidioma production (Wiklund *et al.*, 1995; Brandrud, 1995). In a north-south transect from N Sweden to NE France ectomycorrhizal diversity, i.e. number of fruitbodies as well as species composition, was highest at the northern less polluted sites (Taylor and Read, 1996). Isolates of mycorrhizal fungi differed in their ability to utilize an organic nitrogen source, indicating that species at the most northern site at Umeå in N Sweden are more capable of utilizing organic N sources than species isolated at southern sites. The results supported the author's hypothesis that increasing mineralisation rates or an excess of mineral nitrogen may alter ectomycorrhizal community structure from more specialized protein fungi to less specialized nitrophiles.

During the last decades forest ecosystems of the new German states were influenced by high loads of industrial air pollutants. These deposition loads caused changes in soil chemistry, stand density and forest floor vegetation of impacted ecosystems. Based on an ecosystematic approach within the comprehensive SANA (Regeneration of the Atmosphere above the New Federal States)-project the distribution patterns of ectomycorrhizal morphotypes were investigated in three comparable Scots pine ecosystems located along a deposition gradient of air pollutants. The aim of this investigation was to examine the effects of high nutrient input regimes on the ectomycorrhizal community structure.

2. Materials and methods

2.1. *Experimental sites*

The site Rösa is located about 10 km to the east of the industrial complex Bitterfeld and was influenced by high deposition of sulfur and alkaline dust deposition. The alkaline fly ashes caused a high input of calcium and magnesium to the soil. The site Taura is situated about 50 km north-east of the Halle/Leipzig region near Leipzig, and was moderately impacted by air pollutants, especially by alkaline fly ashes. The reference site Neuglobsow, located in a remote area 75 km north of Berlin, had only received background deposition loads. For stabilization, the site Rösa was treated at least 9 times with 100 kg N/ha between 1970–1985. On the contrary, the site Taura was fertilized only once in 1981 with urea (quantity unknown), and the site Neuglobsow with 100 kg N/ha in 1973 and 1974. The investigations were carried out in 40 to 60 year old Scots pine stands growing on Spodi-dystric Cambisols (Rösa), Cambic Podsols (Taura) and Dystric Cambisols (Neuglobsow) derived from glacial outwash sediments. The mean thickness of the organic layer was 9.7 cm at Rösa, 6.5 cm at Taura, and 6.1 cm at Neuglobsow and the pH_{H_2O} in the A-horizon was 5.11 at Rösa, 3.95 at Taura, and 3.78 at Neuglobsow (Weisdorfer *et al.*, 1995). Mean concentrations of nutrients in the soil solution are shown in Weisdorfer *et al.*, this volume. Nitrification rate was more than 85% at Rösa and about 40% at Taura. There was almost no nitrification in Neuglobsow (Fischer *et al.*, 1995). Forest floor vegetation was dominated by *Calamagrostis epigeios*, *Brachypodium sylvaticum*, *Rubus idaeus*, and *Rubus fruticosus* at Rösa, *Avenella flexuosa* at Taura and *Vaccinium myrtillus*, *A. flexuosa* and mosses at Neuglobsow. Throughfall was 622 mm, 690 mm, and 522 mm from 11.93 to 10.94 and 427 mm, 513 mm, and 491 mm from 11.94 to 10.95 at Rösa, Taura, and Neuglobsow, respectively (see Weisdorfer *et al.*, this volume).

2.2. *Preparation of mycorrhizal types*

Eight to ten soil cores (8 cm i.d.) per site were taken every four to eight weeks during the vegetation periods of 1994 and 1995 (March–November). The cores were divided into organic layer and mineral soil fraction (0–10 cm). Roots were cleaned drily to remove adhering soil. Mycorrhizal types were sorted under a dissecting microscope according to morphological criteria. If possible, the mycorrhizal types were identified with the Colour Atlas of Ectomycorrhizae (Agerer, 1987–1996). The mycorrhizae were stained with toluidine blue for 10 min and the mycorrhizal projection area (vertical topview on mycorrhizae) of each type was measured using colour image analysis (Olympus CUE 3).

3. Results

At all sites projection areas of mycorrhizae related to soil volume were clearly higher in the organic layer than in the mineral soil fraction in both vegetation periods (Figures 1 and 2). The highest mycorrhizal areas were found in the organic layer of the site Neuglobsow. In comparison, mycorrhizal areas of the impacted sites Rösa and Taura were much lower (Figures 1 and 2). In 1994 highest mycorrhizal areas were found in the organic layer of Neuglobsow in autumn (Figure 1), whereas in 1995 they were highest in spring (Figure 2).

During both vegetation periods mycorrhizal areas of the mineral soil fraction were extremely low at Taura, but somewhat higher at Rösa and Neuglobsow. However, in mineral soil projection areas of some types, namely the smooth-yellow types and the white type, were higher at Rösa than at

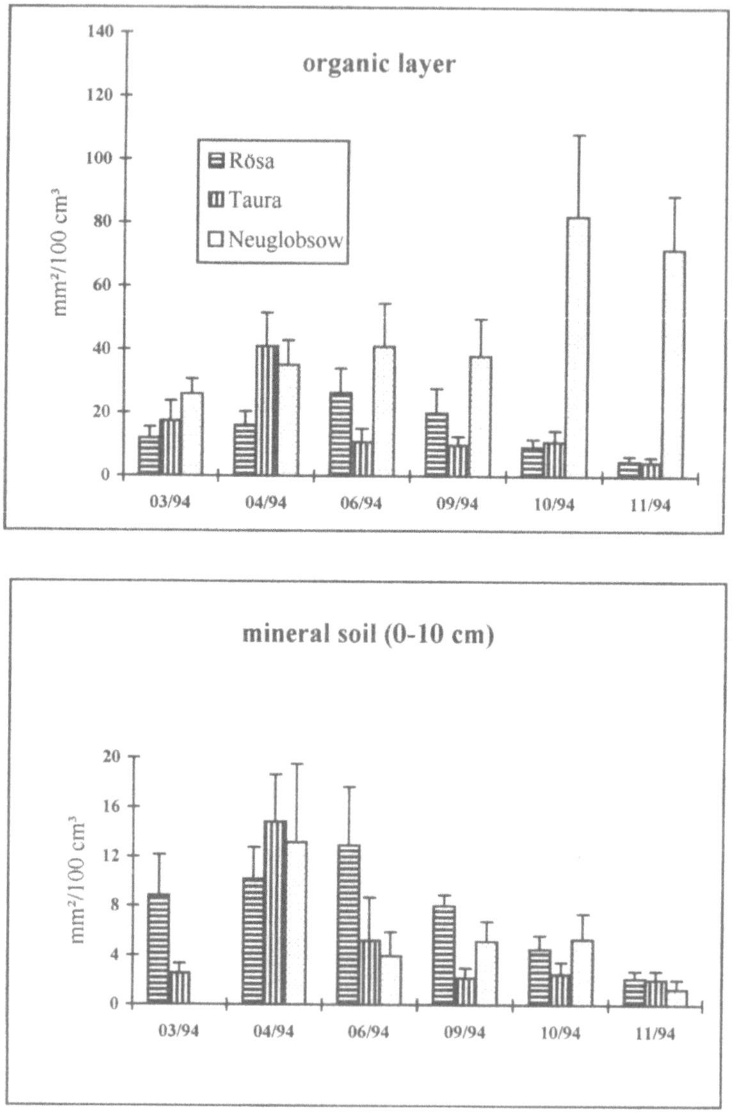

Figure 1. Spatial and temporal distribution of mycorrhizal morphotypes during the investigation period 1994. Mean values of projection areas in $mm^2/100\ cm^3$. Vertical bars indicate standard errors (n = 7–11)

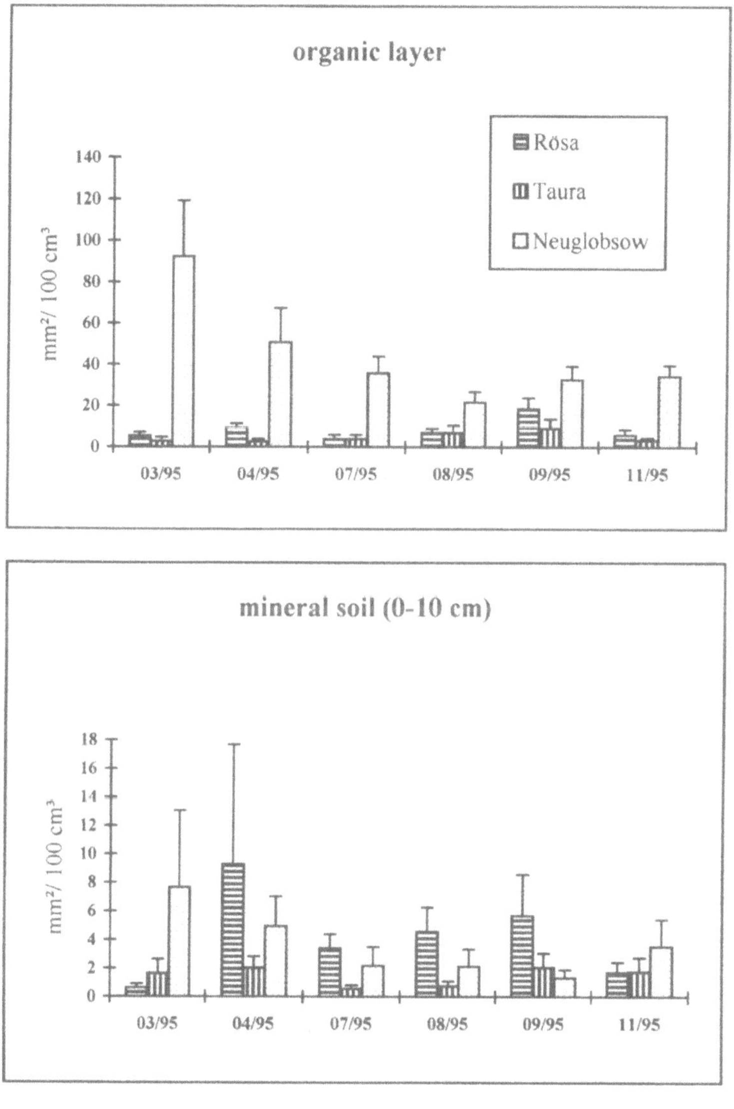

Figure 2. Spatial and temporal distribution of mycorrhizal morphotypes during the investigation period 1995. Mean values of projection areas in $mm^2/100\ cm^3$. Vertical bars indicate standard errors (n = 7–11)

Neuglobsow (Table 1). Whereas mean mycorrhizal area of both types was 42.0 mm/100 cm at Rösa, it only was 11.5 mm/100 cm at Neuglobsow. That means, 35.8% of the mycorrhizal area were found in the mineral soil fraction at Rösa but only 4.3% at Neuglobsow. At Taura the mycorrhizal area of these morphotypes was similar to the area at Neuglobsow (12.3 mm/100 cm).

Table 1. Projection areas of the smooth yellow types and the white type in 1994 and 1995. Mean values in $mm^2/100\ cm^3$ ($n = 13$)

Experimental site	Organic layer	Mineral soil
Rösa	75.4	42.0
Taura	42.1	12.3
Neuglobsow	267.0	11.5

Only few mycorrhizal types dominated in the organic layers of all three sites (Figures 3 and 4). These types were the smooth-yellow types, the woolly-yellow type, the white-fringe type, the white type and *Pinus sylvestris-Cenococcum geophilum*. Species composition differed qualitatively. This could be confirmed for both vegetation periods. Only at Neuglobsow mycorrhizal types forming extensive rhizomorphs were found in the organic layer. These were the white-fringe type and mycorrhizae of *Pinus sylvestris-Dermocybe cinnamomea* and *Pinus sylvestris-Dermocybe semisanguinea*. The mycorrhizal areas of the white-fringe type were 38.9 $mm^2/100\ cm^3$ ($n = 7$) in 1994 and 11.8 $mm^2/100\ cm^3$ ($n = 6$) in 1995 and those of the mycorrhizae of the genus *Dermocybe* were 14.0 $mm^2/100\ cm^3$ ($n = 7$) in 1994 and 11.7 $mm^2/100\ cm^3$ ($n = 6$) in 1995. All three types were lacking at Rösa and Taura.

Whilst the white-fringe type and the *Dermocybe*-mycorrhizae were only found at the reference site Neuglobsow, the grainy-brown type and the dark-brown type were only present at the impacted sites Rösa and Taura. However, their mean mycorrhizal areas were very low: the grainy-brown type had an area of 2.1 $mm^2/100\ cm^3$ at Rösa and 5.9 $mm^2/100\ cm^3$ ($n = 13$) at Taura and the dark-brown type 3.0 $mm^2/100\ cm^3$ at Rösa and 2.5 $mm^2/100\ cm^3$ ($n = 13$) at Taura.

The mycorrhizal areas of *Pinus sylvestris-Paxillus involutus* were 0.8, 1.3, and 64.6 $mm^2/100\ cm^3$ ($n = 13$) in the organic layers of Rösa, Taura, and Neuglobsow, respectively. Hence, this mycorrhizal type was almost lacking at the impacted sites.

Quantitative differences in the abundance of mycorrhizal types were striking. In 1994 mycorrhizal areas of the white type were 16.4, 18.1, and 60.7 $mm^2/100\ cm^3$ ($n = 7$) at Rösa, Taura, and Neuglobsow, respectively. Remarkable were also the differences between the areas of the smooth-yellow types (cf. *Russula*) in both vegetation periods. Whereas the mycorrhizal area of these

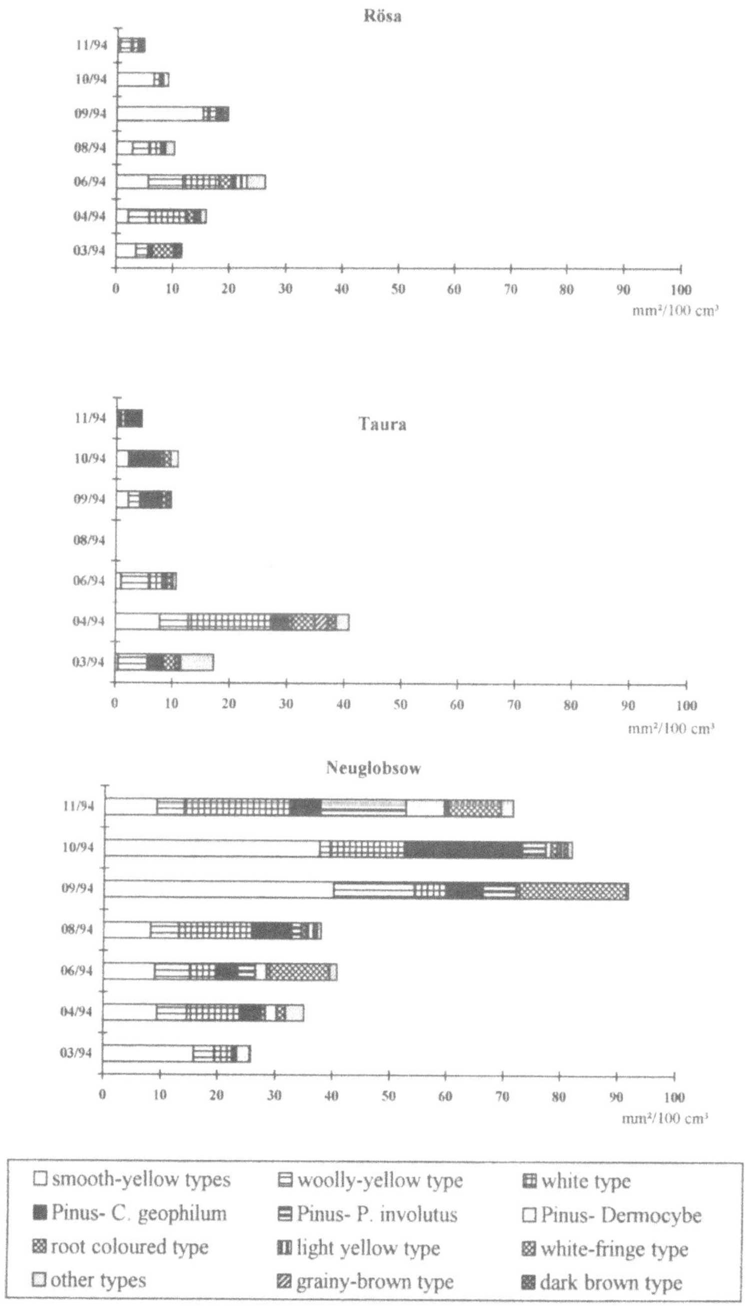

Figure 3. Seasonal distribution of mycorrhizal morphotypes in the organic layer during the investigation period 1994. Mean values of projection areas in $mm^2/100\ cm^3$ (n = 7–11)

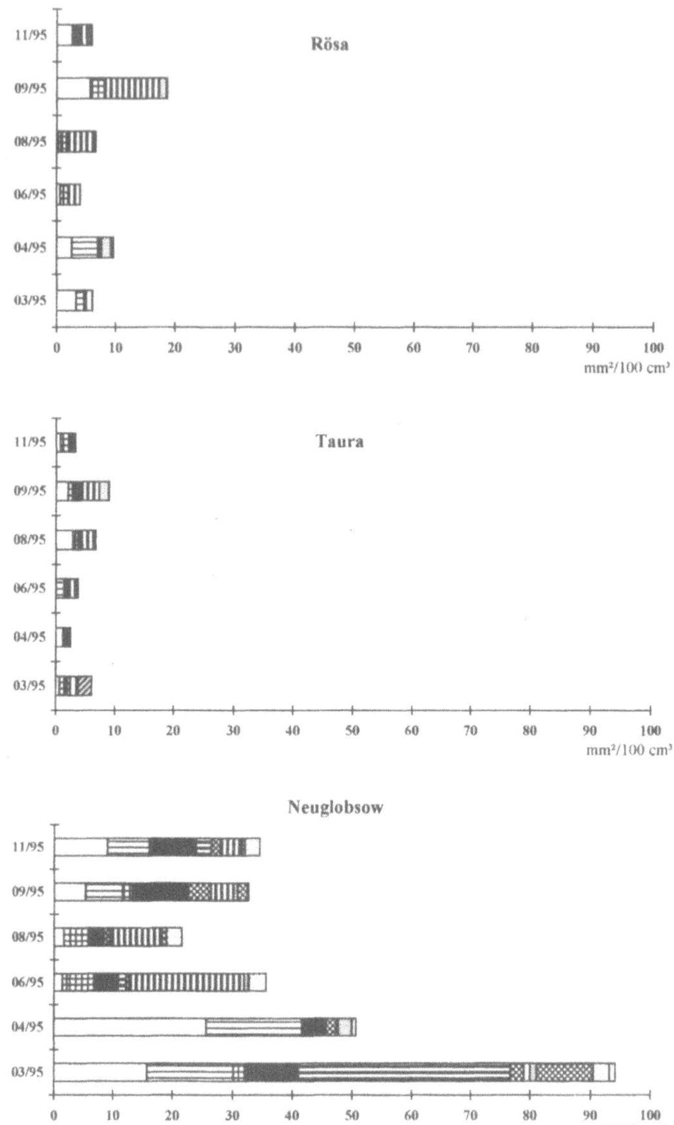

Figure 4. Seasonal distribution of mycorrhizal morphotypes in the organic layer during the investigation period 1995. Mean values of projection areas in $mm^2/100\ cm^3$ (n = 7–11)

types was 188.3 $mm^2/100\ cm^3$ at Neuglobsow, they were only 51.7 and 21.7 $mm^2/100\ cm^3$ (n = 13) at the impacted sites Rösa and Taura. The highest mycorrhizal area of these types was found in autumn 1994 (77.8 $mm^2/100\ cm^3$ for September and October).

4. Discussion

The predominance of ectomycorrhizae in organic layers is well documented (Mikola and Laiho, 1962; Mikola, 1963; Trappe and Fogel, 1977; Harvey *et al.*, 1979) and was confirmed in the present investigation. In 1994 the highest mycorrhizal projection areas were found at the reference site Neuglobsow in autumn. Remarkable were the high mycorrhizal areas of the smooth-yellow types which correlated with number and variety of fruitbodies formed by the genus *Russula*. A positive correlation between fruitbody formation and percentage of mycorrhizal tips was also found by other authors (Laiho, 1970; Ritter and Tölle, 1978; Termorshuizen and Schaffers, 1989; Agerer, 1990; Jansen, 1991). In 1995, the highest mycorrhizal areas were found in spring at the reference site, whereas the impacted sites Rösa and Taura did not show any maxima neither in spring nor in autumn during both vegetation periods. The results indicate that formation of fine root maxima reported by other authors (Haug and Feger, 1990 and citations therein) do not necessarily develop. Haug and Feger (1990) attribute omission of fine root maxima to climatic conditions, e.g. mild winters. As climatic conditions were comparable at all sites the stronger influence of nutritional effects on fine root dynamics caused by air pollutants and fertilization seems more likely. However, small mycorrhizal areas during the summer periods 1994 and 1995 are obviously directly or indirectly related to summer dryness.

Mycorrhizal areas of the smooth-yellow types and the white type were higher in the mineral soil fraction at Rösa than at Neuglobsow. Obviously, the small mycorrhizal area of those types in the organic layer of Rösa was somewhat compensated in the mineral soil fraction.

The small mycorrhizal areas at the pollution impacted sites Rösa and Taura are supposed to be an effect of high nutrient inputs resulting from high atmospheric deposition over decades, fertilization, and altered rates of nitrification due to high pH values and high base saturation in the organic layer. Soil solution chemistry revealed high NO_3 and Ca concentrations that were released from the organic layer of the site Rösa (Weisdorfer *et al.*, 1995; Weisdorfer *et al.*, this volume), where also highest N-content in needles was found. However, P-content in needles was suboptimal at this site (Ende and Hüttl, this volume). $P0_4$-concentrations of the soil solution were not detectable at all sites (Weisdorfer, personal communication). The low phosphate supply in the pine needles at Rösa coincides with the small mycorrhizal areas and the lack of types forming extensive rhizomorphs. As mentioned before, mycorrhizal fungi excrete phosphatases to make organically bound phosphate available for tree nutrition or influence weathering and solubility of P by releasing organic acids (e.g. oxalic acid) into the rhizosphere (Duchesne *et al.*, 1989; Griffiths *et al.*, 1994; McElhinney and Mitchell, 1995). Phosphate may be stored as polyphosphates in the vacuoles of mycorrhizal fungi and is remobilized if required (Ashford *et al.*, 1986; Kottke and Martin, 1994).

Influenced by high atmospheric deposition and fertilization, increased biomass and changes in the composition of forest floor vegetation has caused a change in the C- and N-budget at the site Rösa. C-mineralisation of the necromass of forest floor vegetation was slower in Rösa than in Neuglobsow (Bergmann *et al.*, this volume). This means that in Neuglobsow, where the mycorrhizal area was highest, decomposition of the necromass of forest floor vegetation was faster. This is in agreement with results of Zhu and Ehrenfeld (1996), who found litter decomposition to be faster in the presence of mycorrhizal roots. As thickness of the organic layer was highest at the site Rösa, N-supply was higher at Rösa than at Neuglobsow. This surplus of nitrogen in the necromass of forest floor vegetation at Rösa has led to N-mineralisation by soil microorganisms already at the beginning of decomposition (Bergmann *et al.*, this volume). Whereas in needle litter and necromass of the forest floor vegetation a significant N-immobilization was found at Neuglobsow, no N-immobilization was evident at Rösa. On the contrary, necromass of forest floor vegetation has released significant N-levels at the latter site (Bergmann *et al.*, this volume). Further, alkaline ashes have led to high pH values and high base saturation at the site Rösa resulting in increased microbial activity and thus higher net N-mineralisation than at Neuglobsow. Likewise, nitrification was more than 85% at this impacted site, whilst nitrification was practically lacking at the reference site Neuglobsow (Bergmann *et al.*, this volume). The increased NO_3-N availability at Rösa may have triggered the decrease of mycorrhizal infection. In soils continuously supplied with nutrients either as fertilizers or as deposition, mycelial growth is likely to remain reduced (Arnebrant and Söderström, 1992; Arnebrant, 1994). Especially, reduction of extramatrical mycelium by inorganic nitrogen was found by several authors (Arnebrant, 1994; Wallander and Nylund, 1992). As mentioned by these authors, reduction in the amount of mycelium is a possible explanation for the lowered quantity of ectomycorrhizal root tips.

Species composition differed qualitatively. Only at Neuglobsow mycorrhizal types were found which formed extensive rhizomorphs, such as the white-fringe type and mycorrhizae of the genus *Dermocybe. Dermocybe cinnamomea*, known to colonize nutrient poor substrates with low pH (Høiland, 1983), is a species rejecting lime (Agerer, 1989) and seems to suffer like other types forming rhizomorphs (*Piloderma croceum* and *Hebeloma*) from high nitrogen content in the organic layer (Markkola and Ohtonen, 1988; Termorshuizen and Schaffers, 1991). Similarly, the genus *Cortinarius*, which also forms mycorrhizae with extensive rhizomorphs, reacts very sensitively to high nitrogen levels (Ohenoja, 1988; Brandrud, 1995). Independent of the small areas of all morphotypes, the lack of mycorrhizal types forming extensive rhizomorphs suggests a reduction of extramatrical mycelium at the impacted sites, too.

The white type was observed to form clusters under the surface of mosses at the site Neuglobsow. This site showed a greater cover and thickness of the moss layer than the sites Rösa and Taura, which both exhibited a similar small moss layer. As forest floor vegetation is dominated by *Calamagrostis epigeios* at Rösa and *Avenella flexuosa* at Taura, cluster formation of the white type was never found. Termorshuizen and Schaffers (1989) found coralloid mycorrhizae significantly negatively related to SO_2 concentrations in the Netherlands.

Liming leads to changes in fruitbody formation and altered numbers of mycorrhizae (Kuyper, 1989; Agerer, 1989; Erland and Söderström, 1991; Lehto, 1994). Liming decreased fruitbody production of *Russula ochroleuca* and *Hygrophorus pustulatus* (Agerer, 1989). Erland and Söderström (1991) found a pink mycorrhizal type in higher quantities when *Pinus sylvestris* seedlings were planted in a lime treated humus. However, the classified mycorrhizal types were found in all treatments. Lehto (1994) found the number of mycorrhizae of *Piloderma croceum* and of smooth mycorrhizal types to be reduced by lime. Consequently, the deposition of calcium rich fly ashes can be a further explanation for the strong reduction of the smooth-yellow types (cf. *Russula*) at the impacted sites Rösa and Taura. These types were present in high quantities at the reference site Neuglobsow. This coincided with a high fruitbody production of the genus *Russula* only at this site. Some species of this genus react very sensitively to nitrogen (Brandrud, 1995) and lime (Agerer, 1989).

Two mycorrhizal types, namely the grainy-brown type and the dark-brown type, were only found at the impacted sites Rösa and Taura. However, their mycorrhizal areas were very small. This means that there was no reduction of the number of mycorrhizal morphotypes, but some types are replaced by others. Taylor and Read (1996) found that two mycorrhizal species, *Tylospora* spec. and *Lactarius rufus*, dominated in a spruce forest characterized by high input of mineral nitrogen. Both fungi are known to have limited ability to utilize organic nitrogen sources (Ryan and Alexander, 1992). The nitrophily of Lactarius rufus was recently confirmed by several authors (Brandrud, 1995; Wiklund *et al.*, 1995).

Areas of mycorrhizae formed by *Paxillus involutus* were smaller at the impacted sites Rösa and Taura than at Neuglobsow. This finding differs from the results of other authors who found *Paxillus involutus* to be a nitrophilous key species (Arnebrant, 1994; Brandrud, 1995).

As ectomycorrhizal fungi have different properties, the change in community structure might affect the nutrient uptake of the trees (Arnebrant, 1994). Ammonium, nitrate as well as the complete nutrient solution affected mycelial growth negatively. This is possibly induced by changes in carbon allocation of the trees (Nylund, 1988; Wallander and Nylund, 1991, 1992; Wallander, 1995). This assumption is supported by a higher fine root turnover at the site Neuglobsow than at the impacted sites Rösa and Taura (Strubelt *et al.*, this volume). Likewise, high quantities of mycorrhizae showing reduced vitality were found at the impacted sites indicating a long life-span of mycorrhizae. At the reference site Neuglobsow percentages of vital as well as dead mycorrhizae were enhanced, suggesting higher mycorrhiza turnover rates than at the impacted sites (Münzenberger *et al.*, 1995). This means that there was a higher carbon allocation to the root system in Neuglobsow. Obviously, trees allocate more C to the root system for production of fine roots and mycorrhizae to improve nutrient uptake at this relatively nutrient-poor reference site.

Acknowledgements

This study was supported by the Federal Ministry of Education, Science, Research and Technology (BMBF), Bonn, Germany, as a part of the SANA-project. The authors thank Elisabeth Laska for skillful technical assistance.

References

Abuzinadah RA, Read DJ. 1986a. The role of proteins in the nitrogen nutrition of ectomycorrhizal plants. I. Utilization of peptides and proteins by ectomycorrhizal fungi. New Phytol. 103, 481–493.

Abuzinadah RA, Read DJ. 1986b. The role of proteins in the nitrogen nutrition of ectomycorrhizal plants. III. Protein utilization by *Betula*, *Picea* and *Pinus* in mycorrhizal association with *Hebeloma crustuliniforme*. New Phytol. 103, 507–514.

Abuzinadah RA, Read DJ. 1989. The role of proteins in the nitrogen nutrition of ectomycorrhizal plants. IV. The utilization of peptides by birch (*Betula pendula* L.) infected with different mycorrhizal fungi. New Phytol. 112, 55–60.

Abuzinadah RA, Finlay RD, Read DJ. 1986. The role of proteins in the nitrogen nutrition of ectomycorrhizal plants. II. Utilization of proteins by mycorrhizal plants of *Pinus contorta*. New Phytol. 103, 495–506.

Agerer R. 1987–1996. Colour Atlas of Ectomycorrhizae. 1.-10. delivery. Einhorn, Schwäbisch Gmünd.

Agerer R. 1989. Impacts of artificial acid rain and liming on fruitbody production of ectomycorrhizal fungi. Agric Ecosyst Environ. 28, 3–8.

Agerer R. 1990. Gibt es eine Korrelation zwischen Anzahl der Ektomykorrhizen und Häufigkeit ihrer Fruchtkörper? Z Mykol. 56, 155–158.

Antibus RK, Linkins AE. 1992. Effects of liming a red pine forest floor on mycorrhizal numbers and mycorrhizal and soil acid phosphatase activities. Soil Biol Biochem. 24, 479–487.

Antibus RK, Kroehler CJ, Linkins AE. 1986. The effects of external pH, temperature, and substrate concentration on acid phosphatase activity of ectomycorrhizal fungi. Can J Bot. 64, 2383–2387.

Arnebrant K. 1994. Nitrogen amendments reduce the growth of extramatrical ectomycorrhizal mycelium. Mycorrhiza. 5, 7–15.

Arnebrant K, Söderström B. 1992. Effects of different fertilizer treatments on ectomycorrhizal colonization potential in two Scots pine forests in Sweden. For Ecol Manage. 53, 77–89.

Arnolds E. 1991. Decline of ectomycorrhizal fungi in Europe. Agric Ecosyst Environ. 35, 209–244.

Ashford AE, Peterson RL, Dwarte D, Chilvers GA. 1986. Polyphosphate granules in eucalypt mycorrhizas: determination by energy dispersive x-ray microanalysis. Can J Bot. 64, 677–687.

Bending GD, Read DJ. 1995a. The structure and function of the vegetative mycelium of ectomycorrhizal plants. V. Foraging behaviour and translocation of nutrients from exploited litter. New Phytol. 130, 401–409.

Bending GD, Read DJ. 1995b. The structure and function of the vegetative mycelium of ectomycorrhizal plants. VI. Activities of nutrient mobilizing enzymes in birch litter colonized by *Paxillus involutus* (Fr.) Fr. New Phytol. 130, 411–417.

Brandrud TE. 1995. The effects of experimental nitrogen addition on the ectomycorrhizal fungus flora in an oligotrophic spruce forest at Gårdsjön, Sweden. For Ecol Manage. 71, 111–122.

Dähne J, Klingelhöfer D, Ott M, Rothe GM. 1995. Liming induced stimulation of the amino acid metabolism in mycorrhizal roots of Norway spruce (*Picea abies* [L.] Karst.). Plant Soil. 173, 67–77.

Dighton J. 1983. Phosphatase production by mycorrhizal fungi. Plant Soil. 71, 455–462.

Dighton J. 1995. Nutrient cycling in different terrestrial ecosystems in relation to fungi. Can J Bot. 73 (Suppl. 1), S1349–S1360.
Dighton J, Coleman DC. 1992. Phosphorus relations of roots and mycorrhizas of *Rhododendron maximum* L. in the southern Appalachias, N. Carolina. Mycorrhiza. 1, 175–184.
Duchesne LC, Ellis BE, Peterson RL. 1989. Disease suppression by the ectomycorrhizal fungus *Paxillus involutus*: contribution of oxalic acid. Can J Bot. 67, 2726–2730.
Erland S, Söderström B. 1990. Effects of liming on ectomycorrhizal fungi infecting *Pinus sylvestris* L. I. Mycorrhizal infection in limed humus in the laboratory and isolation of fungi from mycorrhizal roots. New Phytol. 115, 675–682.
Erland S, Söderström B. 1991. Effects of lime and ash treatments on ectomycorrhizal infection of *Pinus sylvestris* L. seedlings planted in a pine forest. Scand J For Res. 6, 519–525.
Fellner R, and Pešková V. 1995. Effects of industrial pollutants on ectomycorrhizal relationships in temperate forests. Can J Bot. 73 (Suppl. 1), S1310– S1315.
Fischer T, Bergmann C, Hüttl RF. 1995. Auswirkungen sich zeitlich ändernder Schadstoffdepositionen auf Prozesse des Kohlenstoff- und Stickstoffumsatzes im Boden. In: Atmosphärensanierung und Waldökosysteme. Eds. RF Hüttl, K Bellmann, W Seiler. pp. 143–160. Eberhard Blottner Verlag, Taunusstein.
Giltrap NJ. 1982. Production of polyphenol oxidases by ectomycorrhizal fungi with special reference to *Lactarius* spp. Trans Br Mycol Soc. 78, 75–81.
Griffiths RP, Caldwell BA. 1992. Mycorrhizal mat communities in forest soils. In: Mycorrhizas in Ecosystems. Eds. DJ Read, DH Lewis, AH Fitter, IJ Alexander. pp 98–105. CAB International, Wallingford, Oxon.
Griffiths RP, Baham JE, Caldwell BA. 1994. Soil solution chemistry of ectomycorrhizal mats in forest soil. Soil Biol Biochem. 26, 331–337.
Gulden G, Høiland K, Bendiksen K *et al.* 1992. Fungi and air pollution. Mycocoenological studies in three oligotrophic spruce forests in Europe. Bibl Mycol. 144, 1–81.
Harvey AE, Larsen MJ, Jurgensen MF. 1979. Comparative distribution of ectomycorrhizae in soils of three western Montana forest types. Forest Sci. 25, 350–360.
Haug I. 1990. Mycorrhization of *Picea abies* with *Pisolithus tinctorius* at different nitrogen levels. Ecosyst Environ. 28, 167.
Haug I, Feger KH. 1990. Effects of fertilization with $MgSO_4$ and $(NH_4)_2SO_4$ on soil solution chemistry, mycorrhiza and nutrient content of fine roots in a Norway spruce stand. Water Air Soil Pollut. 54, 453–467.
Høiland K. 1983. *Cortinarius* subgenus *Dermocybe*. Opera Botanica. 71, 1–113.
Jansen AE. 1991. The mycorrhizal status of Douglas Fir in The Netherlands: its relation with stand age, regional factors, atmospheric pollutants and tree vitality. Agric Ecosyst Environ. 35, 191–208.
Kottke I, Martin F. 1994. Demonstration of aluminium in polyphosphate of *Laccaria amethystea* (Bolt. ex Hooker) Murr. by means of electron energy-loss spectroscopy. J Microsc. 174, 225–232.
Kuyper TW. 1989. Auswirkungen der Walddüngung auf die Mykoflora. Beitr Kennt Pilze Mitteleur. 5, 5–20.
Laiho O. 1970. *Paxillus involutus* as a mycorrhizal symbiont of forest trees. Acta Forest Fenn. 106, 1–72.
Lehto T. 1994. Effects of liming and boron fertilization on mycorrhizas of *Picea abies*. Plant Soil. 163, 65-68.
Markkola AM, Ohtonen R. 1988. The effect of acid deposition on fungi in forest humus. In: Ectomycorrhiza and Acid Rain. Eds. AE Jansen, J Dighton, AHM Bresser. pp 122–126. Bilthoven, The Netherlands.
McElhinney C, Mitchell DT. 1995. Influence of ectomycorrhizal fungi on the response of Sitka spruce and Japanese larch to forms of phosphorus. Mycorrhiza. 5, 409–415.
Menge JA, Grand LF. 1978. Effect of fertilization on production of epigeous basidiocarps by mycorrhizal fungi in loblolly pine plantations. Can J Bot. 56, 2357–2362.

Mikola P. 1963. Beziehungen der Mykorrhizatypen zu forstlichen Humustypen. In: Mykorrhiza. Internationales Mykorrhizasymposium Weimar, 1960. pp. 279-284. VEB G. Fischer, Jena.
Mikola P, Laiho O. 1962. Mycorrhizal relations in the raw humus layer of northern spruce forests. Comm Inst For Fenn. 55, 1–13.
Münzenberger B, Schmincke B, Strubelt F, Hüttl RF. 1995. Reaction of mycorrhizal and non-mycorrhizal Scots pine fine roots along a deposition gradient of air pollutants in eastern Germany. Water Air Soil Pollut. 85, 1191–1196.
Ohenoja E. 1988. Behaviour of mycorrhizal fungi in fertilized forests. Karstenia. 28, 27–30.
Read DJ. 1991. Mycorrhizas in ecosystems. Experientia. 47, 376–391.
Ritter G. 1990. Zur Wirkung von Stickstoffeinträgen auf Feinwurzelsystem und Mykorrhizabildung in Kiefernbeständen. Beitr Forstwirtsch. 24, 100–104.
Ritter G, Tölle H. 1978. Stickstoffdüngung in Kiefernbeständen und ihre Wirkung auf Mykorrhizenbildung und Fruktifikation der Symbiosepilze. Beitr Forstwirtschaft. 4, 162–166.
Rühling Å, Tyler G. 1990. Soil factor influencing the distribution of macrofungi in oak forests of southern Sweden. Holarct Ecol. 13, 11–18.
Ryan EA, Alexander IJ. 1992. Mycorrhizal aspects of improved growth of spruce when grown in mixed stands on heathlands. In: Mycorrhizas in Ecosystems. Eds. DJ Read, DH Lewis, AH Fitter, IJ Alexander. pp 237–245. CAB International, Wallingford, Oxon.
Sastad SM, Jenssen HB. 1993. Interpretation of regional differences in the fungal biota as effects of atmospheric pollution. Mycol Res. 97, 1451–1458.
Shaw PJA, Dighton J, Poskitt J. 1992. Studies on the effect of SO_2 and O_3 on the mycorrhizas of Scots pine by observations above and below ground. In: Mycorrhizas in Ecosystems. Eds. DJ Read, DH Lewis, AH Fitter, IJ Alexander. pp 208–213. CAB International, Wallingford, Oxon.
Taylor AFS, Read DJ. 1996. A European north-south survey of ectomycorrhizal populations on spruce. In: Mycorrhizas in Integrated Systems from Genes to Plant Development. Eds. C Azcon-Aguilar, JM Barea. pp 144–147. European Commission EUR 16728, Luxembourg.
Termorshuizen AJ, Ket PC. 1989. The effects of fertilization with ammonium and nitrate on mycorrhizal seedlings of *Pinus sylvestris*. Agric Ecosyst Environ. 28, 497–501.
Termorshuizen AJ, Schaffers AP. 1989. The relation in the field between fruitbodies of mycorrhizal fungi and their mycorrhizas. Agric Ecosyst Environ. 28, 509–512.
Termorshuizen AJ, Schaffers AP. 1991. The decline of carpophores of ectomycorrhizal fungi in stands of *Pinus sylvestris* L. in The Netherlands: possible causes. Nova Hedwigia. 53, 267–289.
Trappe JM, Fogel RD. 1977. Ecosystematic functions of mycorrhizae. In: The Belowground Ecosystem. Colorado State University, Range Science Department Scientific Series. pp. 205–214. For Collins, Colorado.
van Dijk FFG, Louw MHJ de, Roelofs JGM, Verburgh JJ. 1990. Impact of artificial ammonium-enriched rainwater on soils and joung coniferous trees in a greenhouse. Part II – Effects on the trees. Environ Pollut. 63, 41–59.
Vogt KA, Publicover DA, Vogt DJ. 1991. A critique of the role of ectomycorrhizas in forest ecology. Agric Ecosyst Environ. 35, 171–190.
Wallander H. 1995. A new hypothesis to explain allocation of dry matter between mycorrhizal fungi and pine seedlings in relation to nutrient supply. Plant Soil. 168/169, 243–248.
Wallander H, Nylund J-E. 1992. Effects of excess nitrogen and phosphorus starvation on the extramatrical mycelium of ectomycorrhizas of *Pinus sylvestris* L. New Phytol. 120, 495–503.
Weisdorfer M, Schaaf W, Hüttl RF. 1995. Auswirkungen sich zeitlich ändernder Schadstoffdepositionen auf Stofftransport und -umsetzung im Boden. In: Atmosphärensanierung und Waldökosysteme. Eds. RF Hüttl, K Bellmann, W Seiler. pp. 56–74. Eberhard Blottner Verlag, Taunusstein.
Wiklund K, Nilsson L-O, Jacobsson S. 1995. Effect of irrigation, fertilization, and artificial drought on basidioma production in a Norway spruce stand. Can J Bot. 73, 200–208.
Zhu W, Ehrenfeld JG. 1996. The effect of mycorrhizal roots on litter decomposition, soil biota, and nutrients in a spodosolic soil. Plant Soil. 179, 109–118.

10

Decomposition of needle-, herb-, root-litter, and Of-layer-humus in three Scots pine stands

C. BERGMANN, T. FISCHER and R.F. HÜTTL

1. Introduction

The processes of litter decomposition, mineralization, and humus accumulation have early been recognized to be crucial for nutrient storage and supply to plants (e.g. Bocock and Gilbert, 1957; Falconer *et al.*, 1933; Gustafson, 1943; Lunt 1935; Minderman, 1968; Shanks and Olson, 1961; Witkamp and Olson, 1963). Climatic conditions and litter quality greatly influence the process of litter decomposition (Couteaux *et al.*, 1995; Fog, 1988). Concerning the litter quality, changes in the kind and amount of mineral and organic compounds during decomposition were measured and discussed with respect to their availability or decomposition-enhancing or -retarding effects. The course of decomposition was described mathematically in order to make the turnover- and accumulation-processes of different types of litter and soil humus comparable among each other (e.g. Jenny *et al.*, 1949; Olson, 1963). Several researchers emphasized that the decomposition course should not be represented by simple exponential functions because of the complex composition of organic material (e.g. Howard and Howard, 1974; Minderman, 1968).

Both 'classical' ways, the mathematical and the chemical-analytical, were followed and combined in several ecosystem studies throughout the world to gain a better understanding (e.g. Aber and Melillo, 1980; Berg and Ekbohm, 1983, 1991; Fahey, 1983; Fog, 1988; Wieder and Lang, 1982). Recently, with the studies of Berg and coworkers, the concept of asymptotic decay gained more and more attention (e.g. Berg and Ekbohm, 1991; Berg *et al.*, 1995b and 1996). Aber and Melillo (1980) recognized the inverse-linear relationship between the remaining mass and its nitrogen concentration. Thus, they expressed organic matter- and nitrogen-dynamics of decomposing litter in a single function for the first time and developed a concept of litter transfer to soil organic matter. They proposed to define this 'switch' or 'transfer point' as the moment when net N immobilization switches to net N mineralization (Aber *et al.*, 1990). The amount of decomposed litter present in the moment when this point is reached was supposed to contribute to the long-term organic matter storage (Aber and Melillo, 1980). Similar assumptions were made by Berg *et al.* (1995b) concerning the 'asymptotic remaining mass'. But in this stage, the decomposed material does not yet chemically resemble the material

R.F. Hüttl and K. Bellmann, Changes of Atmospheric Chemistry and Effects on Forest Ecosystems, 151–176.

collected from the Of-horizon. Therefore, it can only form the first step towards long-term organic matter and deserves a closer charcterization with respect to organic matter and N turnover.

In this study, various litter types (pine needles, herbaceous understorey, pine roots) as well as the organic layers in Scots pine stands under different historical air pollution regimes were analysed. The basic hypothesis was that litterfall, i.e. organic mass input and its chemical composition, is affected by atmospheric deposition (Baronius and Fiedler, 1993; Fangmeier *et al.*, 1994; Tamm, 1991; Trautmann *et al.*, 1970) and thus litter decomposition and humus accumulation rates will be changed. This, in turn, would have consequences especially for the N turnover, which characterizes the N status of an ecosystem. The mass loss data sets are first analysed with conventional mathematical models (Wieder and Lang, 1982). Then, the litter dacay models, the transfer-concept (Aber and Melillo, 1980) and the decay rates of the organic layer material (Of) are combined. By this new approach of data analysis the N- and humus-turnover is described in three phases and the humus accumulation is quantified.

2. Material and Methods

2.1. Sites

The experimental sites were chosen by interdisciplinary scientists of the German SANA-Project, conducted to verify the impacts of industrial air pollution on forest ecosystems. They are located along a former air pollution gradient from Rösa (35 km north of Leipzig), to Taura (45 km north-east of Leipzig), and to the background site Neuglobsow (75 km north of Berlin). Apart from atmospheric deposition (Table 2, Hüttl and Bellmann, this volume), the pine stands received different amounts of N fertilizer between 1970 and 1985 (Taura 100 kg ha^{-1}, Neuglobsow 2 times 100 kg ha^{-1}, Rösa 9 times at least 100 kg ha^{-1}). While the background site (Neuglobsow) was dominated by *Avenella flexuosa, Vaccinium myrtillus*, and mosses, the forest floor in Taura was densely covered by *Avenella flexuosa.* In Rösa the vigorous understorey was dominated by *Calamagrostis epigeios*, *Rubus* sp., and *Brachypodium sylvaticum.* Further site and stand characteristics are summarized in Table 1.

2.2. Sampling

The experimental sites were established during spring/summer 1993. Sampling of litter-input, standing herbaceous understorey biomass, and humus material (Of and Oh horizons) started in June/August 1993 and was repeated monthly throughout the vegetation period (April to November) till November 1995.

Pine litter input was sampled from seven 1 × 1 m^2-collectors per site (see also Bergmann *et al.*, this volume). After drying at 70–80°C, samples were separated

Table 1. Selected site and stand characteristics of the three SANA experimental sites in north-east German lowlands

Stand	Annual precipitation (mm)[a]	Annual temp. (°C)[a]	Stand age (1995) (yrs)	Soil types	Humus form	Depth (cm)	pH (KCl)[b]	%base sat.[b]	%C	%N	Dry mass (Mg ha^{-1})
Rösa	566	8.87	61	Spodi-dystric Cambisols, sand	Mor/moder	6/4	4.1/4.5	57/85	34/22	1.7/1.7	106/99
Taura	565	8.87	45	Cambic Podsols, silty sand	Moder	5/2[c]	3.4	37	39	1.5	65
Neuglobsow	586	8.18	65	Dystric Cambisols, sand	Moder	4[c]	3.0	35	44	1.6	55

[a]Mean values: Rösa (station Wittenberg) and Taura (station Torgau) 1954–1990; station Neuglobsow 1996–1990

[b]Values measured in August 1994, pH in 1:5 1 N KCl-extracts, exchangeable bases in 0.5 N NH_4Cl-extracts

[c]In Neuglobsow, Oh-material could not be distinguished as a horizon, in Taura, Oh-material was usually mixed with upper mineral soil and not sampled separately

into a needle-, a bark and fine litter-, a twig- and a cone-fraction. Aliquots from the needle- and the bark- and fine litter-fraction were ground and chemically analysed.

Standing above ground understorey vegetation was harvested bimonthly in five replicates per site from randomly chosen 50×50 cm^2 plots. The material was sorted into a green (living) and a brown (dead) fraction, dried and processed as the pine litter.

From Of-horizons, humus material was collected for decomposition experiments once in early summer 1993.

Pine fine root bio- and necromass measurements and sampling were conducted by Strubelt (Strubelt *et al.*, this volume) as part of another SANA-subproject. Chemical analyses and pine fine root decomposition experiments were conducted by Bergmann and Strubelt, data were not yet published.

2.3. *Mass loss experiments*

For decomposition experiments, litter types sampled during autumn 1993 and 1994 on the respective sites were confined to litterbags. The litterbags, 20 cm × 10 cm, were made of 2 mm mesh size nylon net.

10 g (dry matter) of needle litter per bag was taken from the samples collected in the littertraps, after mixing the stock (from autumns 1993 and 1994) thoroughly. For the ground vegetation litter decomposition, the routine harvests and additional sampling of dead material provided enough tissue to enclose 5 g d.m. in each bag. The rhizobags were filled with 0.5 g fresh tissue of pine roots from the 1–2 mm class. From the humus layer (Of-horizon), field moist substrate, corresponding to 20 g dry material, was filled into the litterbags.

The root- and Of-litterbags were placed in the Of-layer at about 2–5 cm depth and fastend to a plastic clothes-line, thus facilitating the recollection. The incubation of Of-litterbags started in June/July 1993 at five randomly chosen spots per site. One bag per spot was recollected monthly during the vegetation period 1994, in April 1995, and November 1995. From the rhizobags that were distributed in the Of-horizon at the same spots in late July 1995, 8–12 per site were recollected monthly until November 1995. In November 1994, the needle- and herb layer-litterbags were distributed to four spots at each experimental site. They were placed in the litter layer, among mosses where present, and fastend to clothes-lines to prevent movement and loss by wind and/or small animals. These litterbags were recollected monthly (one per spot) from April to November 1995. After five months (April 1995) they were already completely incorporated into the litter layer, overgrown by mosses (Neuglobsow and Rösa) and grass (Taura). All recollected litterbags were transported in polyethylene bags to the laboratory and cleaned of moss, grass and other remnants. The confined litter was dried at 70–80°C to constant weight, weighed individually per bag, and ground in a mill for chemical analyses. All rhizobags sampled were pooled per site and date to provide enough material for the analyses.

For the determination of initial element concentrations unconfined material from the harvests was used.

Chemical analyses
Dried samples were finely ground in a swinging mill (MM 2000, Retsch). Total carbon, nitrogen and sulphur were measured on individual samples by dry combustion (CNS-Analyser Vario EL, Heraeus), total P, K, Ca, Mn, Fe, Zn and Pb were determined on pooled samples (per site and date) by X-ray-fluorescence (Spectrace 5000, Noran Instruments). Unfortunately, the Mg-determination by the X-ray method showed not to be reliable. Replicate element analyses, performed on approx. 10% of the samples, indicated satisfactory reproducibility, and NBS tissue standards (NIST 1575, pine needle) and laboratory-intern pine needle standards were within 1–6% of the known concentrations for C, N, S, K, Ca, Mn, Fe, and within 10–12% for P, Zn, Pb.

2.4. *Calculations and statistics*

Input of pine litter was calculated from the monthly yields to annual input mass per hectare. For understorey input we assumed that maximum primary production was approximately reached in July when no dead material was sortet out from the harvests. Further, all tissue produced should return to the litter layer within one year, as the understorey was dominated by herbaceous plants (for detailed calculation cf. Bergmann *et al.*, this volume). Pine root litter input was estimated from fine root production and turnover data obtained by Strubelt *et al.* (this volume). All litter inputs were expressed as annual dry mass (d.m.) per area.

The mass loss model used to describe pine needle-, root- and herb-litter decomposition courses is based on the idea of a 'two-fractions-material': an easily decomposable, 'labile' fraction (m_l) which is decomposed with the rate konstant k, following an exponential curve, and the recalcitrant, almost 'inert' fraction (m_i), which in the model is represented by the asymptote.

$$m_r = m_i + m_l * \exp(-k * t), \qquad \text{(eq. 1)}$$

where m_r is the remaining mass in grams and t is time in days. By definition, $m_i + m_l$ equals the initial amount of litter enclosed in the bags (0.5 g, 5 g or 10 g), thus simplifying the model to a two-parameter model (m_l and k). Alternatively, all masses can be expressed as percent of the initially enclosed mass (100%), then $m_i + m_l = 100$.

The model parameters were estimated from the individual data sets by nonlinear least-square approximation. For graphic presentation, mean values and standard deviations were calculated from individual masses or element concentrations when measured in individual samples.

The concept of transfer or 'switch' from litter to humus proposed by Aber and Melillo (1980) uses the inverse-linear relationship

$$m_r = a * \%N + b. \qquad \text{(eq. 2)}$$

The parameters a and b were estimated by least-square approximation. From this function the absolute content of accumulated nitrogen in percent of the original mass of nitrogen can be derived:

$$N_{abs}\,(\%) = (a * \%N + b) * \%N \,/\, \%N_0. \qquad \text{(eq. 3)}$$

$\%N_0$ is the initial concentration of N in the litter material. At maximum N_{abs}, net nitrogen immobilization switches to net nitrogen release. Equation (3) then gives the N concentration of the remaining material at this point ($\%N_{max}$). This, in turn, determines in equation (2) the remaining mass at the 'switch' point, $m_{r,max}$. As discussed by Aber and Melillo (1980), N_{abs} may include not only N in the original litter (= resource-N) but also considerable amounts of microbial N. Then, the ratio of resource-N to microbial N at a given N concentration is approximated by m_r:($N_{abs} - m_r$).

The Student's *t* statistic, when applicable, was used to compare dry mass and nutrient differences. ANOVA comparisons were run with the Student-Newman-Keuls-test. Significance always refers to α = 5 %, unless indicated otherwise.

3. Results and discussion

3.1. Input of organic matter

Annual pine needle input was about 2.8 Mg ha^{-1}, i.e. 58% from total pine litter, and did not differ significantly between sites (Table 2). Of total pine litter, 16–21% consisted of bark and very fine litter material, 23% of twigs and female cones. Total pine litter mass is in the upper range of values reported e.g. by Hoffmann and Krauß (1988) for 60 to 73 yr old Scots pine stands in eastern Germany. It seems that there is no significant effect of pollution loads on annual above ground pine litter mass. This might probably be due to the generally sufficient, although site-specifically different, supply with nitrogen and thus biomass production at all three stands (cf. Hoffmann and Krauß, 1988).

The annual input of understorey litter differed significantly between sites, ranging from 0.75 Mg ha^{-1} at Neuglobsow, 1.41 Mg ha^{-1} at Rösa to 2.29 Mg ha^{-1} at Taura. This is 15, 21 and 32% of total above ground litter at the respective sites. Similar biomass production was documented e.g. by Hofmann (1994) for forest communities with comparable understorey species spectra. For pine and understorey litter input see also Bergmann *et al.* (this volume).

Root litter was estimated separately for finest and fine pine roots and for

Table 2. Chemical composition of needle-, herb-, and root-litter at the three SANA experimental sites

	Input ($Mg\ ha^{-1}\ yr^{-1}$)	%C	N	S	P	K	Ca	Mn	Fe	Pb	Zn
			Concentration of elements ($mg\ g^{-1}$)[c]								
Needle litter[a]											
Rösa	2.8	50	6.4	1.5	0.6	1.7	7.6	0.54	0.34	nd	0.03
Taura	2.8	51	6.0	1.4	0.5	1.5	5.7	0.41	0.21	nd	0.02
Neuglobsow	2.4	51	5.8	1.1	0.7	1.5	5.2	0.92	0.12	nd	0.03
Understorey litter[a]											
Rösa	1.41	43	10.1	2.2	2.8	3.5	3.1	0.14	0.14	nd	0.04
Taura	2.29	45	8.2	1.4	3.6	1.4	1.0	0.10	0.09	nd	0.02.
Neuglobsow	0.75	46	4.3	1.0	4.6	2.2	0.9	0.39	0.04	nd	(0.01)
Root litter[b]	($g\ m^{-2}\ yr^{-1}$)										
Rösa	32/13[b]	48/49	9.0/7.0	2.4/2.4	1.0/0.6	2.6/2.3	5.2/4.5	0.13/0.17	1.72/0.93	0.02/0.01	0.08/0.05
Taura	5/1	48/48	9.0/6.7	2.3/2.5	1.0/0.9	1.6/2.0	4.3/2.8	0.04/0.05	2.86/1.57	0.05/0.05	0.04/0.04
Neuglobsow	27/8	49/49	9.0/6.0	2.1/2.1	0.9/1.3	2.1/2.3	5.0/3.2	0.12/0.14	0.48/0.22	0.02/nd	0.05/0.03

nd, not detected

[a]Values are dry mass

[b]Values are mean annual values for 0–1/1–2 mm input to the forest floor only in (g dry matter $m^{-2}\ yr^{-1}$)

[c]Values for autumn-litter (pine needles and understorey) and July-root biomass

different soil horizons. In this paper, only pine root litter contribution to the organic layers is accounted for. Highest annual input was calculated for Neuglobsow: 0.45 Mg ha^{-1}, which is about 8% of total above and below ground input. At Rösa, pine root litter input was approximately 0.35 Mg ha^{-1} or 5% of total and at Taura where the humus layer was hardly rooted input was only about 0.07 Mg ha^{-1} or 1% of total litter input to the forest floor (Table 2).

3.2. *Chemical composition of the litter types*

The litter types of the three sites show a relatively wide range of chemical composition (Table 2). Obviously, there is a difference in above and below ground litter with respect to site-specific concentrations in macro- and micronutrients/heavy metals.

In pine needle litter, the concentrations of macronutrients are very similar at all sites, only Ca reaches higher values in needles from Rösa. The concentration of Fe increases in the order Neuglobsow > Taura > Rösa, whereas Mn reaches about twice the concentration in Neuglobsow compared to Taura and Rösa. In the understorey litter, the different species spectra obviously cause wider ranges in macronutrient concentrations. Values of Mn and Fe are considerably lower than in pine needle litter. Highest N and K concentrations were measured in understorey litter at Rösa (10.1 mg N g^{-1}, 3.5 mg K g^{-1}), highest P concentrations in the understorey litter at Neuglobsow (4.6 mg g^{-1}). Ca accumulation is stronger in pine needle litter than in understorey necromass. Ca concentrations are highest at Rösa (7.6 mg g^{-1} and 3.1 mg g^{-1}, respectively).

Below ground, concentrations of N and P in the roots do not show any trend between sites and are even higher than in needle litter (9 mg N g^{-1}, 1 mg P g^{-1}). S concentrations also reach higher values in roots than in needle litter, increasing in the order Neuglobsow < Taura < Rösa (2.4 mg g^{-1} at Rösa). K, Ca, Mn, Fe, and Pb concentrations differ between Neuglobsow and Rösa at the one hand and Taura at the other. The concentrations of K, Ca, and Mn are considerably lower in pine roots at Taura (1.6, 4.3 and 0.04 mg g^{-1}, respectively) than at the other sites (> 2.1, > 5.0 and > 0.12, respectively). The reverse is found for Fe, which in addition reaches 10-fold higher concentrations in roots than in needle litter at Taura (2.86 mg g^{-1} in roots).

The chemical composition of the pine needle litter is well within the range reported by Krauß and Hoffmann (1991) for Scots pine in this region. The N-, S-, and Ca-concentrations reflect the different deposition loads of these elements at the experimental sites. A similar even more pronounced pattern is observed in understorey tissues. Compared to Swedish Scots pine stands (e.g. Johansson *et al.*, 1995), needle litter from our stands has higher concentrations of N, P, S, and K, but Ca concentrations of the same level and about half to one third the concentration of Mn. Berg *et al.* (1995a) related nutrient concentrations in Scots pine needle to latitude. The comparison of our data with these relationships reveals that only the S concentrations in needle litter at Taura and Rösa are above the range of the documented pine ecosystems. This may support the assumption of elevated S concentration in needle litter due to air pollution.

3.3. Changes in chemical composition during decomposition

The changes in macronutrient contents, calculated as total amount of the element in mg per g *initial* dry mass, are quite similar both for sites, litter types and elements (Figure 1). Generally, the nutrients were quickly released from the incubated litterbags, as illustrated e.g. for K (Figure 1c). The S release is most pronounced from S-rich materials as pine roots and all litter at Rösa (Figure 1b). An exception is N which was immobilized and even accumulated to various extents in the different litter types (Figure 1a). The micronutrient contents changed corresponding to their site-specifically different initial concentrations (data not shown).

Nitrogen was efficiently accumulated from 3–5 mg g^{-1} to about 9 mg g^{-1} at Neuglobsow in above ground litter. In needle litter at Taura N also accumulated, whereas in needle litter at Rösa the initial amounts of about 6 mg N g^{-1} were only just retained by immobilization. This was also observed in understorey litter at Taura. From understorey litter at Rösa, N was clearly released from the onset of the decomposition experiment. Decomposing pine roots retained nitrogen at the initial level of 6-9 mg g^{-1}.

At all three sites and in both above ground litter types, the P, S, and K contents reached quite stable levels after five to six months of incubation. In needle litter, P decreased only slightly from about 0.6 to 0.4 mg per g initial mass, indicating that P was efficiently retained in the decomposing material. In Of material, the P concentrations are 0.4 mg per g dry matter at Rösa and 0.7 mg per g at Taura and Neuglobsow. A similar pattern is observed for S and K with final levels of about 1 mg S g^{-1} and 0.6 mg K per g initial litter mass (Figure 1b,c). In pine root litter, the release patterns are similar. The Ca-poor understorey litter did not release Ca (level of about 1 mg g^{-1} at all sites), while in needle litter the content decreased from 5–7 mg g^{-1} to 3–5 mg g^{-1} without yet reaching a final level. The Mn content decreased in pine needle and understorey litter at all three sites, while in root litter Mn was retained at the low initial levels. Fe was accumulated in above ground litter. In root litter, this is also observed at Neuglobsow, but at Taura and Rösa the high initial amounts (1.6 mg g^{-1}, 0.9 mg g^{-1}, respectively) decreased to 0.8 and 0.6 mg g^{-1}, respectively.

The observed patterns of nutrient release from decomposing pine needle litter correspond well with those documented in other projects (e.g. Berg and Staaf, 1980a, b; Fahey, 1983). The efficient immobilization of N, P, and S is reported from these researchers, only the concentration level may be lower than at Neuglobsow, Taura and Rösa because of lower initial values. K and Ca seem to be released more rapidly from the litter at Neuglobsow, Taura and Rösa than at sites studied by those authors.

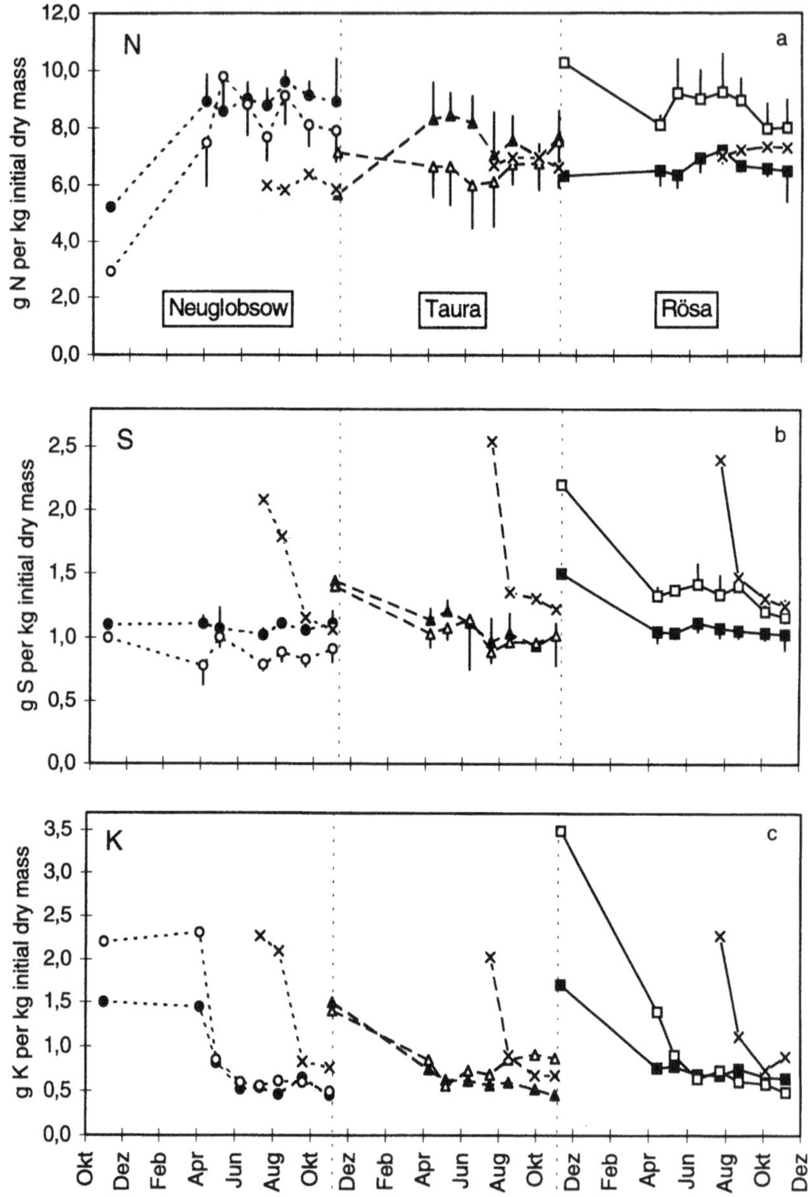

Figure 1a–c. Changes in chemical composition of remaining mass in pine needle, understorey, and root litter during one-year litterbag experiments in the field. closed symbols = needle litter, open symbols = understorey litter, crosses = 1–2 mm root litter. (Nov. 1994 – Nov. 1995)

4. Mass loss experiments

4.1. Above ground litter

The course of pine needle litter decay is similar at the three sites, reaching values about 65% remaining mass after the first six months (May 1995). During the next six months (till November) needle litterbags lost another 20% of their initial mass. A tendency for increasing mass loss in the order Neuglobsow < Rösa < Taura can be observed but not statistically proved. Courses of pine litter mass loss are illustrated in Figure 2a (closed symbols).

The course of understorey litter decay differs, both from pine needle litter decomposition and between the experimental sites (Figure 2a, open symbols). Understorey litter at Neuglobsow decomposed very quickly to less than 45% remaining mass after five months. After that, the remnants only slowly decreased in weight about another 10% of initial mass until November. Completely different, the understorey litter at Rösa, slowly but steadily lost its mass linearly, reaching about 50% remaining mass at the end of the observation period. In Taura, understorey litter followed the decomposition course of pine needle litter, disappearing slightly slower than needle litter in Neuglobsow during these first months of decay.

Classical approach of data analysis

For both litter types, the asymptotic model was fitted to the data sets (Figure 2b). Decomposition rate and proportional amount of the labile fraction were estimated (eq. 1) and parameters compared statistically (Table 3). The results indicate that the 'final' level (asymptote) of mass loss should be around 35% for pine needle litter and around 32% for understorey litter at Neuglobsow and Taura. In contrast, according to the asymptotic model, the understorey litter in Rösa is expected to disappear completely without leaving recalcitrant remnants (m_i=0%). By setting remaining mass, m_r, in the model (eq. 1) to m_i + 1%, it is possible to estimate the time it will take until the enclosed litter would have lost 99% of its initially 'decomposable' fraction. Thus, the labile fraction of pine needle litter (about 65%) should be disappeared after 2 years in Rösa and Taura and after some more than two and a half years in Neuglobsow. Annually, about 35% should thus accumulate in the recalcitrant fraction, i.e. the humus layer. The labile fraction of the understorey litter should rest much longer time on the forest floor in Rösa (>6 yrs.), about half this long in Taura (2.9 yrs) and only some more than a year (1.2 yrs) in Neuglobsow.

In order to characterize the coupled process of litter decay and N turnover in more detail, mass loss and nitrogen concentration data were combined in the inverse-linear function (eq. 2) recognized by Aber and Melillo (1980). The 'switch point' from net nitrogen immobilization to mineralization at maximum absolute N-accumulation (N_{abs}) was determined. From the linear functions and $\%N_{max}$-values the proportion of litter that is transferred to the Of-layer ($m_{r,max}$) was calculated as well as the ratio of resource-N to microbial-N in the

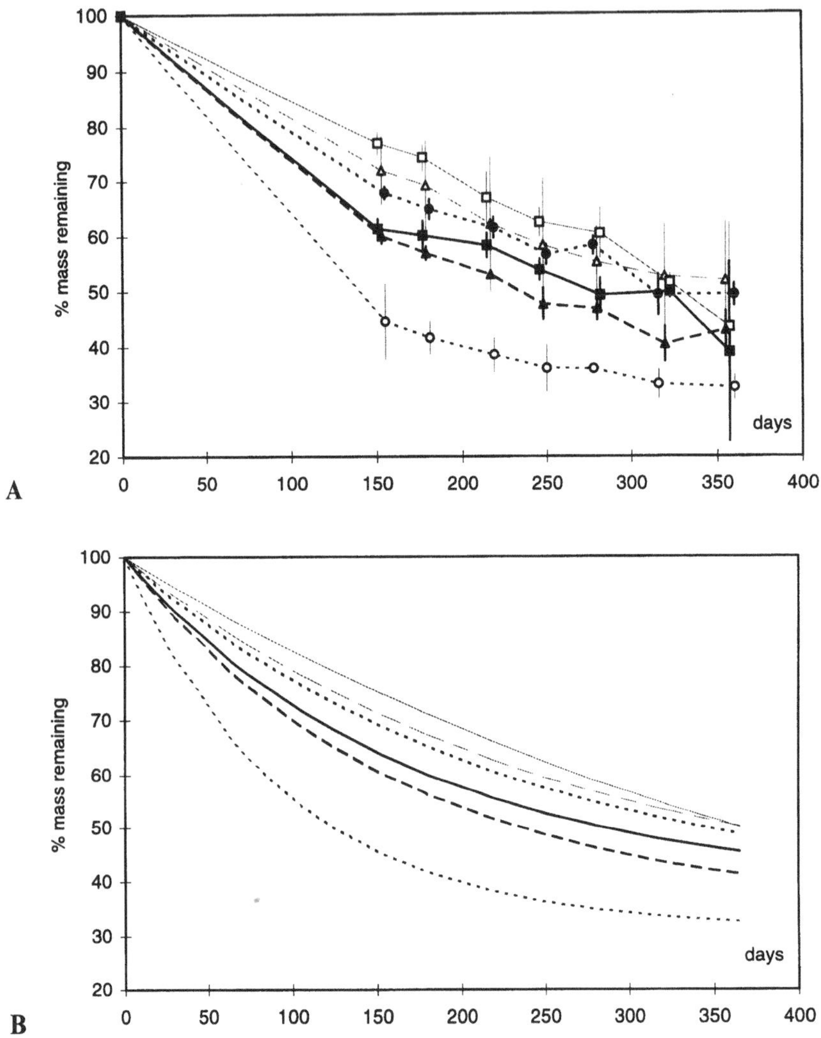

Figure 2. **A**: Decomposition course of pine needle litter (closed symbols) and understorey litter (open symbols) during one-year litterbag experiments at Neuglobsow (dotted lines, circles), Taura (dashed lines, triangles), and Rösa (solid lines, squares). Standard deviation bars. **B**: Idealized curve fit (without symbols)

whole remaining material. (cf. 'Calculations and statistics', text and eq. 3). Results are summarized in Table 4. Figure 3 presents the calculated functions and mean data values per sampling date at Neuglobsow. The slopes of the inverse-linear function are steepest at Rösa (–66, understorey, and –59,

Table 3. Estimated parameters of the asymptotic decomposition model for the different litter types incubated. Significant differences between sites are indicated for the different litter types

	'Labile' fraction, m_l in (%)	Initial konstant, k in ($days^{-1}$)	'Inert' fraction, m_i = 100%–m_l	Time, $t(m_i+1\%)$ in (days)
Needle litter				
Rösa	62	5.8 E-03ab	38a	720
Taura	66	6.0 E-03b	34a	700
Neuglobsow	65	4.3 E-03a	35a	970
Understorey				
Rösa	100	1.9 E-03a	0b	2 300
Taura	67	3.7 E-03b	33a	1 050
Neuglobsow	69	10.4 E-03c	31a	420
Fine roots				
Rösa	25	20.6 E-03	75	160
Taura	28	39.7 E-03	72	80
Neuglobsow	32	12.9 E-03	68	260
Of material				
Rösa	100	2.6 E-04	0	17 712
Taura	100	2.8 E-04	0	16 506
Neuglobsow	100	2.4 E-04	0	19 765

needles) and flattest at Neuglobsow (–31 and -40, respectively). As already shown in Figure 1a, N-accumulation is most pronounced at Neuglobsow, reaching values of 158% of the original N mass in needle litter and 222% in understorey litter. At Taura up to 138%, at Rösa 113% of original N mass has accumulated in needle litter while no accumulation could be observed in understorey litter (theoretical value of 89% N_{abs} at Taura and 92% N_{abs} at Rösa). According to these calculations, about 60 to 65% of needle litter ($m_{r,max}$) is transferred to the Of-layer at the experimental sites. From the understorey litter, only 54% of the original mass of litter input is transferred to the Of-layer at Neuglobsow, but almost all material at Taura and Rösa. At the latter sites, calculated transfer-values of 66 and 78% $m_{r,max}$ are probably too small because the linear fit underestimates the initial nitrogen concentrations. The resource-N to microbial N ratios were below 1 at Neuglobsow and in needle litter at Taura, and above 1 in understorey litter at Taura and all litter at Rösa, probably indicating different microbial population dynamics and substrate-use strategies.

Table 4. Estimated parameters of the inverse-linear relationship between remaining mass and its nitrogen concentration (cf. Aber and Melillo, 1980), illustrated in Figure 3

	Slope	y-intercept	r	Maximum at $\%N_{max}$	Max %N abs[b]	% m_r at N_{max}	N-res.: N-mic	% N end[c]
Needle litter								
Rösa	–59	130	0.90	1.10	112	65	1.4	1.6/1.5
Taura	–51	130	0.98	1.28	138	65	0.9	1.9/1.8
Neuglobsow	–40	121	0.98	1.52	158	60	0.6	2.2/1.8
Understorey								
Rösa	–66	156	0.88	(1.20)[a]	(92)	(78)	5.6	2.4/1.7
Taura	–60	133	0.71	(1.10)	(89)	(66)	2.7	1.7/1.5
Neuglobsow	–31	109	0.95	1.75	222	54	0.3	2.6/2.4

[a]Calculated values are of theoretical character, initial values are underestimated and in reality no maximum was observed

[b]Total nitrogen content in the remaining mass as percentage of original at $\%N_{max}$

[c]Theoretical N concentration in remaining substrate at asymptotic mass level and observed N concentration at the end of the incubation experiment (Nov. 1995)

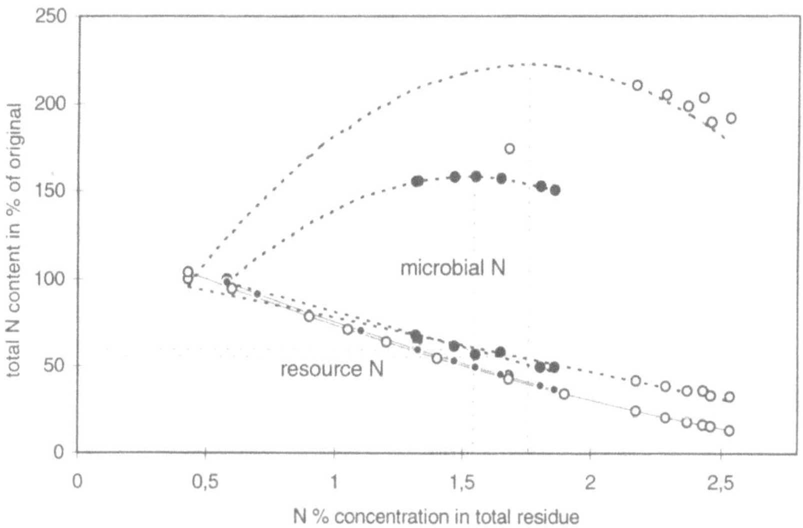

Figure 3. Organic matter and nitrogen dynamics during above ground litter decomposition at Neuglobsow. Inverse-linear relationship between N concentration and remaining mass (dotted line), which approximates remaining resource N (Aber and Melillo, 1980) and total N content in the residue as a percentage of original (dotted line). The calculation of remaining residue N (see eq. 4 and explanation in the text) is given with the solid line. closed circles = needle litter, open circles = understorey litter

4.2. *Below ground litter*

Classical approach of data analysis

The total mass loss in Of-litterbags was not significant, except for the last sampling date in Taura and Neuglobsow, after approx. 900 days of exposure, compared to the first months of incubation. However, if the methodological problem of root-ingrowth, reflected by error bars, especially during the vegetation period, is taken into account, a still lasting tendency of declining total mass is documented. This becomes more obvious, if the decrease of percent organic matter (loss on ignition) from L to Of and from Of in Neuglobsow to Of in Rösa is considered when comparing decomposition processes among sites and litter types. If expressed on organic matter basis, the mass loss of Of-litterbags is evident (Figure 4b). The asymptotic model was fitted to individually calculated remaining ash free matter and resulted in 100% 'degradable' Of material (m_l) which should disappear within 45 to 54 years (Table 3).

Individual rhizobags did not lose more than 30% weight during the incubation time (Figure 4a). Those in Taura lost slightly more than in Rösa and Neuglobsow, but on the whole, the variability between individual bags was relatively high at all sites and differences are not significant (Figure 4a).

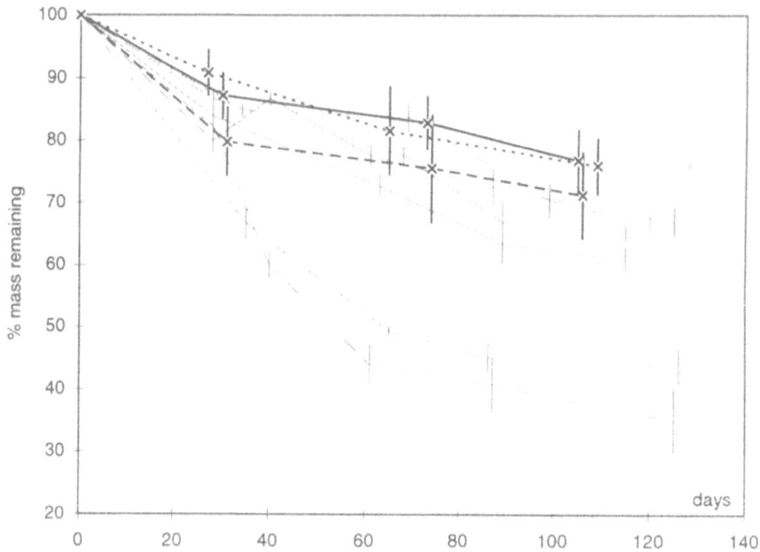

A

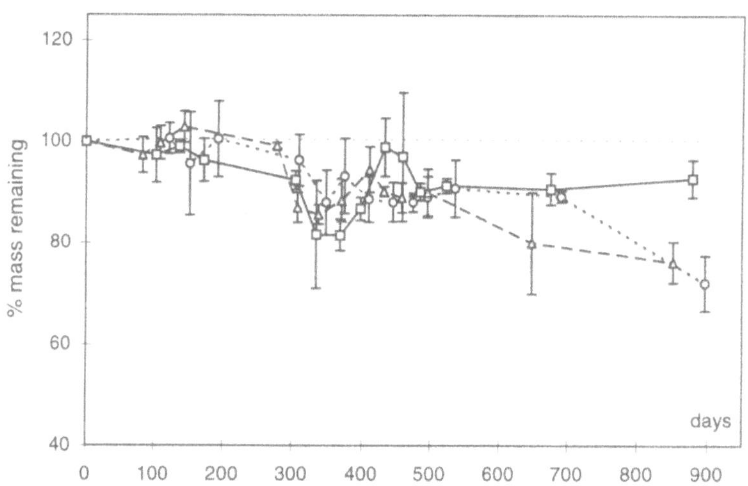

B

Figure 4. **A**: Decomposition course of root litter (crosses) compared to short-term experiments with above ground litter types at the experimental sites. **B**: Decomposition course of ash free Of-material. Symbols and lines see Figure 2

Asymptote estimates indicate no more than 25% (Rösa) to 32% (Neuglobsow) mass loss (Table 3). This small decomposable fraction though should be decomposed within about five months.

Transfer percentages according to Aber and Melillo (1980) were not calculated for pine root litter as it did not show significant nitrogen immobilization (Figure 1a). Of-material, per definition of the inverse-linear and transfer point calculations, is beyond net N immobilization.

Discussion of the classical approach of decomposition data analysis

The model of asymptotic decay was applied to decomposition data of several leaf litter types by e.g. Berg and Ekbohm (1991), Berg *et al.* (1996), who correlated the calculated 'limit values' (asymptotes) to various litter element concentrations. These reported limit- or maximum mass loss values, corresponding to the labile fraction, range from 35 to 100%, those for Scots pine from 75 to 100%. Although the asymptotic mass loss values for needle litter from our stands (62 to 66%) fall within the range for coniferous needle litter from different latitudes, they are significantly lower than those for the Swedish Scots pine stands. This may be due to the higher nutrient concentrations in our litter material (cf. Berg *et al.*, 1993). Usually, deciduous leaf litter have higher maximum mass loss values (Berg *et al.*, 1996) and it might be supposed that herbaceous understorey litter behaves similarly. Correspondingly, the understorey litter at Neuglobsow, Taura and Rösa seems to decompose to greater proportions (67 to 100%) than the needle litter.

The process of humus accumulation, not only depends on the litter mass input and the extent of its decay, but also on the duration of this decomposition process. The exposed needle litter at Neuglobsow, Taura and Rösa lost about 55% of its initial weight within the first year. This is a much faster weight loss than is reported for needle litter from Swedish Scots pine stands, which reaches 27% in the first year (e.g. Berg and Staaf, 1981). Even compared to stands at a similar latitude, the first year loss at our stands reaches highest values, only attained by decomposing Scots pine needle litter in stands of *Pinus radiata* and *Pinus nigra* at the european atlantic coast (Berg *et al.*, 1993). Obviously, climatic conditions play an important role for the decomposition rate (Berg *et al.*, 1993). As can be supposed from data given by Berg *et al.* (1993), the understorey vegetation is sparse at the respective sites, dominated by *Vaccinium myrtillus* and some grasses. At Neuglobsow, the microclimate relevant for decomposition, i.e. substrate moisture, is greatly influenced by a dense bottom layer of mosses which holds much of the incomming rain over long periods. At Taura and Rösa, the dense understorey vegetation cover causes a more balanced and moister forest floor microclimate compared to sparsely grown forest floor (own data, not published; also cf. Lüttschwager *et al.*, this volume). Apart from the litter quality (chemical composition), substrate moisture exerts some control on the course of the decompositon process.

Further approach to organic matter decomposition, humus accumulation and N turnover

With respect to the important role which the organic layer plays concerning nutrient storage and supply to plants, the approach of Aber and Melillo (1980) who combined C and N dynamics within one function gains great significance. In relation to the N-dynamics during the litter decomposition process, three phases can be defined as described in the following discussion.

The first phase is characterized by the invasion of litter with soil fauna and microorganisms, indicated by mass loss from CO_2-respiration and a relative accumulation of nitrogen, expressed as net N immobilization. According to the ideas of Aber and Melillo (1980), the litter becomes soil organic matter, or Of-material at forested sites, at the 'switch' point from net N immobilization to net N mineralization. This is defined to be the end of the first step.

In needle litter at Neuglobsow, the 'switch' point was reached after about 8 months of exposure in July 1995, when only 60% of the original mass remained and the absolute N content had reached 158% of the initial N content. Similar values were reported from a Swedish pine stand by Berg and McClaugherty (1989) ($m_{r,max} \approx 60\%$, $N_{abs} \approx 180\%$), but there the net N release did not start until 18 months had passed. This corresponds to the generally slower decompositon rates at those sites. During this first phase, the needle litter at Neuglobsow accumulated Fe and N and released K, Ca, and small amounts of P. Thereby the C:N ratio decreased from 88 to 33. The imported nitrogen was immobilized in microbial tissue which increased considerably. The slope of the inverse-linear relation between remaining mass and total N concentration gives an estimation of the substrate use efficiency of the microbial population (cf. Aber and Melillo, 1980). If the synthesis of secondary products, apart from microbial biomass, with higher N concentrations than those of the original needle litter (= resource) is neglected for this first phase of decomposition, the ratio of resource N to microbial N is about 0.6 in needle litter in Neuglobsow at the end of this phase (Table 4, calculations from Figure 3). Assuming microbial N to be about 6.5% of microbial biomass (Smith *et al.*, 1993), these quite high amounts of microbial N in needle litter at Neuglobsow (5.6 mg g^{-1} from total N_{abs} of 9.2 mg g^{-1}) clearly lead to high amounts of microbial tissue. This should be considered when calculating the remaining amount of resource N (n_r) or resource mass (m_r). The approximation made by Aber and Melillo (1980) that m_r (resource) $\cong m_r$ (total residue) must be omitted. The percentage amount of remaining litter, m_r (resource), or nitrogen, n_r (resource), which equal each other if its N concentration, $\%N_{resource}$, is assumed constant, must then be calculated from the N concentration of the whole remaining mass, %N. From the equations for m_r and $N_{abs}(\%)$ (eq. 1 and eq. 3) and the following assumption of

$$m_r \text{ (resource)} = m_r \text{ (total)} - m_r \text{ (mic), or } n_r \text{ (resource)} = n_r \text{ (total)} - n_r \text{ (mic)}$$

the following function was developed ($c = \%N_{resource} - \%N_{mic}$):

$$n_r \text{ res.} = \%N^2 (a/c) - \%N (a\ \%N_{mic} - b)/c - b\ \%N_{mic}/c \qquad \text{(eq. 4)}$$

This function of resource remaining mass or nitrogen is included in Figure 3. From the initial N concentration of 5.8 mg N g^{-1}, which is assumed constant, the absolute amount of resource N and, by subtraction, of microbial N can then be calculated. Thus, the remaining mass becomes less if only resource mass is considered (51% instead of 60% in needle litter at Neuglobsow) and the proportion of microbial tissue can be estimated. At the end of phase one, from the remaining 60% of needle litter at Neuglobsow, approximately 16% was microbial tissue.

In the understorey litter at Neuglobsow, the processes are similar but of another magnitude. The slope of the inverse-linear function is somewhat less steep, the remaining mass at the maximum N accumulation of 222% is only 42%. About five months passed until net N immobilization switched to net N mineralization. At this time, the C:N ratio had decreased from above 100 to 28. The microbial N was about five times the resource N. This rapid increase in microbial biomass and the fast mass loss of the understorey litter may partly be due to the relatively low initial litter N concentration (e.g. Berg *et al.*, 1995b, 1996). Usually, however, a positive correlation between mass loss and litter N concentration is observed at the beginning of the decomposition process (e.g. Swift *et al.*, 1979). In this context, it has to be taken into account that annually about 10–20 kg N ha^{-1} enter the forest floor with throughfall (cf. Weisdorfer *et al.*, this volume). Effects of added N to decomposing material is reviewed in detail by Fog (1988) who reports higher decay rates after N addition to easily degradable, N-poor substrate.

The next phase of substrate decomposition is characterized by relative accumulation of carbon and nitrogen, but the C:N ratio still decreases slowly. Secondary products will be synthesized and microbial biomass decreases as the amount of recalcitrant compounds increases relatively, because this signifies a decline in available energy resources. As N release begins to dominate over N immobilization, roots start to penetrate into the substrate and thus help mecanically to bind the now degraded litter material into the Of and make it indistinguishable from the Of horizon. The end of this step should correspond to the end of phase one documented by Aber *et al.* (1990), which coincides with the observed end of the inverse-linear relationship.

From the litterbag experiment with needle litter at Neuglobsow, the end of the inverse-linear relation cannot be defined. Aber *et al.* (1990) proposed to fit the function even to data sets that are not conducted long enough because they assumed that the relation holds until about 20% of initial mass remains. This does not seem reasonable for any of the needle litter samples incubated at our sites, because the fraction of 'inert' substrate was calculated to be about 35%. The asymptotic mass loss calculations done by Berg *et al.* (1996) for various decomposition experiments make a comparison possible with observed 'end point' values from the same incubations conducted by Aber *et al.* (1990). If experiments with leaf litter from White pine, White oak, Aspen, Sugar maple and Red oak are concidered, the mean asymptotic remaining mass is 16 ± 3% (Berg *et al.*, 1996), while the 'end point' of the inverse-linear relation is reached at a mean remaining mass of 22 ± 3% (Aber *et al.*, 1990). The relative loss from 22% to 16% m_r would correspond to a theoretical 'end point' in the needle litter decomposing at Neuglobsow at about 50% m_r, as the asymptotic remaining mass is 35% (Table 3). This approximation suggests that the 'end point' should be reached at the end of our experiment after one year of needle litter decay at Neuglobsow.

After one year, the C and N concentrations in the decomposing needle litter closely attained those of the Of material, if calculated on an ash free basis: 1.8% N, 54% C, C:N 30 in needle litter at Neuglobsow and 1.9% N, 52% C, C:N 27 in the Of-material. The studies of Aber *et al.* (1990), Berg and McClaugherty (1989), and others show that the different characteristic phases during the decomposition process are strongly related to the formation of recalcitrant compounds, as can be expressed e.g. by the concentration of lignin or acid-insoluble-substances or the lignin-cellulose-index. Complex chemical and biological interactions between available nitrogen, initial nitrogen concentrations, degradation products and microflora regulate the decomposition process (e.g. Berg and Ekbohm, 1991; Fog, 1988; Söderström *et al.*, 1983).

The third phase of substrate decomposition is characterized by the stabilization of C and N dynamics, microbial biomass and root growth, and a relative decrease of organic matter mass (ash free material). Aber *et al.* (1990) discussed the concept of a 'decay filter' that converts various types of litter into soil organic matter of a relative uniform composition of carbon fractions. Initially different above ground litter types will not be further distinguishable and thus the decomposition process is expected to follow the course described for Of-material. Within the Of matrix, root litter decomposes.

At Neuglobsow, the root decomposition followed the same pattern as the above ground litter decomposition, but hardly any N immobilization could be observed. The inverse-linear function resulted in a rather steep slope, probably indicating a change in the kind of microbial population. From the comparison of resource-N and microbial N after 2–3 months, little more than 0.1% of microbial biomass can be found in decomposing root litter. This approximated value has the same order as values determined by Substrate Induced Respiration- and Chloroform Fumigation Extraction-methods in the Of-material (3.2 mg C_{mic} per g dry matter; own data, unpublished). The concept of chemically relatively uniform material is supported by the similarity of the exponential functions that result from the Of-litterbag experiments from all three sites (Table 3).

Figure 5 presents the combined data from the litter and humus decomposition at Neuglobsow in the discussed three-phases model. A double exponential curve was fitted to the means of all litter types. The fit to a double exponential model was not reasonable for above ground litter only, because it does not consider any transfer of labile to recalcitrant material for the original substrate (Wieder and Lang, 1982). In this final model, all organic substrate present in the forest floor at Neuglobsow is included. At this site, no Oh-layer could yet be separated which should constitute the next step in the litter decomposition and humus accumulation process. Data from mineral soil horizons were not included in this study.

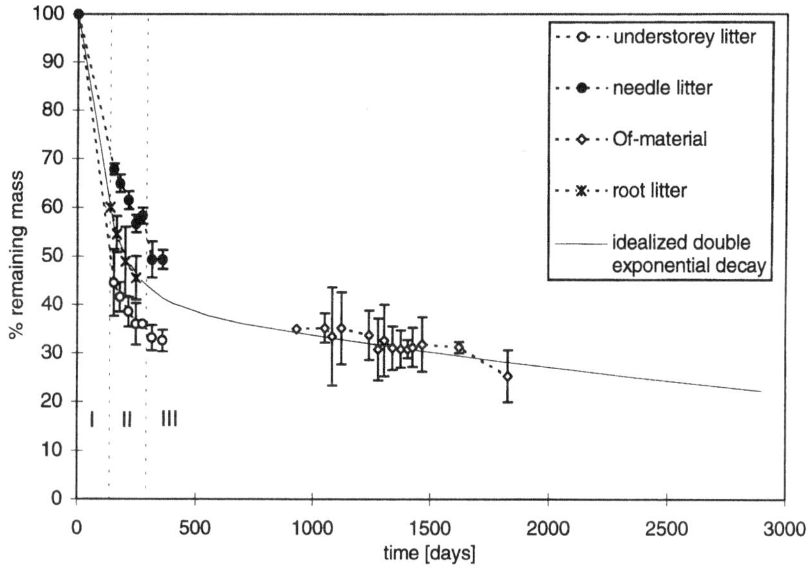

Figure 5. Three-phases-model of litter decomposition at Neuglobsow. Courses of measured mass loss and idealized double exponential curve fit. For explanations see text

The special management history of the Neuglobsow site, which was afforested at the beginning of this century after agricultural use and now is planted with the second generation of pine, allows to check our model of litter input – decomposition – accumulation against the observed stock of forest floor organic matter. The approximated calculation of litter accumulation was based on the assumption of almost linear steady litter input of the stand during its 62 years of age (in 1995). The understorey was assumed to reach todays primary production within 20 years. Decay rates of bark, twigs and cones were approximated to be somewhat higher (bark) or lower (twigs and cones) than that of needle litter (for cones cf. Berg and Staaf, 1981; for bark, twigs, cones e.g. cf. Fahey, 1983). Data on pine fine root production were taken from Strubelt *et al.* (this volume). These calculations result in about 33 Mg organic matter per hectare in the forest floor at Neuglobsow, while the determined stock was about 46 Mg organic matter per hectare. Considering all the assumptions made and the orders of magnitude from stock determination in centimetres and grams to tons per hectare calculation, these results show a good agreement. Apart from this, it must be taken into account, that the decomposition studies could not be repeated during other years to cover a higher variation of climatic conditions. However, it is assumed from moisture and temperature records of a 2.5-year-period that microclimatic conditions in the organic layers do not differ strongly between years.

For the calculation of annual N turnover, the data of remaining mass (resource, microbial and total) at the different decomposition phases and the respective N concentrations were used. Table 5 summarizes the N input with litterfall (beginning of phase one) and the remaining amounts of N at the beginning of phase two ($L \rightarrow Of_1$) and three ($Of_1 \rightarrow Of_2$) on an annual area basis (kg ha^{-1} yr^{-1}). It is supposed, that the main amount of N is imported and immobilized into microbial tissues during phase one and released during phase two. Thus, there would be an annual net N release of 2 kg N ha^{-1} from needle litter that passed the first two decomposition phases at Neuglobsow and, simultaneously, a net immobilization of 2.6 kg N ha^{-1} in understorey litter. Summed up, this results in a net immobilization of about 0.6 kg N ha^{-1} yr^{-1} at Neuglobsow.

At Taura and Rösa, litter and Of decomposition follows the same general pattern as discussed for Neuglobsow litter. The 'switch point', transfer from L to Of, is reached somewhat earlier than at Neuglobsow for pine litter and about two months later for the understorey litter. The amount of litter transferred to Of are higher, 65–78% compared to 54–60% at Neuglobsow. The N-dynamics differ among sites, as is indicated by the results of the mineralization-immobilization calculations (Table 5). The slopes of the inverse-linear relationships are steeper at Taura and Rösa, less nitrogen is immobilized in less microbial biomass and the rates of understorey litter decomposition do not differ considerably from the needle decomposition rates. This may be due to higher initial N concentrations at Taura and Rösa as is discussed above for the comparison of needle and understorey decay rates at Neuglobsow. This is supported by a significant negative relationship between the initial N concentration and k (E-03 day^{-1}) which is obtained from all the above ground litter decomposition experiments:

$$k = -12.4 * (\%N) + 13.8, \ r = 0.77^{**}.$$

If all data per site are combined to fit idealized double exponential curves, again the concept of the 'decay-filter' is supported by the results:

Neuglobsow:	$m_r = 58\% \exp(-8.58 \text{ E-03} * t) + 42\% \exp(-2.17 \text{ E-04} * t)$; $R^2 = 0.78^{***}$ (Figure 5)
Taura:	$m_r = 70\% \exp(-4.25 \text{ E-03} * t) + 30\% \exp(-2.17 \text{ E-04} * t)$; $R^2 = 0.96^{***}$
Rösa:	$m_r = 67\% \exp(-4.31 \text{ E-03} * t) + 34\% \exp(-2.17 \text{ E-04} * t)$; $R^2 = 0.95^{***}$

The approximation of humus accumulation can be calculated only for Of-horizons from the conducted experiments as discussed for Neuglobsow (see above). This is not possible for Taura and Rösa because of the site history (long-term forestry sites) and the presence of an Oh horizon. The formation of Oh-material is not simply a subsequent step in the litter decomposition but a combination with humification processes and therefore not considered in the decomposition model.

Table 5. Summarized data for N input with litterfall (L, beginning of phase one) and for the remaining amounts of organic matter and N (in resource, microbes and total) at the beginning of phase two ($L \rightarrow Of_1$) and three ($Of_1 \rightarrow Of_2$) on an annual area basis (kg N ha^{-1} yr^{-1})

	$L \rightarrow Of_1$ m_r resource	mic. biomass	L **Input**	**Total**	$L \rightarrow Of_1$ Resource	Microbial	**Total**	$Of_1 \rightarrow Of_2$ Resource	Microbial	$L–Of_2$ **Release**
	%					kg N ha^{-1} yr^{-1}				
Neuglobsow										
needle litter	51	16	**26.9**	**38.5**	17.2	21.3	**25.3**	24.9	0.4	**2**
understorey litter	42	22	**3.2**	**7.1**	1.4	5.7	**5.8**	1.1	4.7	**–2.6**
										–0.6
Taura										
needle litter	57	12	**30.7**	**39.8**	20.5	19.3	**27.6**	27.2	0.4	**3.1**
understorey litter	63	5	**18.8**	**16.7**	11.8	4.9	**11.7**	11.5	0.2	**7.1**
										10.2
Rösa										
needle litter	60	8	**35.5**	**38.8**	24.3	14.5	**28.3**	28.0	0.3	**7.2**
understorey litter	75	3	**14.2**	**13.2**	10.7	2.5	**0.2**	0.2	0.0	**14.0**
										21.2

The calculation of annual net N release from the datasets for Taura and Rösa (10 and 21 kg N ha^{-1} yr^{-1}, respectively; Table 5) reveals that at both sites N is not net retained in the uppermost humus layer but transferred to deeper humus horizons and probably to the mineral soil. Further movement cannot be calculated from litterbag experiments and must be assessed by other methods (Bergmann, 1998; Fischer *et al.*, 1996; Weisdorfer *et al.*, this volume).

Conclusions

Different rates of organic matter turnover in the forest floor are mainly due to the understorey growth and species spectra at the three test sites. The actual amount of above ground pine litter input does not seem to be affected by historical air pollution or fertilizer application. The chemical composition of the needle litter only indicates a slightly higher nitrogen, sulphur and calcium uptake at Rösa compared to Taura and Neuglobsow, which can be explained by the higher Ca- and S-deposition and N-fertilization at Rösa in the past. The decomposition rates of the needle litter do not differ significantly between the sites. In contrast, the type, amount, and chemical composition of the understorey litter, contributing 15 to 32% of total above ground litter, is a sensitive indicator for effects of historical deposition. Decomposition rates obviously depend on the spectrum of understorey species, N availability from the litter material itself and from input with precipitation and on microclimatic conditions, particularly forest floor moisture. The transfer pattern from fresh litter to Of-material is similar for all three sites. This is in accordance with the 'decay-filter' concept discussed by Aber *et al.* (1990). Also in agreement with this concept, the nitrogen dynamics may differ considerably among sites although the carbon chemistry is similar. At the experimental sites, N stocks increase with the historic deposition load gradient in the order Neuglobsow < Taura < Rösa. This results from different amounts of forest floor dry mass while C/N ratios decrease conversely. Additionally, the understorey necromass C/N ratio is highest at Neuglobsow (115 compared to 55 and 43 at Taura and Rösa, respectively). These results indicate that a great part of the nitrogen from deposition and fertilization is accumulated in the forest floor and understorey vegetation. For the upper humus layers (L, Of_1) the litterbag experiments allow to calculate the amount of net annual N release. At Taura and even more pronounced at Rösa, an annual N surplus in the freshly decaying and one to two year old litter is not efficiently immobilized neither in soil fauna (e.g. microbial tissue) nor in vegetation via plant uptake but transferred to deeper (humus) horizons.

References

Aber JD, Melillo JM. 1980. Litter decomposition: measuring relative contributions of organic matter and nitrogen to forest soils. Can J Bot. 58, 416–421.

Aber JD, Melillo JM, McClaugherty C. 1990. Predicting long-term patterns of mass loss, nitrogen dynamics, and soil organic matter formation from initial fine litter chemistry in temperate forest ecosystems. Can J Bot. 68, 2201–2208.

Baronius G, Fiedler HJ. 1993. Zur monatlichen Veränderung der Nährelementgehalte in Nadeln von Kiefernbeständen im Immissionsgebiet Dübener Heide. Wiss Z Techn Univers Dresden. 42(1), 93–97.

Berg B, Ekbohm G. 1983. Nitrogen immobilization in decomposing needle litter at variable carbon: nitrogen ratios. Ecology. 64(1), 63–67.

Berg B, Ekbohm G. 1991. Litter mass-loss rates and decomposition patterns in some needle and leaf litter types. Long-term decomposition in a Scots pine forest. VII. Can J Bot. 69, 1449–1456.

Berg B, McClaugherty C. 1989. Nitrogen and phosphorus release from decomposing litter in relation to the desappearance of lignin. Can J Bot. 67, 1148–1156.

Berg B, Staaf H. 1980a. Decomposition rate and chemical changes of Scots pine needle litter. II. Influence of chemical composition. In: Structure and Function of Northern Coniferous Forests – An Ecosystem Study. Ed. T Persson. Ecol Bull (Stockholm). 32, 373–390.

Berg B, Staaf H. 1980b. Decomposition rate and chemical changes of Scots pine needle litter. I. Influence of stand age. In: Structure and Function of Northern Coniferous Forests – An Ecosystem Study. Ed. T Persson. Ecol Bull (Stockholm). 32, 363–372

Berg B, Staaf H. 1981. Leaching, accumulation, and release of nitrogen in decomposing forest litter. In: Terrestrial Nitrogen Cycles. Eds. FE Clark, T Rosswall. Processes, Ecosystem Strategies, and Management Impacts. Ecol Bull (Stockholm). 33, 163–178.

Berg B, Calvo de Anta R, Escudero A *et al.* 1995a. The chemical composition of newly shed needle litter of Scots pine and some other pine species in a climatic transect. Long-term decomposition in a Scots pine forest. X. Can J Bot. 73, 1423–1435.

Berg B, Ekbohm G, Johansson M-B, McClaugherty C, Rutigliano F, Virzo De Santo A. 1996. Maximum decomposition limits of forest litter types: a synthesis. Can J Bot. 74, 659–672.

Berg B, McClaugherty C, Virzo de Santo A, Johansson M, Ekbohm G. 1995b. Decomposition of litter and soil organic matter – can we distinguish a mecanism for soil organic matter buildup? Scand J For Res. 10, 108–119.

Berg, B, Berg MP, Bottner P *et al.* 1993 Litter mass loss rates in pine forests in Europe and Eastern United States: some relationships with climate and litter quality. Biogeochemistry. 20, 127–159.

Bergmann C. 1998. Stickstoff-Umsätze in der Humusauflage unterschiedlich immissionsbelasteter Kiefernbestände (*Pinus sylvestris* L.) im nordostdeutschen Tiefland mit besonderer Berücksichtigung des gelösten organischen Stickstoffs. Dissertation, BTU Cottbus. Cottbuser Schriften zu Bodenschutz und Rekultivierung Bd 1.

Bocock KL, Gilbert OJW. 1957. The disappearance of leaf litter under different woodland conditions. Plant Soil. 9, 179–185.

Couteaux M-M, Bottner P, Berg B. 1995. Litter decomposition, climate and litter quality. TREE. 10(2), 63-66.

Fahey TJ. 1983. Nutrient dynamics of aboveground detritus in lodgepole pine (*Pinus contorta ssp. latifolia*) ecosystems, southeastern Wyoming. Ecol Monographs. 53, 51–72.

Falconer GJ, Wright JW, Beall HW. 1933. The decomposition of certain types of fresh litter under field conditions. Am J Bot. 20, 196–203.

Fangmeier A, Hadwiger-Fangmeier A, Van der Eerden L, Jäger H-J. 1994. Effects of atmospheric ammonia on vegetation – a review. Environ Pollut. 86, 43–82.

Fischer T, Bergmann C, Hüttl RF. 1996. Auswirkungen sich zeitlich ändernder Schadstoffdepositionen auf Prozesse des Kohlenstoff- und Stickstoffumsatzes im Boden. Abschlußbericht, SANA-Projekt.

Fog K. 1988. The effect of added nitrogen on the rate of decomposition of organic matter. Biol Rev. 63, 433–462.

Gustafson FG. 1943. Decomposition of the leaves of some forest trees under field conditions. Plant Physiol. 18, 704–707.

Hoffmann H, Krauss HH. 1988. Streufallmessungen in gedüngten und ungedüngten mittelalten Kiefernbeständen auf Tieflandstandorten der DDR. Beitr Forstwirtschaft. 22(3), 97–100.

Hofmann G. 1994. Wälder und Forsten – Mitteleuropäische Wald- und Forstökosystemtypen in Wort und Bild. Der Wald, Sonderheft Waldökosystem-Katalog. 52 S.

Howard PJA, Howard DM. 1974. Microbial decomposition of tree and shrub leaf litter. I. Weight loss and chemical composition of decomposing litter. Oikos. 25, 341–352.

Jenny H, Gessel SP, Bingham FT. 1949. Comparative study of decomposition rates of organic matter in temperate and tropical regions. Soil Sci. 68, 419–432.

Johansson M-B, Berg B, Meentemeyer V. 1995. Litter mass-loss rates in late stages of decomposition in a climatic transect of pine forests. Long-term decomposition in a Scots pine forest. IX. Can J Bot. 73, 1509–1521.

Krauss HH, Hoffmann H. 1991. Streufallmessungen in gedüngten und ungedüngten mittelalten Kiefernbeständen auf Tieflandstandorten der ehemaligen DDR. Teil II: Gehalte und Umlauf an den Nährstoffen N, P, K, Ca, Mg. Beitr Forstwirtschaft. 25(3), 119–130.

Lunt HA. 1935. Effects of weathering upon dry matter and composition of hardwood leaves. J For. 33, 607–609.

Minderman G. 1968. Addition, decomposition and accumulation of organic matter in forests. J Ecol. 56, 355–362.

Olson, JS. 1963. Energy storage and the balance of producers and decomposers in ecological systems. Ecology. 44(2), 322–331.

Shanks RE, Olson JS. 1961. First-year breakdown of leaf litter in Southern Appalachian forests. Science. 134, 194–195.

Smith JL, Papendick RI, Bezdicek DF, Lynch JM. 1993. Soil organic matter dynamics and crop residue management In: Soil Microbial Ecology. Ed. FB Metting Jr. Marcel Dekker, New York pp. 65–98.

Söderström B, Baath E, Lundgren B, 1983, Decrease in soil microbial activity and biomasses due to nitrogen amendments. Can J Microbiol. 29, 1500–1506.

Swift MJ, Heal OW, Anderson JM. 1979. Decomposition in Terrestrial Ecosystems. Studies in Ecology, Vol. 5. Blackwell Scientific, University of California Press, Berkeley and Los Angeles, 1979.

Tamm CO. 1991. Nitrogen in Terrestrial Ecosystems. Ecological Studies Vol. 81. Springer Verlag, Berlin Heidelberg, 1991.

Trautmann W, Krause A, Wolff-Straub R. 1970. Veränderungen der Bodenvegetation in Kiefernforsten als Folge industrieller Luftverunreinigung im Raum Mannheim – Ludwigshafen. Schriftenreihe Vegetationskunde, 5.

Wieder RK, Lang GE. 1982. A critique of the analytical methods used in examining decomposition data obtained from litter bags. Ecology. 63(6), 1636–1642.

Witkamp M, Olson J. 1963. Breakdown of confined and non-confined oak litter. Oikos. 14(2), 138–147.

11
Seasonal variability of organic matter and N input with litterfall in Scots pine stands

C. BERGMANN, T. FISCHER and R.F. HÜTTL

1. Introduction

As part of the SANA subproject on organic matter and N turnover, the seasonal variability of mass input and N concentration of Scots pine litter was recorded by monthly collections at three sites of historically different air pollution and fertilization impacts. Furthermore, seasonal growth dynamics, biomass production and litter (= necromass) input to the forest floor from the ground vegetation was analysed in detail. The ground vegetation biomass differs obviously at the chosen sites according to their deposition history, being dominated by *Vaccinium myrtillus* and *Avenella flexuosa*, only *Avenella flexuosa* or *Calamagrostis epigeios*, *Rubus* sp., and *Brachypodium sylvaticum*. It was hypothesised that pine litter mass would also differ between the three sites. In addition, N fertilization and deposition of alkaline dusts and SO_2 should exert an effect on the chemical composition of living and dead plant tissues and thus on the amount of N returned to the forest floor via litterfall. In this paper the seasonal variability of organic matter and N input from Scots pine and ground vegetation at the SANA-sites is described and data are compared among sites. Further nutrient element concentrations in litter from pines and ground vegetation are published by Bergmann *et al.* (this volume).

2. Materials and methods

2.1. Sampling

The experimental sites, located along an air pollution gradient from Rösa to Taura and Neuglobsow, were established in spring 1993. For a more detailed site description see elsewhere in this volume (e.g. Bergmann *et al.*). The sampling of pine litter input and standing herbaceous ground vegetation biomass started in June/August 1993 and was repeated throughout the vegetation period (April to November) until November 1995.

Pine litter input was sampled monthly from seven 1×1 m^2-collectors per site. The collectors were constructed from steel posts to which four aluminium tubes were attached to produce a rectangular frame, about one metre above the

R.F. Hüttl and K. Bellmann, Changes of Atmospheric Chemistry and Effects on Forest Ecosystems, 177–186.

ground. Nylon net (2 mm mesh size) was hung in the frame, closed at the bottom end, and fastened to the ground to prevent it being loosened or opened by wind and storm. For sampling, the net was opened and litter material collected on a tarpaulin, also hung in the frame, to which a funnel was connected that led the material into a bucket. After drying at 70–80°C, samples were separated into needles, bark and fine litter, twigs and cones. Aliquots from the needle- and the bark and fine litter-fraction were milled and chemically analysed.

Bio- and necromass of the standing ground vegetation was harvested bimonthly in five replicates per site from randomly chosen $50 \times 50\ cm^2$ quadrats. Mosses were not sampled, nor were individual understorey phanerophytes harvested which were present occasionally at the Rösa site. The material (phytomass, P) was sorted into a green (living, i.e. phytobiomass, B) and a brown (dead, i.e. phytonecromass, N) fraction (Janetschek, 1982), dried and processed as described for the pine litter.

Dried samples were finely ground in a swinging mill (MM 2000, Retsch). Total nitrogen was measured on individual samples by dry combustion (CNS-Analyser Vario EL, Heraeus). Laboratory-intern pine needle standards were within 1–6% of the known concentrations for N.

2.2. *Calculations and statistics*

Input of pine litter per site was calculated from the monthly replicate yields per square metre to mean monthly and annual input of dry mass per hectare.

For the calculation of ground vegetation litter input some assumptions had to be made:

- the sampled species are all herbaceous in character and their entire above ground biomass will die at the end of every vegetation period and return to the litter layer until next spring,
- the maximum primary production (biomass maximum) will approximately be reached at or between the summer harvests in June and August. Until June there will be no necromass production and after August biomass will not increase any more.

These assumptions do not, in fact, take the perennial nature of the (dwarf) shrubs *Vaccinium myrtillus* and *Rubus* sp. into account.

Figure 1 illustrates the assumptions made above from which the following equations (equation 1–4) were deduced. They allow for the calculation of bimonthly litter input (L), i.e. all necromass which is no longer 'standing' phytonecromass, to the forest floor.

August to October	$L =$	$\Delta P = P_{Aug} - P_{Oct}$	(eq. 1)
October to April	$L =$	$P_{Oct} - N_{Apr}$	(eq. 2)
April to June	$L =$	$\Delta N = N_{Apr} - N_{Jun} = N_{Apr} - 0$	(eq. 3)
June to August	$L =$	0	(eq. 4)

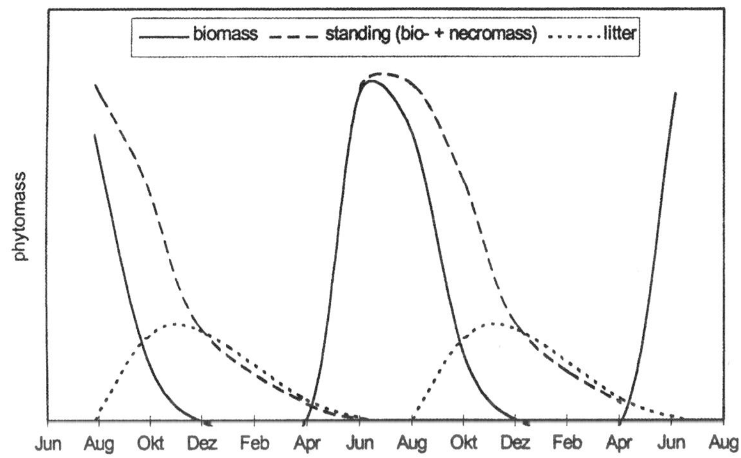

Figure 1. Model of seasonal course of phytobiomass, phytonecromass and litter development of the ground vegetation

The Student's *t* statistic, when applicable, was used to compare dry mass and nutrient differences. ANOVA comparisons were run with the Student-Newman-Keuls-test. Significance always refers to $\alpha = 5\%$, unless indicated otherwise.

3. Results

3.1. Pine litter input

The mean annual input of pine litter was between 4.3 and 5.3 Mg ha^{-1}. Needle input accounted for 2.4 to 2.8 Mg ha^{-1}, i.e. 56 to 61% from total pine litter, and did not differ significantly between the sites. Of the total pine litter, 16–21% consisted of bark and very fine litter material, 23% of twigs and female cones. There is a tendency for higher proportions of needle, bark and fine material input at Neuglobsow compared to Taura and Rösa. At the latter sites the percentage of cones from total litter input is higher than at Neuglobsow. Differences in mean annual amounts between sites are not significant, either for individual fractions or for total pine litter. The seasonal variation of total litter input and proportions of individual fractions are illustrated in Figure 2, data are given in Table 1.

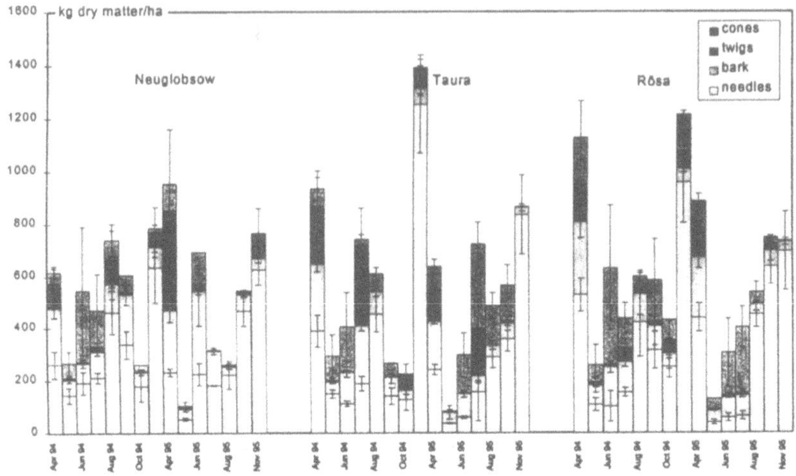

Figure 2. Seasonal variability of pine litter input, separately recorded for the needle-, bark and fine litter-, twig- and cone-fraction

The N concentration was measured in the individual samples of the needle- and the bark and fine litter-fraction. For each of the sampling years (1993–1995) they show a marked seasonal variability but do not differ among sites (Figure 3a,b; Table 1). Highest N concentrations are usually reached in those small litter amounts falling during May (1.4 to 2.5% N in bark and fine litter and 1.0 to 1.3% N in needles, depending on year and site). Lowest concentrations, usually measured in late summer- or autumn-litter, range from 0.5 to 0.9% N in bark and fine litter and from 0.5 to 0.6% in needles, depending on year and site. The differences between highest and lowest mean N concentrations in needle litter are significant at all three sites. The combination of the summed litter mass input (Figure 4a) and the respective monthly mean N concentrations (Figure 3) gives the mean curve of cumlative N input with litterfall which is illustrated in Figure 4b. The mean annual input of N with pine litter is 27 kg N ha^{-1} at Neuglobsow, 31 kg N ha^{-1} at Taura and 36 kg N ha^{-1} at Rösa.

3.2. *Ground vegetation litter input*

The bimonthly sampled phytomass of the ground vegetation differed significantly among sites. Highest phytomass production was always recorded at Taura, while the yields were always lowest at Neuglobsow. In August, when the phytomass production reached a maximum, the differences between sites were most pronounced and highly significant: 2.3 Mg d.m. ha^{-1} standing phytobiomass at Taura, 1.4 Mg ha^{-1} at Rösa and 0.7 Mg ha^{-1} at Neuglobsow. The

Table 1. Input of litter (kg dry matter ha^{-1} yr^{-1}) from pine stands and ground vegetation and corresponding N concentrations (%N); data are monthly or bimonthly mean values for the sampling period 1993–1995

	April		May		June		July	
	dm	%N	dm	%N	dm	%N	dm	%N
Neuglobsow								
Needles	246	0.98	99	1.01	208	0.87	197	0.80
Bark and fine litter	227	0.86	49	2.19	197	1.49	116	1.06
Twigs	240		10		17		12	
Cones	71		29		197		69	
Understorey litter			65	1.00			0	
Taura								
Needles	316	1.14	93	1.17	87	1.08	173	1.06
Bark and fine litter	221	0.65	44	1.04	105	0.95	142	0.82
Twigs	174		6		3		246	
Cones	75		46		157		173	
Understorey litter			210	1.47			0	
Rösa								
Needles	485	0.91	74	1.23	80	0.99	110	1.07
Bark and fine litter	256	0.73	60	1.95	113	1.43	98	1.02
Twigs	115		9		10		31	
Cones	152		52		266		182	
Understorey litter			317	0.98			0	
	Aug		**Sep**		**Oct**		**Nov**	
	dm	%N	dm	%N	dm	%N	dm	%N
Neuglobsow								
Needles	342	0.66	286	0.67	206	0.54	631	0.56
Bark and fine litter	70	1.09	112	0.88	43	0.83	62	0.96
Twigs	53		10		6		76	
Cones	36		32		12		7	
Understorey litter			321	0.54			360	0.43
Taura								
Needles	372	0.82	160	0.71	153	0.58	1044	0.62
Bark and fine litter	65	0.71	51	0.78	33	0.77	43	0.73
Twigs	4		6		33		25	
Cones	107		56		34		17	
Understorey litter			796	0.76			1286	0.65
Rösa								
Needles	439	0.81	318	0.76	286	0.59	827	0.64
Bark and fine litter	72	0.74	63	0.67	40	0.82	43	0.86
Twigs	15		14		32		37	
Cones	44		85		45		67	
Understorey litter			195	1.14			896	1.01

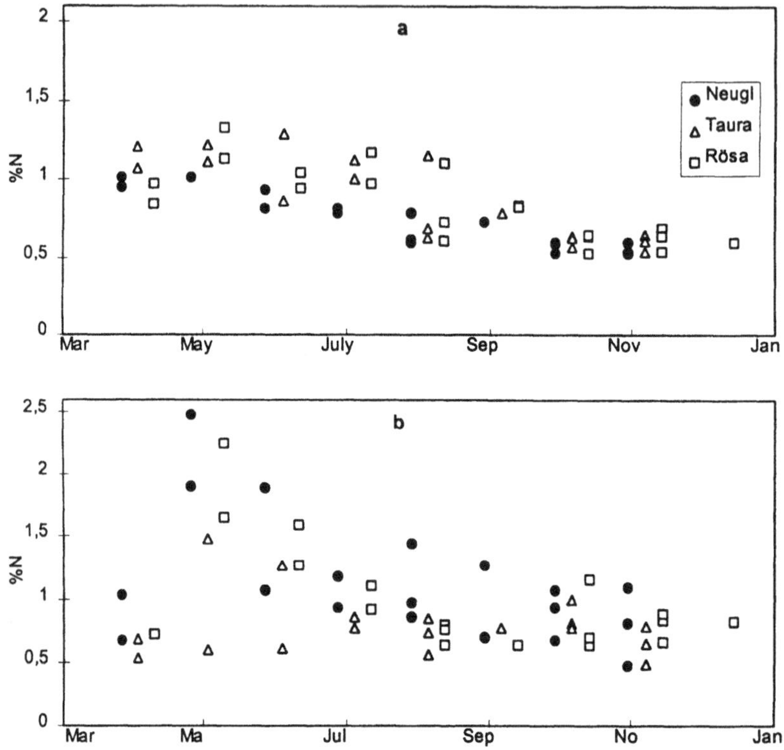

Figure 3. Seasonal variability of N concentrations in (a) the bark and fine litter-fraction and (b) the needle-fraction of the pine litter; data are monthly means for 1993, 1994 and 1995

seasonal phytomass development is illustrated in Figure 5. Positive columns indicate standing phytobio- and -necromass, while negative columns give the amount of fallen necromass, i.e. litter input to the forest floor. Generally, the course of phytomass production follows the curve shown in Figure 1. Some phenological differences between Taura and Neuglobsow on the one hand, and Rösa on the other can be observed. At Rösa the time of highest litterfall from the ground vegetation is delayed to late autumn, winter and early spring if compared to Neuglobsow and Taura where most of the necromass is shed in late summer, autumn and winter (cf. Table1). The annual input of ground vegetation litter, by calculation comes to the amount of maximum biomass production (0.75 Mg ha^{-1} at Neuglobsow, 1.41 Mg ha^{-1} at Rösa and 2.29 Mg ha^{-1} at Taura). This accounts for 15, 21 and 32% of total above ground litter at the respective sites.

The mean annual N concentrations in the ground vegetation litter range from 4.3 mg g^{-1} at Neuglobsow to 10.1 mg g^{-1} at Rösa. As for pine litter, they show a marked seasonality (Figure 6a,b; Table 1). In contrast to the similarity

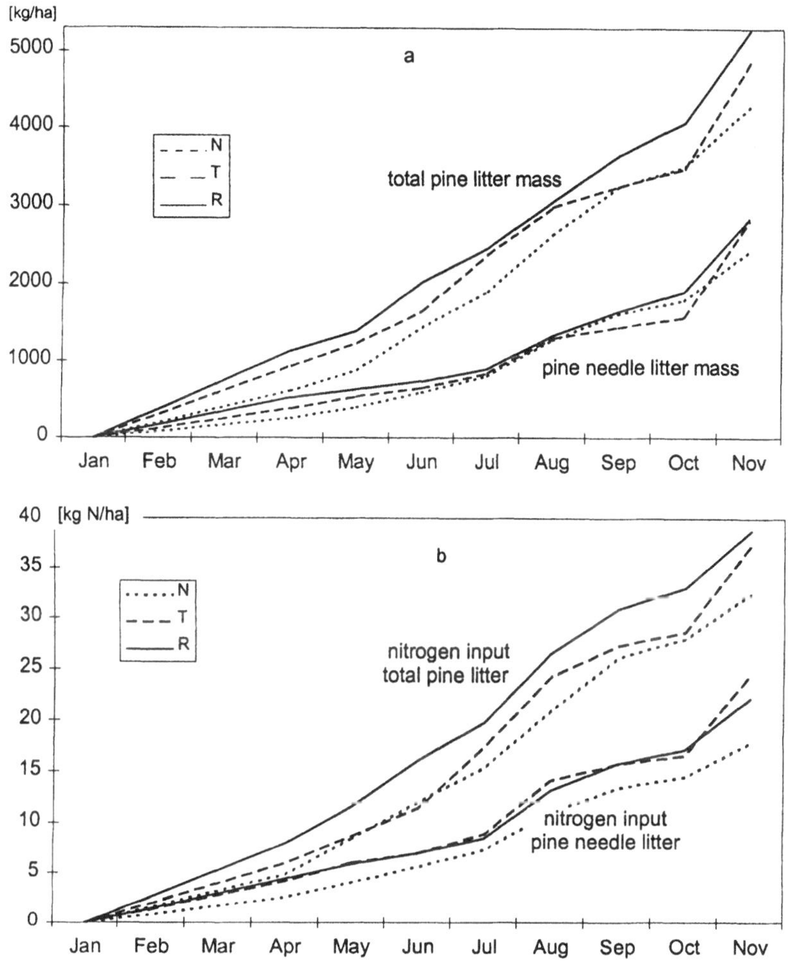

Figure 4. Cummulative input of (a) total pine litter mass and pine needle litter and (b) nitrogen with litterfall at Neuglobsow, Taura and Rösa

of N concentrations in pine litter at the three sites, there is a clear differentiation in phytomass N concentrations between Neuglobsow at the one hand and Taura and Rösa at the other. Mean biomass N concentrations are about 1.3% at Neuglobsow and 2.0 to 2.3 at the other sites. Mean necromass N concentrations are 0.7%, 1.0%, and 1.2% at Neuglobsow, Rösa, and Taura, respectively. In the biomass, highest N concentrations were measured in early spring (1.9%, 2.5% and 3% at Neuglobsow, Rösa and Taura, respectively), which were diluted during biomass production in summer to 1.1%, 1.7% and 1.6% at the respective sites. Maximum necromass N concentrations were also reached in

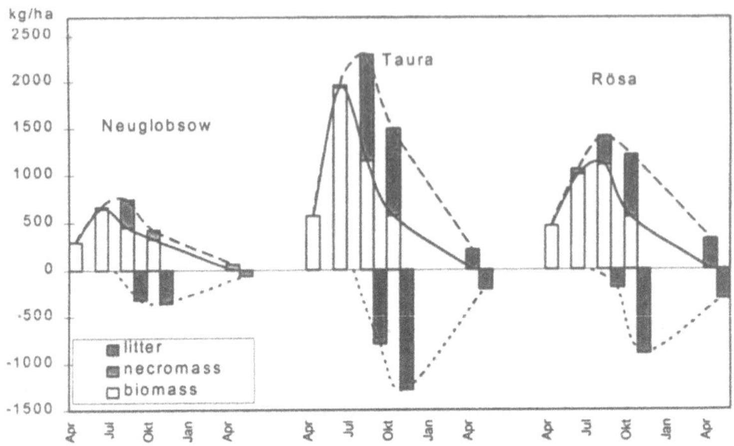

Figure 5. Seasonal development of phytomass (positive columns) and litter input (negative columns) in the ground vegetation communities at Neuglobsow, Taura and Rösa

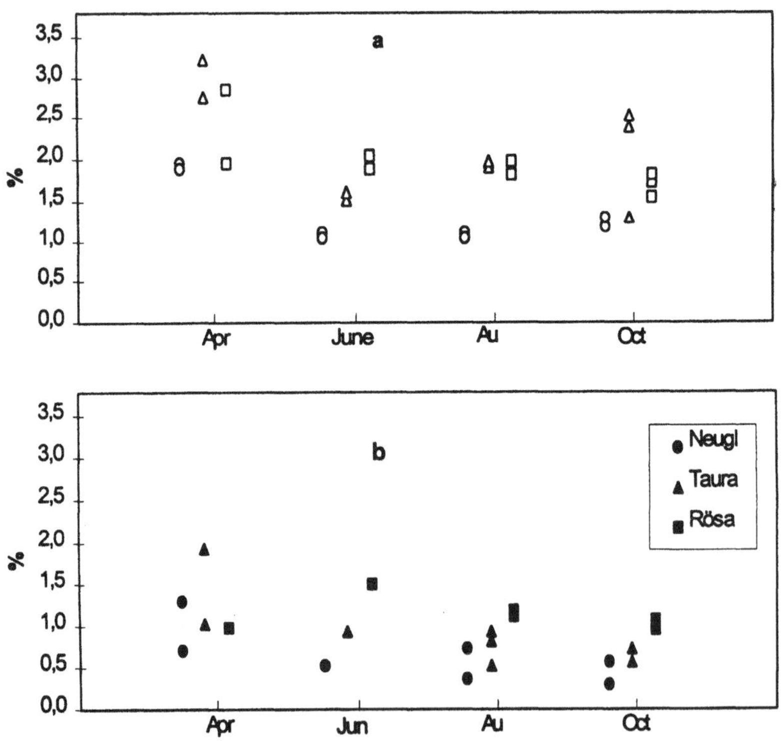

Figure 6a–b. (a) Seasonal variability of N concentrations in the phytobiomass; data are bimonthly means for 1993, 1994 and 1995; (b) Seasonal variability of N concentrations in the phytonecromass of the ground vegetation; data are bimonthly means for 1993, 1994 and 1995

spring (1.0 to 1.5%) and decreased until autumn to values from 0.5 to 1.0%. Combined with litter mass input, the following mean annual N inputs with understorey litter are calculated: 3 kg N ha^{-1} at Neuglobsow, 19 kg N ha^{-1} at Taura and 14 kg N ha^{-1} at Rösa.

4. Discussion

4.1. Litter input

Total pine litter mass is in the upper range of values reported e.g. by Hoffmann and Krauß (1988) for 60 to 73-yr-old Scots pine stands in eastern Germany. Compared to data from Swedish Scots pine ecosystems (100%), needle litter at Neuglobsow, Taura and Rösa is about 175% and perfectly fits the extrapolation of a linear regression from Berg *et al.* (1995b) which relates needle litterfall to latitude. No significant effect of historical pollution loads or fertilizations on annual above ground pine litter mass is detected. This may well be due to the relatively high foliage biomass, which in turn may result from the generally sufficient, although site-specifically different, supply with nitrogen (cf. Hoffmann and Krauß, 1988). At Neuglobsow, the higher stand density might cause relatively high litter mass and thus hide differences that could be due to air pollution or fertilization treatments.

The alteration of species spectra of ground vegetation due to increased amounts of acids or nutrients is a phenomenon that was already documented by Fangmeier *et al.* (1994), Tamm (1991) and Trautmann *et al.* (1970). At Neuglobsow, Taura and Rösa the observed differentiation follows the known pattern: *Vaccinium mytillus*, restricted to low pollution areas, cannot compete with *Avenella flexuosa* at poor sites or with *Calamagrostis epigeios* and *Rubus* sp. at nutrient enriched sites. The height and density of the understorey obviously correlates with litter mass which was documented e.g. by Hofmann (1994).

4.2. N concentration and N input

The N concentrations of the pine needle litter is well within the range reported by Krauß and Hoffmann (1991) for Scots pine in this region. There is a tendency of higher N concentrations in the needle litter from Rösa where the N availability was raised during the last decades by continuous fertilization. A similar even more pronounced pattern is observed in ground vegetation tissues. Compared to Swedish Scots pine stands (e.g. Johansson *et al.*, 1995), needle litter from Neuglobsow, Taura and Rösa has higher N concentrations. But they can be fitted perfectly into a regression from Berg *et al.* (1995a), who related N concentrations in Scots pine needle litter to latitude. As far as the annual N input with litterfall is concerned, the stands at the three experimental sites are certainly comparable to coniferous ecosystems with approximately 30 kg

N ha^{-1} yr^{-1} in 4 Mg litter ha^{-1} yr^{-1} which were documented by Vitousek (1982). The high annual biomass production and N input via understorey litter at Taura and Rösa compared to Neuglobsow shows that at these ecosystems considerable 'surplus' amounts of N are annually immobilized in plant tissues and thus stored in a 'one-year-N-cycle'.

References

Berg B, Calvo de Anta R, Escudero A *et al.* 1995a. The chemical composition of newly shed needle litter of Scots pine and some other pine species in a climatic transect. Long-term decomposition in a Scots pine forest. X Can J Bot. 73, 1423–1435.

Berg B, McClaugherty C, Virzo de Santo A, Johansson M, Ekbohm G. 1995b. Decomposition of litter and soil organic matter – can we distinguish a mechanism for soil organic matter buildup? Scand J For Res. 10, 108–119.

Fangmeier A, Hadwiger-Fangmeier A, Van der Eerden L, Jäger H-J. 1994. Effects of atmospheric ammonia on vegetation – a review. Environ Pollut. 86, 43–82.

Hofmann G. 1994. Wälder und Forsten – Mitteleuropäische Wald- und Forstökosystemtypen in Wort und Bild. Der Wald, Sonderheft Waldökosystem-Katalog. 52 S.

Hoffmann H, Krauß HH. 1988. Streufallmessungen in gedüngten und ungedüngten mittelalten Kiefernbeständen auf Tieflandstandorten der DDR. Beitr Forstwirtschaft. 22(3), 97–100.

Janetschek H. (Hrsg.) 1982. Ökologische Feldmethoden. Ulmer Stuttgart.

Johansson M-B, Berg B, Meentemeyer V. 1995. Litter mass-loss rates in late stages of decomposition in a climatic transect of pine forests. Long-term decomposition in a Scots pine forest. IX Can J Bot. 73, 1509–1521.

Krauß HH, Hoffmann H. 1991. Streufallmessungen in gedüngten und ungedüngten mittelalten Kiefernbeständen auf Tieflandstandorten der ehemaligen DDR. Teil II: Gehalte und Umlauf an den Nährstoffen N, P, K, Ca, Mg. Beitr Forstwirtschaft. 25(3), 119–130.

Tamm CO. 1991. Nitrogen in Terrestrial Ecosystems. Ecological Studies Vol. 81. Springer Verlag, Berlin Heidelberg, 1991.

Trautmann W, Krause A. Wolff-Straub R. 1970. Veränderungen der Bodenvegetation in Kiefernforsten als Folge industrieller Luftverunreinigung im Raum Mannheim – Ludwigshafen. Schriftenreihe Vegetationskunde, 5.

Vitousek PM. 1982. Nutrient cycling and nutrient use efficiency. Am Naturalist. 119, 353–372.

12
Soil chemical response to drastic reductions in deposition and its effects on the element budgets of three Scots pine ecosystems

M. WEISDORFER, W. SCHAAF, R. BLECHSCHMIDT, J. SCHÜTZE and R.F. HÜTTL

1. Introduction

During the last 15–20 years the environmental effects of acidic deposition, in particular soil acidification, were studied with different research approaches, ranging from the evaluation of element budgets of whole watersheds to experiments with soil samples in the laboratory, both with and without manipulation (Matzner, 1989; Marschner, 1990; Johnson and Lindberg, 1992; Schaaf, 1992; Beier *et al.*, 1993; Koopmans *et al.*, 1995; Bredemeier *et al.*, 1995; Matzner and Murach, 1995; Moldan *et al.*, 1995; Visser and van Breemen, 1995). In our study, the experimental sites are located along a gradient of atmospheric pollutant deposition, especially with respect to sulfur and alkaline dust. Additionally, the present soil chemical properties of the experimental sites reflect different periods of pollutant deposition with corresponding soil chemical properties of forest ecosytems in Western Europe and thus, can be regarded as the result of a temporal gradient. This gradient ranges from the early 1970s, influenced by high deposition loads of sulfur and dust (site 'Rösa') to the early 1980s where the soil chemistry was affected by less alkaline dust but still high sulfur immissions (site 'Taura'). Since presently the three sites are characterized by relatively low deposition loads, the project can be seen as a 'roof experiment without roof'. Therefore, we are able to study simultaneously the development of the soil chemistry of three soils representing different situations of deposition loads.

The main objective of our project was to study the element budgets and the soil chemical properties to identify dominating processes in the soil and to evaluate the stability and sensitivity of ecosystems with respect to environmental changes. Another important question is whether the Scots pine ecosystems, especially the compartment 'soil' will recover from long-term pollution and what are the governing processes.

Because of high emissions of sulfur in the past we focus on the S dynamics in the ecosystems and in particular on the SO_4^{2-} retention/release characteristics of the soils. Among the effects of elevated pollutant deposition (H^+, SO_4^{2-}) on mineral soil chemical reactions and solution chemistry, inorganic sulfate

R.F. Hüttl and K. Bellmann, Changes of Atmospheric Chemistry and Effects on Forest Ecosystems, 187–225.

adsorption capacity as well as S mineralization and immobilization processes have been considered to be important parameters by affecting cation leaching and pH of the soil solution. (Singh, 1984; Nodvin *et al.*, 1986; Marsh *et al.*, 1987; Sposito, 1989; Courchesne and Hendershot, 1990; David *et al.*, 1991; Alewell and Matzner, 1993; MacDonald *et al.*, 1994; Ajwa and Tabatabai, 1995).

Specific and/or nonspecific inorganic adsorption of SO_4^{2-} may result in an increase in pH due to protonation of OH-groups and formation of 'outer-sphere complexes' or the release of OH^- by ligand exchange reaction and formation of 'inner-sphere complexes'. Furthermore, SO_4^{2-} adsorption at surface sites of hydrous oxides of Al and Fe create additional negative charge and increase soil effective cation exchange capacity (ECEC). As a result, SO_4^{2-} adsorption may be able to decrease the leaching of H^+, base cations, and Al from the mineral soil to the groundwater. On the other hand, once input concentrations of SO_4^{2-} are reduced, SO_4^{2-} desorption may prevent or delay de-acidification due to an increasing transfer of adsorbed SO_4, H^+, and Al^{n+} into solution. Both the amounts and the bonding form of SO_4^{2-} in a mineral soil depend on pH, ionic strength, DOC concentration and fractions, composition of the soil solution and contents of hydrous oxides of Al and Fe (Ajwa and Tabatabai, 1995). A further SO_4^{2-} retention/release mechanism may be the precipitation/dissolution of aluminium-sulfate minerals such as alunite, basalunite, and jurbanit (Reuss and Walthall 1989).

2. Materials and methods

2.1. *Experimental sites and soils*

The three investigated Scots pine ecosystems are located along a deposition gradient. The site 'Rösa' (12°26′ E 51°38′ N) is located about 10 km to the east of the industrial complex Bitterfeld and represents the most severely affected area that was heavily impacted mainly by sulfur and alkaline dust depositions in the past. The site 'Taura' (13°2′ E 51°28′ N) is situated about 50 km north-east of the Halle/Leipzig region and was also influenced by high deposition of sulfur, but less alkaline dust deposition. 'Neuglobsow' (13°2′ E 53°8′ N) represents the 'background' site and is located in northern Brandenburg. The soils at the sites are Spodi-dystric Cambisols (Rösa), Cambic Podzols (Taura) and Dystric Cambisols (Neuglobsow) derived from glacial outwash sediments.

2.2. *Field measurements and soil samples*

At all investigation sites soil solution was continuously collected at three measurement plots per site using ceramic plates and suction cups (P80 material) with two replicates in four soil depths (0 cm = mineral soil input, 20 cm, 50 cm, 100 cm). Three mixed samples were composed from every soil depth. Soil solutions were sampled every two weeks from September 1993 to October 1995. Soil tensions were recorded using five continuously registering pressure transducer tensiometers per depth (20 cm, 50 cm, 100 cm, 180 cm). Throughfall was collected with ten bulk samplers and one recording sampler. Climatic parameters like precipitation, temperature, humidity, wind velocity and global radiation were recorded at a nearby clear-cut at each site. Soil samples were collected from all humus layers and mineral soil horizons at three small pits per site in October 1993. To investigate SO_4^{2-} retention mechanisms, we collected additional soil samples in April 1995. The soil samples were air dried and passed through a 2 mm mesh prior to chemical analyses.

2.3. *Analytical methods*

Soil solution, bulk precipitation, and throughfall were analyzed for pH, the cations calcium (Ca^{2+}), magnesium (Mg^{2+}), potassium (K^+), sodium (Na^+), manganese (Mn^{2+}), aluminium (Al^{n+}), iron ($Fe^{2+/3+}$) by AAS (Unicam 939) and ICP-AES (Unicam 701), chloride (Cl^-), sulfate (SO_4^{2-}), and nitrate (NO_3^-) by ionic chromatography (IC Dionex DX 500) and dissolved organic carbon (DOC) with TOC-analyzer after 0.45 μm filtration (Shimadzu TOC 500). A rapid flow analyzer (Flow solution Peerstop Alpkem) was used for NH_4 analyses.

The pH of the soil samples was measured in 1:2.5 water, 0.01 M $CaCl_2$-, and 1 M KCl-extract (Meiwes *et al.*, 1984). The effective cation exchange capacity (ECEC) was determined using 0.5 N NH_4Cl (Trüby and Aldinger, 1986). Ca^{2+}, Mg^{2+}, K^+, Na^+, Fe^{3+}, Al^{n+}, and Mn^{2+} were determined by ICP-AES (Unicam 701). The Ca/Al, Mg/Al and Ca/H-ratios were determined in the 1:2 water-extract (Meiwes *et al.*, 1984). Total carbon (C), nitrogen (N), and sulfur (S) were determined with CHN-analyzer (Leco CHN 1000) and S-analyzer (Leco SC432), respectively. Amorphous Al- and Fe-Oxides (Alo/Feo) of the soils were determined by oxalate extraction, total Al- and Fe-oxides (Ald/Fed) by dithionite extraction (Blume and Schwertmann, 1969).

Different sulfate fractions were obtained for the Bhs, AB and Bw (separated into Bw1 and Bw2) horizons by continuous extraction with bidistilled water, 0.02 N K_2HPO_4, and 0.02 N NaOH (soil:solution ratio 1:5) (Kaiser, 1992). As shown by Alewell and Matzner (1995), K_2HPO_4 extraction was not suitable for organically influenced horizons. From A horizons, they extracted more SO_4^{2-} with NaCl-solutions compared to the K_2HPO_4-extract. Our results correspond to this, since for the E horizons we also found more SO_4^{2-} in the $MgCl_2$ than in the K_2HPO_4-extract (data not shown). Sulfate in the extracts was measured by ion chromatography (IC Dionex DX 500). H_2O-extractable SO_4^{2-} is regarded as representing soluble SO_4^{2-}-salts, originating partly during soil sample treatment (air-drying). K_2HPO_4-extractable SO_4^{2-} characterizes the sum of specifically and nonspecifically adsorbed SO_4^{2-}, especially in B horizons. To obtain information about nonspecifically adsorbed SO_4, we conducted a further extraction with 0.02 M $MgCl_2$. Cl^- is only in the position to exchange non-specifically adsorbed SO_4^{2-} (McBride, 1994). The difference between K_2HPO_4- and $MgCl_2$-extractable SO_4^{2-} is considered to represent the specifically adsorbed SO_4^{2-} fraction, and the difference of $MgCl_2$ extractable and H_2O soluble SO_4^{2-} to characterize the non-specifically bound SO_4^{2-} fraction. NaOH extraction gives information about the presence of $AlOHSO_4$ minerals. Because inorganic SO_4^{2-} retention plays a minor role in the E-horizons due to both the dominating organic S-storage and extensive DOC-adsorption on positively charged sites, for the determination of the SO_4^{2-} fractions we only took the Bs/AB and Bw horizons into consideration.

Soil water fluxes were calculated on a daily basis using the SOIL-Model (Janssen, 1991) and calibrated with measured tension data.

3. Results

3.1. *Water fluxes*

Table 1 shows measured data and model results of the water fluxes through the three ecosystems for the two-year period. The data reveal considerably lower precipitation amounts in the second year at the sites Rösa and Taura, whereas at Neuglobsow the values are similar. Interception losses in the canopy vary from 22% to 27% in the first year and increase to 30–35% in the drier period. The values for the actual transpiration show only small differences between the two years and vary between 200 and 270 mm/yr. These model results are well in accordance with studies in comparable ecosystems (Marschner, 1990) and to transpiration measurements of the stand and ground vegetation by Luettschwager *et al.*, (1997, this volume).

Table 1. Water fluxes at three sites

	Rösa		Taura		Neuglobsow	
	P1	P2	P1	P2	P1	P2
	mm		mm		mm	
BP	801	614	945	738	714	762
TF	622	427	690	513	522	491
% Interception	22	30	27	30	27	36
Act. Transp.	254	220	271	248	199	211
Evapotrans.	328	282	352	330	256	270
MSI	482	272	576	400	431	380
20 cm	370	190	462	292	364	307
50 cm	291	130	347	213	327	273
100 cm	256	126	288	206	305	269

P1: 11.93–10.94; TF: throughfall; P2: 11.94–10.95; MSI: mineral soil input; BP: bulk precipitation

Amounts and distribution of precipitation and throughfall as well as climatic conditions result in generally lower soil water fluxes in the second year. The site Rösa is affected most by a 50% reduction in leaching losses in the second period. The water fluxes in at a depth of 100 cm (= 'ecosystem output') reach 30–40% of the throughfall (TF) input at Rösa and Taura. At Neuglobsow the corresponding amounts are clearly higher (55–58% of TF), which can be attributed to ground vegetation differences as well as to the unfavourable soil physical properties (Weisdorfer *et al.*, 1995).

3.2. *General soil physical and chemical properties*

Selected soil chemical properties at the site Rösa (Table 2) reveal high pH values of the podsolic sandy soil, especially in the humus layers. The pH(H_2O) ranges from 5.2 to 5.7 in the Oa and Oe horizon, respectively. The ECEC is quite low due to the sandy texture. Because of the low clay content, organic exchange sites play the most important role. Most cation exchange sites, particulary those in the Oa, Oe, and E horizon are occupied by Ca^{2+}, resulting in high base saturation (BS $>70\%$). Although Al^{3+} dominates the exchangeable cations in the lower mineral soil, the BS generally exceeds 20%. The close C/N ratios, high pH, and BS especially in the Oe horizon, indicate favourable conditions for mineralization. Total S contents amount to about 2900 mg/kg in the Oa and 2078 mg/kg in the Oe horizon. In the mineral soil, the Bw horizon shows the highest S content.

Table 2. Chemical and physical properties of soil at Rösa

Horizon	Depth (cm)	pH H_20	pH $CaCl_2$	C/N	C t/ha	Total S mg/kg
Oa	10–4	5.2	4.4	22	37	2906.00
Oe	4–0	5.7	4.8	13	27	2078.00
E	0–7	5.1	4.2	27	17	66.15
Bhs	7-16	5.0	4.3	18	11	60.32
Bw	16–50	4.7	4.3	8	21	72.65
BC	50–60	4.7	4.3	3	2	59.28
C	60+	5.0	4.3	1	3	22.40

Horizon	ECEC mmolc./kg	BS %	Al^{3+} %	Clay %	Silt %	Sand %	K_w mm/d
Oa	250.56	67.75	24.56				2846
Oe	290.84	92.01	6.15				
E	41.90	73.81	19.39	3.4	8.6	88.0	898
Bhs	21.07	36.11	54.39	4.7	10.4	84.9	490
Bw	15.34	25.13	66.88	2.2	10.2	87.6	426
BC	8.19	23.64	62.83	0.9	6.1	92.7	660
C	5.25	23.96	62.18	0.8	5.1	94.1	840

ECEC: effective cation exchange capacity BS: base saturation K_w: saturated hydraulic conductivity

The molar Ca/Al ratios as well as the Mg/Al ratios determined in the 1:2 water extract (Table 3) show favourable soil chemical conditions throughout the whole soil profile. The Ca^{2+} concentrations are very high, especially in the Bw horizon. The Al concentrations are extremely low corresponding to the high pH.

Table 3. Ca/Al-, Mg/Al-, and Ca/H- ratios in the 1:2 water extract at Rösa

Horizon	Depth (cm)	H^+	Ca^{2+}	Mg^{2+} mmol/l	Al^{n+}	Ca/Al	Mg/Al mol/mol	Ca/H
E	0–7	0.013	0.271	0.030	0.106	2.66	0.29	25.03
Bhs	7–16	0.013	0.211	0.028	0.049	4.62	0.63	25.09
Bw	16–50	0.020	0.344	0.027	0.038	8.92	0.74	20.75
BC	50–60	0.023	0.247	0.025	0.021	11.16	1.06	12.10
C	60+	0.013	0.110	0.010	0.007	15.92	5.49	9.23

At Taura, the $pH(H_2O)$ of the humus layers and the mineral soil is distinctly lower in comparison to Rösa ranging from 4.1 in the Oa horizon and 3.8 in the Oe horizon to 4.6 in the C horizon (Table 4). With the exception of the Oa horizon, the low BS throughout the whole profile indicates progressed soil acidification. The BS ranges from 43.4 to 66.1% and from 5.6 to 24.2% in the humus layer and in the mineral soil, respectively. In contrast to the Oe horizon at Rösa, the C/N ratio of this layer at Taura points to more unfavorable mineralization and nitrification conditions also due to the low pH and BS. The percentage of Al^{3+} at the ECEC increases from 22% in the Oa horizon to 74–85% in the lower mineral soil. By comparing the total S content of Taura with that of Rösa, distinctly lower amounts (Oa: 2405 mg/kg; Oe: 993 mg/kg) are conspicuous. On the other hand, in the mineral soil we determined substantially higher amounts of total S, especially in the Bw horizon (Bw: 124 mg/kg).

At Taura, the Al^{n+} concentrations in the water extract are clearly higher compared to Rösa, especially in the lower mineral soil. But the Ca^{2+} concentrations are 3–4 times lower than at Rösa. Consequently, the Ca/Al and Mg/Al ratios (Table 5) are considerably lower compared to Rösa. In addition, low pH values lead to low Ca/H ratios in the upper mineral soil.

Soil pH and ECEC at Neuglobsow and Taura show similar levels. The $pH(H_2O)$ ranges from 3.8 in the humus layer to 4.9 in the mineral soil at Neuglobsow. BS is generally below 10% and the exchange sites are occupied mainly by Al^{3+} except in the E and the C horizon (Table 6). The AB horizon reveals a relatively high ECEC due to its higher content in organic matter as a consequence of former agricultural landuse at that site. The BS of the Oa layer is clearly lower compared to Taura and Rösa. The humus layer at Neuglobsow shows lower S contents (1665 mg/kg) than at Rösa and Taura.

Table 4. Chemical and physical properties of soil at Taura

Horizon	depth (cm)	pH H_20	pH $CaCl_2$	C/N	C t/ha	Total S mg/kg
Oa	7–2	4.1	3.3	21	48	2405.00
Oe	2–0	3.8	3.2	30	23	993.00
E	0–6	4.0	3.3	13	6	143.25
Bhs	6–13	4.2	3.6	10	6	49.35
Bw	13–47	4.4	4.2	5	13	124.00
BC	47–63	4.4	4.3	2	1	62.80
C	63+	4.6	4.3	1	2	44.65

Horizon	ECEC mmol/kg	BS %	Al^{3+} %	Clay %	Silt %	Sand %	K_w mm/d
Oa	260.44	66.11	22.50				633
Oe	167.50	43.40	38.82				
E	24.59	24.20	38.80	4.0	12.8	83.2	241
Bhs	24.93	11.27	50.23	3.4	11.6	85.0	390
Bw	14.80	5.61	84.88	3.6	14.8	81.6	504
BC	6.98	6.11	79.61	2.9	7.2	89.9	275
C	3.33	6.48	73.53	0.9	6.4	92.7	525

ECEC: effective cation exchange capacity BS: base saturation Kw: saturated hydraulic conductivity

Table 5. Ca/Al-. Mg/Al-. and Ca/H - ratios in the 1:2 water extract at Taura

Horizon	Depth (cm)	H^+	Ca^{2+}	Mg^{2+} mmol/l	Al^{n+}	Ca/Al	Mg/Al Mol/mol	Ca/H
E	0–6	0.130	0.060	0.020	0.140	0.46	0.13	0.45
Bhs	6–13	0.120	0.090	0.020	0.050	1.85	0.34	0.75
Bw	13–47	0.070	0.150	0.020	0.140	1.04	0.13	2.15
BC	47–63	0.050	0.070	0.010	0.060	1.08	0.16	1.50
C	63+	0.030	0.040	0.010	0.020	4.23	1.02	1.11

The water extracts show similar Al^{n+} concentrations in the mineral soil as at Rösa, but clearly lower Ca^{2+} concentrations (Table 7). Comparing the Al^{n+} concentrations of the water extract at Neuglobsow with those at Taura, distinctly lower concentrations in the lower mineral soil are obvious.

The Fe_o and Fe_d contents (Table 8a) of the Rösa soil decrease with increasing soil depth from 1159 and 1739 mg/kg in the E horizon to 629 and 1141 mg/kg in the Bw2 horizon, respectively. The Al_o maximum in the Bw horizon can be explained by both downward migration of Al due to initial podsolization processes in the upper mineral soil in the past and precipitation of amorphous Al-minerals in the Bw-horizon at higher pH.

Table 6. Chemical and physical properties of soil at Neuglobsow

Horizon	Depth (cm)	pH H_20	pH $CaCl_2$	C/N	C t/ha	Total S mg/kg
Oa	5–0	3.8	2.9	28	47	1665.00
E	0–3	3.9	3.2	18	6	194.50
AB	3–19	4.5	4.1	11	23	58.55
BW	19–48	4.6	4.3	2	7	81.35
BC	48–65	4.5	4.3	n.d.	1	42.35
C	65+	4.9	4.5	n.d.	1	34.15

Horizon	ECEC mmol(c)/kg	BS %	Al^{3+} %	Clay %	Silt %	Sand %	K_w mm/d
Oa	188.42	48.04	17.59				751
E	42.95	15.53	51.63	2.7	7.8	89.5	551
AB	30.26	5.56	81.72	4.2	5.8	90.0	789
BW	19.71	3.90	88.70	<0.1	12.1	87.9	1132
BC	14.71	7.04	83.16	2.1	5.2	92.7	716
C	13.19	47.22	44.30	1.9	5.0	93.1	1630

ECEC: effective cation exchange capacity
BS: basesaturation saturation
K_w: saturated hydraulic conductivity

Table 7. Ca/Al-, Mg/Al-, and Ca/H- ratios in the 1:2 water extract at Neuglobsow

Horizon	Depth (cm)	H^+	Ca^{2+}	Mg^{2+} mmol/l	Al^{n+}	Ca/Al	Mg/Al mol/mol	Ca/H
E	0–3	0.194	0.076	0.040	0.184	0.40	0.22	0.39
AB	3–19	0.042	0.049	0.027	0.063	0.82	0.40	1.16
Bw	19–48	0.043	0.055	0.019	0.043	1.72	0.68	1.36
BC	48–65	0.028	0.050	0.018	0.017	2.78	1.11	1.79
C	65+	0.023	0.096	0.023	0.007	28.60	5.50	4.13

At Taura, oxalate and dithionite soluble Fe and Al levels decrease with increasing soil depth (Table 8b) below the E horizon. Podsolic soils are characterized by the movement of Fe and Al from the eluvic (E) horizon. Amounts of oxalate extractable Fe from the Taura soil reflect this obviously by well pronounced minima in the E horizon and maxima in the Bhs horizon. As at Rösa, we found increasing extractable Al_o down to the Bw horizon, also showing substantial evidence of podsolization processes.

Initial podsolization processes can be recognized at Neuglobsow by both a substantially higher Al_o amount in the AB and the Bw horizon and the Fe_d maximum in the E horizon (Table 8c). Furthermore, the Al_o maximum together with the relatively high content in organic matter in the AB horizon points to a higher buffer capacity of the upper mineral soil than at Taura.

Table 8a–c. Oxalate and dithionite extractable Fe and Al at the three sites

	Fe_o	Fe_d mg/kg	Al_o
a. Rösa			
E	1158.50	1738.50	520.75
Bhs	1055.00	1490.36	1277.75
Bw1	847.50	1289.44	2204.50
Bw2	628.50	1141.71	1989.25
BC	248.00	435.53	889.50
C	19.50	77.19	279.50
b. Taura			
E	778.50	1580.61	460.75
Bhs	1573.50	2359.09	1003.50
Bw1	930.50	2006.69	2299.75
Bw2	536.00	1383.49	1455.25
BC	127.50	388.65	311.50
C	85.00	310.45	187.50
c. Neuglobsow			
E	543.50	3166.06	838.50
AB	619.00	1706.89	2146.25
Bw1	818.00	1150.69	2185.00
Bw2	307.50	696.46	711.00
BC	448.00	1358.71	476.75
C	242.50	472.23	232.50

3.3. Inorganic sulfate fractions

At Rösa, water soluble SO_4^{2-} amounts are highest in the Bw horizons (52 mg/kg) (Table 9a). The non-specifically adsorbed SO_4^{2-} contents are low and range from 21 (Bw horizons) to 31 mg/kg (Bsh horizon). The Bw horizons show significantly higher quantities of specifically bound SO_4^{2-} (60 mg/kg).

In comparison to the Rösa soil, the Bw horizons at Taura show similar amounts of water soluble and non-specifically bound SO_4^{2-}, but considerably more specifically sorbed SO_4^{2-} (Table 9b). 285 and 106 mg/kg SO_4^{2-} are specifically bound in the Bw1 and Bw2 horizon, respectively. The amounts of non-specifically bound SO_4^{2-} are very small and range from 17 to 27 mg/kg. As at Rösa, the Bhs horizon reveals the highest level of nonspecifically, electrostatically bound SO_4^{2-}.

Only at Neuglobsow, we extracted considerable amounts of SO_4^{2-} with $MgCl_2$ from the soil (Table 9c). The amounts of non-specifically adsorbed SO_4^{2-} range from 32 mg/kg in the AB to 78 mg/kg in the Bw1 horizon. We determined elevated levels of specifically bound SO_4^{2-} in the AB horizon (64 mg/kg) and the Bw1 horizon (171 mg/kg). On comparing the AB horizon with the Bhsv/Bhs horizons of Rösa and Taura, the amounts of specifically adsorbed SO_4^{2-} are found to be much higher. The NaOH extractable SO_4^{2-} amounts are low at all investigation sites.

Table 9a–c. Inorganic sulfate contents (mg/kg)

	H_2O-extract	$MgCl_2$-extract	K_2PO_4-extract	NaOH-extract
a. Rösa				
Bhs	32.4	31.2	7.2	16.3
Bw1	51.1	22.9	59.8	20.7
Bw2	52.3	21.1	59.2	20.3
b. Taura				
Bhs	32.6	27.1	21.7	17.8
Bw1	86.0	17.4	285.1	43.8
Bw2	77.6	21.9	106.4	22.2
c. Neuglobsow				
AB	29.5	32.3	62.8	21.2
Bw1	40.9	75.4	170.6	28.8
Bw2	20.1	39.7	12.0	6.3

3.4. Deposition and canopy interactions

Bulk precipitation and throughfall concentrations

The mean pH values in bulk precipitation (BP) and throughfall (TF) are similar for the three sites and range from 4.3 to 4.4 and from 4.1 to 4.2, respectively (Table 10). Generally, the TF concentrations are on a higher level compared to BP due to concentration effects by interception losses and due to leaching processes, including both removal of material that is physically and chemically bound to plant surfaces or interiors and washoff of dry deposition. Mean SO_4^{2-} concentrations are at the same level at Rösa and Taura and lower at Neuglobsow (Table 10) The elevated mean SO_4^{2-} concentrations in the BP and TF point to a presently higher S-load at Rösa and Taura. The mean Ca concentrations of the BP are low (1.3–1.8 mg/L) indicating that presently there is no substantial influence of alkaline dust deposition at the three sites. The

Table 10. Mean element concentrations in bulk precipitation (BP) and throughfall (TF) (1.11.1993–31.10.1995)

		pH	Ca^{2+}	Mg^{2+}	K^+	Na^+	Fe^{3+}	Mn^{2+} mg/L	Al^{n+}	NH_4^+	SO_4^{2-}	NO_3^-	Cl^-	DOC
R	BP	4.42	1.78	0.21	0.62	0.97	0.06	0.01	0.15	1.53	6.79	4.58	1.93	4.76
	TF	4.13	4.76	0.75	2.97	1.88	0.20	0.13	0.37	2.77	17.81	9.03	3.41	18.67
T	BP	4.34	1.36	0.17	0.64	0.96	0.07	0.01	0.11	1.40	6.52	4.33	1.76	4.47
	TF	4.11	3.31	0.51	2.24	1.67	0.15	0.07	0.28	2.87	15.66	8.61	3.08	14.55
N	BP	4.39	1.25	0.21	0.61	1.42	0.05	0.02	0.07	1.51	5.22	4.93	2.21	5.71
	TF	4.18	2.63	0.60	3.10	2.66	0.09	0.35	0.17	1.69	10.56	7.15	4.46	25.76

R = Rösa; T = Taura; N = Neuglobsow

mean Ca concentrations in TF decrease in the order Rösa > Taura > Neuglobsow. The NH_4^+ and NO_3^- concentrations in BP are similar at all sites, but in TF Rösa and Taura show clearly higher values (Table 10). The mean K, Mn, and DOC concentrations are considerably higher in the TF compared to the BP.

Time courses of SO_4^{2-} concentrations (Figure 1a–c) show pronounced seasonal patterns, especially in TF at Rösa and Taura with higher SO_4^{2-} concentrations during the winter months. The differences between the SO_4^{2-} concentration in BP and TF at Neuglobsow are distinctly lower compared to Rösa and Taura, particularly in winter.

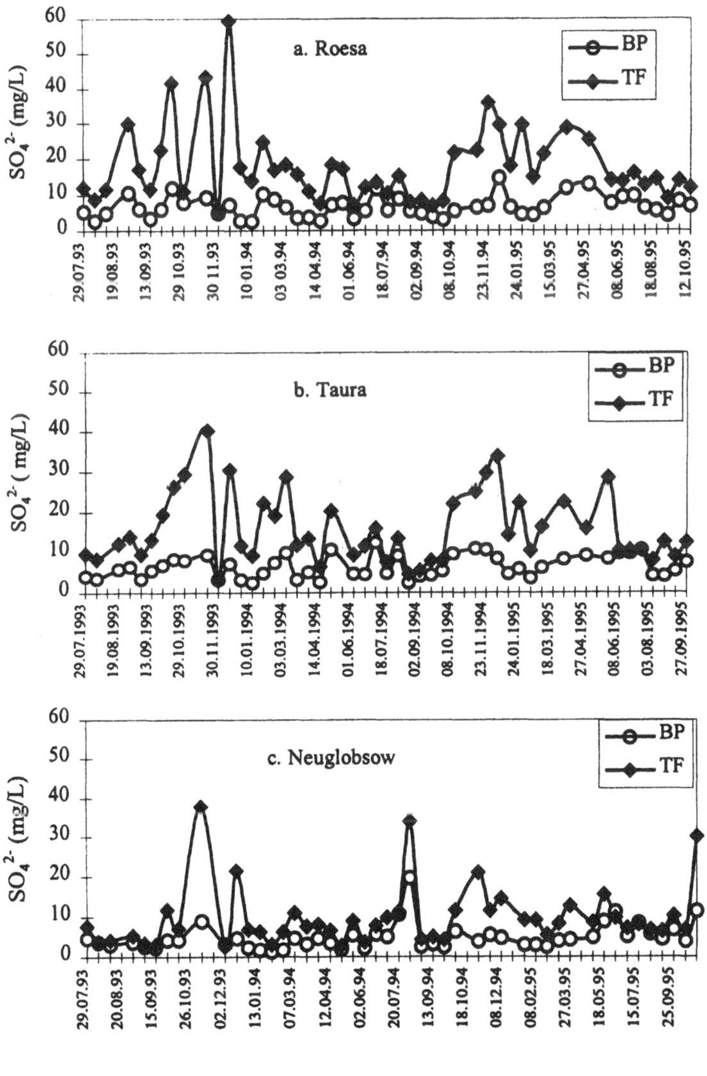

Figure 1. (a) Time courses of the SO_4^{2-} concentrations of the bulk precipitation (BP) and the throughfall (TF) at Rösa. (b–c) Time courses of the SO_4^{2-} concentrations of the bulk precipitation (BP) and the throughfall (TF) at Taura and Neuglobsow

In winter 1993/1994, the peak concentrations of SO_4^{2-} are paralleled by elevated H^+ concentrations in TF (Figure 2a–c). Highly gaseous S deposition leads to corresponding higher H^+ deposition rates due to the formation of sulfuric acid. The lack of H^+ concentration peaks in the second winter period mainly resulted from lower SO_2 emissions.

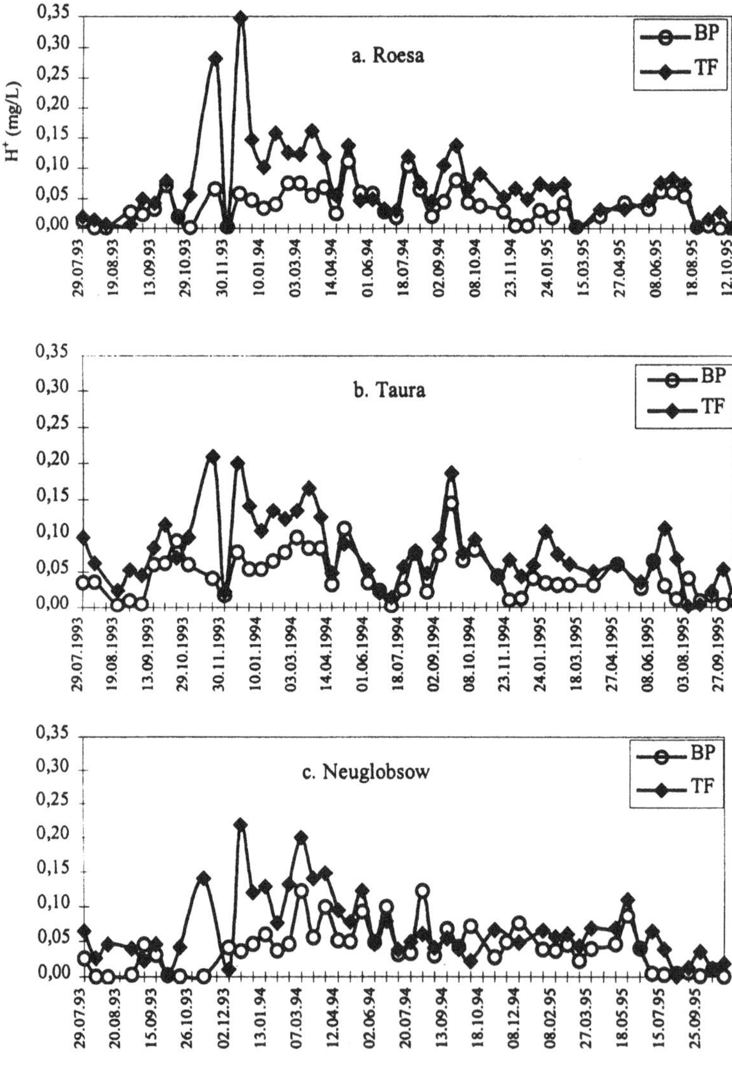

Figure 2. (a–c.) Time courses of the H^+ concentration of the bulk precipitation (BP) and the throughfall (TF) at the three sites

Presented time courses (Figure 2a–c) exceed the period that was taken into consideration for calculating mean concentrations and element budgets (1.11.1993–31.10.1995). The exclusive view of that period may give the impression of decreasing pH values in the BP and the TF over time. Taking the whole presented time course of H^+ concentrations into consideration (1.9.1993–31.10.95) a different result is visible and emphazises the importance of extended observation periods for characterizing trends concerning the deposition load.

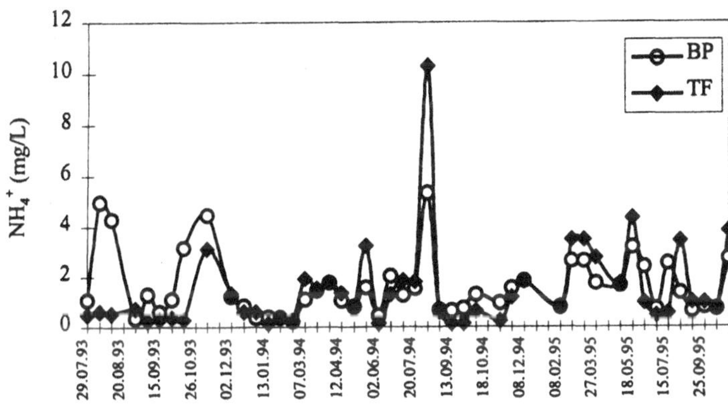

Figure 3. Time courses of the NH_4^+ concentrations of the bulk precipitation (BP) and the throughfall (TF) at the site Neuglobsow

The NH_4^+ concentrations in the BP and TF at Neuglobsow (Figure 3) are similar. Especially in autumn, BP concentrations exceed those of the TF, indicating N uptake by canopy surfaces (3.4.2). At Rösa and Taura, NH_4^+ concentrations are generally higher in the TF than in the BP.

Total deposition and canopy interactions

To quantify the rates of deposition and element turnover in the canopy from measured BP and TF we applied the approach by Ulrich (1981). Na^+ (or Cl^-) is considered to be a conservative element with respect to canopy interactions. The particulate interception deposition IDp is calculated from the ID of Na according to equation (1) and (2):

$$\text{Eq. (1)}: ID_{Na} = FD_{Na} - PD_{Na}$$
$$\text{Eq. (2)}: ID/PD_{Nap} = ID/PD_{xp}$$

- ID = interception deposition (IDg (gaseous deposition) + IDp (particulate deposition))
- PD = bulk precipitation deposition
- FD = throughfall deposition
- x = Ca^{2+}, Mg^{2} , K, Mn^{2}, Fe^{3+}, Al^{3+}, SO_4^{2-}, (NH_4^+, NO_3^-)

Gaseous deposition of SO_2 (IDg) can be calculated from the difference of total deposition (TD) and IDp. The model contains uncertainties with regard to the N deposition, and may underestimate actual N input.

The deposition rates (kg/ha) and canopy sink/source functions for a period of two years are presented in Table 11. Total proton deposition is distinctly higher at Rösa and Taura compared to Neuglobsow due to substantially higher gaseous inputs of SO_2. Proton buffering in the canopy is clearly higher at Rösa and Taura resulting in almost equal FD rates. The TD of Ca^{2+} decreases in the order Rösa > Taura > Neuglobsow. The same sequence applies also for the Ca^{2+} leaching rates. During the two years, SO_4-S input is highest at Rösa (52 kg/ha) and Taura (51 kg/ha), but distinctly lower at the 'background'-site (29 kg/ha). The large differences can partly be attributed to elevated ID_p- and ID_g-S depositions at Rösa and Taura, especially in winter 1993/1994. At all sites the mobile cation K^+ is leached in the highest amounts. At Neuglobsow, Mn^{2+} also contributes considerably to canopy leaching.

Table 11. Element deposition and canopy budgets in kg/ha (1.11.1993–31.10.1995)

	H^+	Ca^{2+}	Mg^{2+}	K^+	Fe^{3+}	Mn^{2+}	Al^{n+}	NH_4^+-N	SO_4^{2-}-S	NO_3^--N
Rösa										
PD	0.58	21.61	2.55	7.62	0.67	0.15	1.87	15.35	29.69	12.79
Idp	0.23	8.64	1.02	3.05	0.27	0.06	0.75	6.14	11.88	5.12
Idg	0.68								10.09	
TD	1.49	30.25	3.58	10.66	0.94	0.21	2.62	21.49	52.47	17.91
LE	-0.65	10.76	2.86	15.02	0.78	0.83	0.48	-0.85		-0.05
FD	0.84	41.01	6.44	25.68	1.73	1.04	3.10	20.64	52.47	17.86
Taura										
PD	0.85	21.62	2.69	9.76	0.96	0.14	1.59	16.80	33.21	14.28
Idp	0.17	4.32	0.54	1.95	0.19	0.03	0.32	3.36	6.64	2.86
Idg	0.72								11.54	
TD	1.74	25.95	3.23	11.71	1.15	0.17	1.91	20.16	51.39	17.14
LE	-0.77	6.81	1.81	10.57	0.59	0.59	0.78	1.24		1.46
FD	0.97	32.76	5.04	22.28	1.74	0.76	2.69	21.40	51.39	18.60
Neuglobsow										
PD	0.65	14.19	2.56	6.39	0.58	0.23	0.73	12.89	21.37	13.39
IDp	0.22	4.82	0.87	2.17	0.20	0.08	0.25	4.38	7.27	4.55
IDg										
TD	0.87	19.01	3.43	8.57	0.78	0.31	0.98	17.27	28.54	17.94
LE	-0.14	1.71	1.06	13.98	0.12	1.78	0.36	-8.37		-6.18
FD	0.73	20.72	4.49	22.55	0.90	2.09	1.34	8.90	28.29	11.76

PD: Precipitation deposition; TD: Total deposition
IDp: Particulate interception deposition
LE: leaching
IDg: Gaseous deposition
FD: Throughfall deposition

The total rate of N deposition calculated by this approach is only a conservative guess because a considerable amount of N may be directly assimilated by the foliage. Nevertheless, whereas the TD of N are similar at all sites, N uptake in the canopy can be assumed only for the 'background'-site, indicated also by lower FD compared to PD. Total inorganic N deposition is calculated to be 25–39 kg/ha for the two-year period.

3.5. *Soil solution composition*

The mean composition of the soil solutions collected from 4 soil depths are presented in Table 12. The three sites show distinct differences in the pH values. The mean pH of 4.7 in the mineral soil input (MSI) at Rösa is 0.4 units higher than at Taura and Neuglobsow. This is in accordance with the pH values presented in chapter 3.2. At Rösa, the pH of the soil solution decreases with soil depth to about 4.4 in 100 cm. The Ca^{2+} concentrations in the MSI are

Table 12. Mean element concentrations in the soil solutions (1.11.1993–31.10.1995)

	pH	Ca^{2+}	Mg^{2+}	K^+	Na^+	Fe^{3+}	Mn^{2+}	Al^{n+}	NH_4^+	SO_4^{2-}	NO_3^-	Cl^-	DOC
							mg/L						
Rösa													
MSI	4.72	32.17	2.37	1.94	3.24	0.26	0.09	1.01	0.38	52.47	28.69	3.80	35.42
20 cm	4.74	27.93	1.78	1.97	2.49	0.20	0.08	1.24	0.25	54.64	17.82	5.09	28.70
50 cm	4.43	35.57	1.54	1.78	2.99	0.09	0.09	1.61	0.25	86.16	8.58	5.41	17.71
100 cm	4.37	54.43	2.35	2.88	3.24	0.11	0.17	1.94	0.29	132.50	9.87	5.88	17.55
Taura													
MSI	4.27	10.07	1.32	1.67	2.52	0.29	0.10	3.24	0.44	33.49	4.69	2.13	38.82
20 cm	4.30	9.85	1.09	1.61	2.00	0.26	0.09	3.41	0.22	34.70	4.31	3.87	26.94
50 cm	4.05	16.41	1.31	1.34	3.01	0.17	0.08	9.76	0.32	95.10	2.70	4.69	16.10
100 cm	4.06	16.35	1.41	2.47	2.55	0.10	0.11	8.13	0.46	80.50	7.03	5.68	14.05
Neuglobsow													
MSI	4.28	3.88	0.94	1.25	3.45	0.31	0.44	2.15	0.19	9.02	0.78	4.25	48.49
20 cm	4.51	3.79	0.77	0.86	3.15	0.14	0.68	1.41	0.17	14.78	0.64	4.74	23.00
50 cm	4.33	4.57	0.86	1.57	3.33	0.16	0.85	1.42	0.15	25.37	0.47	4.80	7.57
100 cm	4.34	8.72	1.42	2.02	4.79	0.11	1.00	1.70	0.13	39.88	0.73	6.26	8.20

MSI: mineral soil input

highest at Rösa (32 mg Ca/L) decreasing clearly in the order Taura (10 mg Ca/L) > Neuglobsow (4 mg Ca/L). The same trend is found for Mg concentrations, but on a much lower level. MSI at Rösa shows elevated SO_4^{2-} and NO_3^- concentrations, corresponding to high S and N contents of the humus layer and favourable conditions for mineralization and nitrification. The SO_4^{2-} concentrations increase with soil depth not only due to concentration effects but obviously also due to SO_4^{2-} release from the Bw and the BC horizon. At Rösa, the NO_3^- concentrations in leachate decrease strongly with soil depth indicating plant and/or microbial uptake in the upper soil horizons. DOC concentrations also decrease with soil depth, but remain at considerably high levels (17–18 mg/L) in the lower mineral soil. Generally, average NH_4^+ concentrations are extremely low and do not exceed 0.4 mg/L.

At Taura, soil solution from the organic horizons is lower in pH, in mean Ca^{2+} and Mg^{2+} concentrations, but higher in Al^{n+} concentrations compared to Rösa. Al values increase from 3.2 mg/L in MSI to about 10 mg/L at 50 cm depth. The strong increase below 20 cm depth confirms the results of the 1:2 water extract (Table 5) revealing high water extractable Al^{n+} contents in the Bw horizon. Mean SO_4^{2-} concentration is also high in the MSI and increases strongly from 35 to 95 mg/L below the Bw horizon (50 cm depth). Mean NO_3^- concentration in seepage water from the humus layer is very low due to unfavourable conditions for nitrification. Whereas Ca^{2+} is the main counter ion of SO_4^{2-} at Rösa, Al^{n+} is of increasing importance at Taura. The mean soil solution composition at Taura is in accordance with the moderate alkaline dust but high S depositions in the past.

At Neuglobsow, the mean pH value is highest in 20 cm soil depth due to the high buffer capacity of the overlaying AB horizon. Compared to the other sites, the mean Ca^{2+} concentrations throughout the whole profile are very low and range from about 4 mg/L in the MSI to about 9 mg/L in 100 cm soil depth. Mean SO_4^{2-} concentration in the MSI is on a distinctly lower level than at Rösa and Taura. The NH_4^+ and NO_3^- concentrations are also extremely low throughout the entire profile. In conformity with both a high percentage of Mn at the ECEC and a high Mn leaching from the canopy, the Mn concentrations in seepage water are clearly higher than at Rösa and Taura. Despite a low pH and BS of the organic layer, the DOC concentration in the MSI is highest at Neuglobsow, pointing to extensive mineralization. In contrast to Rösa and Taura, the DOC concentrations decrease more strongly below 20 cm soil depth.

Ca/Al-, Mg/Al- and Ca/H-ratios in the soil solution (Table 13) are similar to those of the 1:2 water extracts (Table 3, 5, and 7). Taking into account the activity coefficients, especially at Rösa and Taura, the clearly lower coefficients lead to even higher Ca/Al- and Mg/Al-ratios.

At Rösa and Taura, we find a highly significant correlation between Ca^{2+} and NO_3^- concentration in MSI (Table 14). In addition we find significant correlation between Al^{n+} and NO_3^- at Taura. Whereas at Rösa Ca^{2+} is significantly negative correlated with DOC we find the same correlation between Al^{3+} and DOC at Taura. Whereas at Rösa, there is no correlation between H^+ and NO_3^-, we find a weakly positive correlation ($r = 0.32$; $p = 0.023$) at Taura. Correlation analysis only for the two winter periods (10.93–3.94 and 10.94–3.95: data not shown) reveals a more significant correlation ($r = 0.49$; $p = 0.018$) between H^+ and NO_3^- at Taura probably due to the lack of NO_3^- uptake/immobilization. Only at Neuglobsow, we find strongly positive correlations between DOC and almost all main cations.

In soil solutions from 50 cm soil depth (Table 15) at Rösa there is a highly

Table 13. Molar Ca/Al-, Mg/Al-, and Ca/H- concentration ratios in the soil solutions and corresponding activity ratios

	Activity coef. (Al)	Ca/Al-mmol	Ca/Al (activity)	Mg/Al-mmol	Mg/Al (activity	Ca/H (activity)
Rösa						
MSI	0.59	21.44	28.38	2.61	3.48	33.02
20 cm	0.60	15.12	19.87	1.59	2.11	30.48
50 cm	0.56	14.87	20.18	1.06	1.45	18.30
100 cm	0.51	18.88	26.97	1.34	1.95	22.88
Taura						
MSI	0.66	2.09	2.62	0.45	0.57	3.83
20 cm	0.65	1.95	2.44	0.36	0.45	4.04
50 cm	0.55	1.13	1.56	0.15	0.21	3.46
100 cm	0.56	1.35	1.84	0.19	0.26	3.57
Neuglobsow						
MSI	0.73	1.22	1.44	0.49	0.58	1.62
20 cm	0.74	1.81	2.13	0.61	0.72	2.65
50 cm	0.71	2.17	2.61	0.68	0.82	2.07
100 cm	0.66	3.46	4.32	0.93	1.16	3.90

MSI: mineral soil input

positive correlation between SO_4^{2-} and Ca^{2+} and Mg^{2+} (Ca^{2+}–SO_4^{2-}: r = 0.86, r = 0.92). At Neuglobsow there is only a significant correlation between SO_4^{2-} and Mn^{2+}.

Temporal patterns of SO_4^{2-} concentration in the MSI and 50 cm depth reveal higher SO_4^{2-} concentrations at Rösa and Taura compared to Neuglobsow during the whole oberservation period (Figures 4 and 5).

The elevated values at the beginning of the first investigation period are partly caused by the high deposition impact (especially gaseous and particle deposition) during winter 1993/94 resulting in peak concentrations of 100–130 mg SO_4^{2-}/L in MSI and 200–350 mg SO_4^{2-}/L at 50 cm depth. MSI concentrations of NO_3^- (Figure 6) show elevated values during the autum and winter months for Rösa and Taura, but on a clearly lower level for the latter site. The high NO_3^- concentration peaks at Rösa are also reflected in increased concentrations at 50 cm soil depth (Figure 7).

3.6. *Element fluxes*

Annual element fluxes were calculated for two periods (01.11.1993–31.10.1994 = P1; 01.11.1994–31.10 1995 = P2) based on measured and simulated water fluxes, and element concentrations of the BP, TF, and the soil solutions (Tables 16–18)

Table 14. Pearson correlation coefficients for mean solution concentrations in the mineral soil input (bold: $p<0.005$)

Rösa

H^+	Ca^{2+}	Mg^{2+}	K^+	Fe^{3+}	Mn^{2+}	Al^{n+}	NH_4^+	SO_4^{2-}	NO_3^-	DOC	
1.000	–0.032	–0.235	–0.121	–0.051	0.029	0.063	0.026	0.089	–0.094	–0.175	H
	1.000	**0.627**	0.197	–0.238	0.222	–0.031	–0.180	**0.713**	**0.664**	**–0.485**	Ca
		1.000	0.135	–0.161	0.070	–0.333	–0.001	**0.425**	0.387	–0.307	Mg
			1.000	–0.062	0.000	0.027	**0.447**	0.205	0.061	0.037	K
				1.000	**0.717**	–0.220	–0.075	–0.242	–0.193	0.175	Fe
					1.000	–0.190	–0.142	0.084	0.269	–0.195	Mn
						1.000	–0.114	–0.075	–0.037	–0.030	Al
							1.000	–0.171	–0.076	–0.010	NH_4
								1.000	**0.496**	–0.256	SO_4
									1.000	**–0.447**	NO_3
										1.000	DOC

Taura

H^+	Ca^{2+}	Mg^{2+}	K^+	Fe^{3+}	Mn^{2+}	Al^{n+}	NH_4^+	SO_4^{2-}	NO_3^-	DOC	
1.000	0.285	–0.054	0.197	0.012	0.068	**0.408**	–0.014	0.385	0.315	–0.395	H
	1.000	**0.473**	0.299	–0.040	0.140	**0.514**	0.084	**0.621**	**0.743**	–0.157	Ca
		1.000	0.330	0.017	0.157	0.105	0.167	0.222	0.279	0.210	Mg
			1.000	–0.017	0.331	0.262	0.055	0.212	**0.574**	0.005	K
				1.000	–0.217	–0.130	–0.009	–0.270	–0.128	0.003	Fe
					1.000	0.062	0.117	0.071	0.151	0.298	Mn
						1.000	–0.071	**0.858**	0.356	**–0.476**	Al
							1.000	0.038	–0.060	0.219	NH_4
								1.000	0.412	–0.369	SO_4
									1.000	–0.244	NO_3
										1.000	DOC

Neuglobsow

H^+	Ca^{2+}	Mg^{2+}	K^+	Fe^{3+}	Mn^{2+}	Al^{n+}	NH_4^+	SO_4^{2-}	NO_3^-	DOC	
1.000	0.224	0.049	0.036	0.230	–0.003	0.164	**0.403**	–0.046	0.124	0.113	H
	1.000	0.806	0.038	**0.659**	**0.530**	**0.614**	**0.453**	**0.439**	–0.021	**0.825**	Ca
		1.000	0.158	**0.578**	**0.559**	**0.628**	**0.425**	**0.606**	–0.033	**0.754**	Mg
			1.000	–0.066	0.113	0.268	0.277	0.124	0.102	–0.025	K
				1.000	0.345	**0.524**	0.383	0.159	–0.150	**0.704**	Fe
					1.000	**0.624**	0.193	**0.487**	–0.188	**0.416**	Mn
						1.000	**0.455**	**0.438**	0.111	**0.549**	Al
							1.000	0.165	0.197	**0.475**	NH_4
								1.000	–0.059	0.210	SO_4
									1.000	–0.174	NO_3
										1.000	DOC

At Rösa, proton fluxes decrease from 0.64 (TF) to 0.07 kg ha $^{-1}$ yr^{-1} (MSI) in P1 and from 0.19 to 0.05 kg ha $^{-1}$ yr^{-1} in P2, respectively, after percolation through the humus layer (Table 16). Ca^{2+}, Mg^{2+}, SO_4^{2-}-S, and DOC fluxes in particular increase strongly from TF to MSI. In P1, 147 kg Ca^{2+} ha^{-1} yr^{-1} and 68 kg SO_4^{2-}-S ha^{-1} yr^{-1} are released from the organic surface layer. Whereas the DOC deposition via TF is similar for the two years (86 and 89 kg DOC ha^{-1} yr^{-1}, respectively), the calculations show lower mobilization in P2 compared to

Table 15. Pearson correlation coefficients for mean solution concentrations at a soil depth of 50 cm (bold: $p < 0.005$)

H^+	Ca^{2+}	Mg^{2+}	K^+	Fe^{3+}	Mn^{2+}	Al^{n+}	NH_4^+	SO_4^{2-}	NO_3^-	DOC	
					Rösa						
1.000	–0.096	–0.186	–0.317	–0.127	–0.092	0.067	–0.239	–0.072	0.162	–0.247	H^+
	1.000	**0.899**	0.036	–0.165	**0.662**	0.042	–0.193	**0.863**	–0.203	–0.054	Ca^{2+}
		1.000	0.159	–0.160	**0.657**	0.035	–0.115	**0.826**	–0.196	0.071	Mg^{2+}
			1.000	0.111	0.040	–0.193	0.212	0.105	–0.068	**0.662**	K^+
				1.000	0.127	–0.111	0.124	–0.216	0.144	0.156	Fe^{3+}
					1.000	0.098	–0.220	**0.618**	–0.028	–0.091	Mn^{2+}
						1.000	–0.307	0.124	0.103	–0.380	Al^{n+}
							1.000	–0.167	–0.261	0.389	NH_4^+
								1.000	–0.298	0.100	SO_4^{2-}
									1.000	–0.058	NO_3^-
										1.000	DOC
					Taura						
1.000	–0.053	–0.054	0.045	0.175	–0.005	–0.021	–0.143	–0.060	–0.161	–0.182	H^+
	1.000	**0.872**	**0.426**	0.346	0.355	**0.816**	0.307	**0.927**	–0.097	**0.778**	Ca^{2+}
		1.000	**0.391**	0.232	0.382	**0.687**	0.092	**0.800**	–0.099	**0.662**	Mg^{2+}
			1.000	0.231	0.223	0.371	**0.429**	**0.405**	0.109	**0.448**	K^+
				1.000	0.236	**0.424**	–0.001	0.328	–0.071	0.243	Fe^{3+}
					1.000	0.255	0.161	0.346	–0.117	0.351	Mn^{2+}
						1.000	0.229	**0.917**	0.062	**0.702**	Al^{n+}
							1.000	0.301	–0.030	**0.581**	NH_4^+
								1.000	–0.044	**0.805**	SO_4^{2-}
									1.000	–0.107	NO_3^-
										1.000	DOC
					Neuglobsow						
1.000	–0.138	–0.325	–0.105	–0.157	–0.300	–0.083	–0.088	–0.379	0.054	0.014	H^+
	1.000	**0.610**	0.305	–0.008	0.083	–0.222	0.153	0.144	–0.099	0.352	Ca^{2+}
		1.000	0.301	0.214	0.334	–0.147	0.129	0.192	–0.230	0.208	Mg^{2+}
			1.000	0.174	–0.141	–0.246	**0.426**	–0.155	0.277	**0.407**	K^+
				1.000	0.033	–0.070	0.249	–0.081	0.110	0.084	Fe^{3+}
					1.000	0.282	–0.307	**0.640**	–0.352	**–0.445**	Mn^{2+}
						1.000	–0.278	0.158	–0.281	**–0.482**	Al^{n+}
							1.000	–0.161	0.324	**0.411**	NH_4^+
								1.000	–0.083	–0.248	SO_4^{2-}
									1.000	0.298	NO_3^-
										1.000	DOC

P1 (102 and 174 kg DOC ha^{-1} yr^{-1}). This is probably due to a combined effect of considerably higher water fluxes as well as more intensive mineralization in P1. In both periods, high amounts of NO_3^--N are leached from the humus layer due to high nitrification rates. Deposited NH_4^+-N and mineralized N seems to be almost completely nitrified since only 0.80–1.8 kg ha^{-1} yr^{-1} are leached from the humus layer. In P1 high Ca^{2+} and SO_4^{2-}-S fluxes in MSI and further increasing fluxes, in particular below the Bw horizon (50 cm soil depth) lead

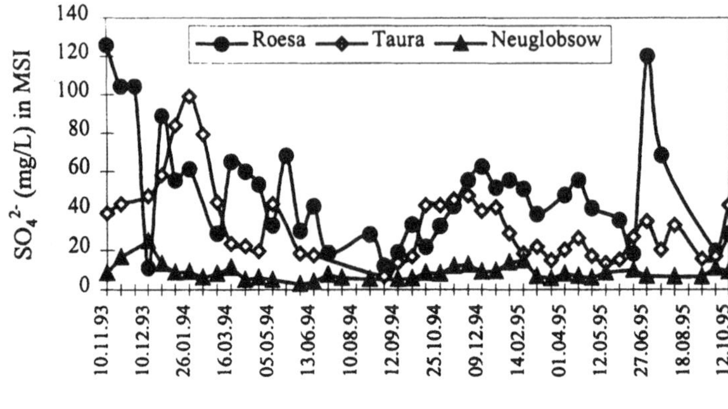

Figure 4. Time courses of the SO_4^{2-} concentrations in the mineral soil input (MSI) at the three sites

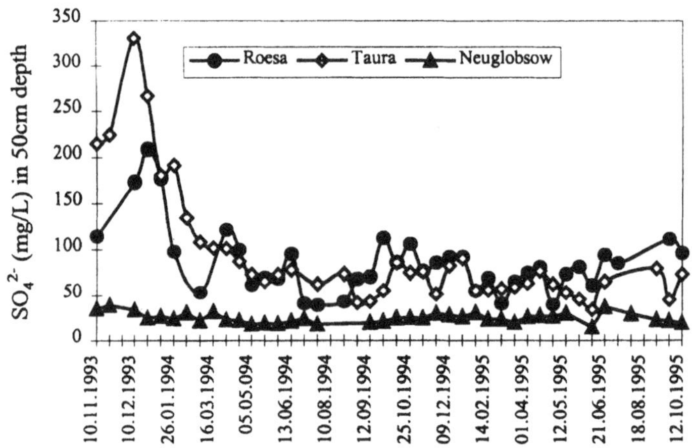

Figure 5. Time courses of the SO_4^{2-} concentrations at 50 cm soil depth at the three sites

to enormous outputs from the mineral soil in 100 cm depth (168 kg Ca^{2+} ha^{-1} yr^{-1} and 112 SO_4^{2-}-S kg ha^{-1} yr^{-1}). Whereas the soil budgets for Ca^{2+}, Mg^{2+}, Al^{n+}, and SO_4^{2-}-S are negative, the soil acts as a sink for N and H^+. In general, element fluxes from the humus layer are clearly lower in the second period. But Ca^{2+} and SO_4^{2-}-S are removed from the humus layer in still substantial amounts (Table 16). These results are in accordance with the lower DOC mobilization from the humus layer indicating lower cation and S releases by mineralization. Correspondingly, percolation losses at a depth of 100 cm are lower and amount to 63 kg Ca^{2+} ha^{-1} yr^{-1} and 54 kg SO_4^{2-}-S ha^{-1} yr^{-1}. The total

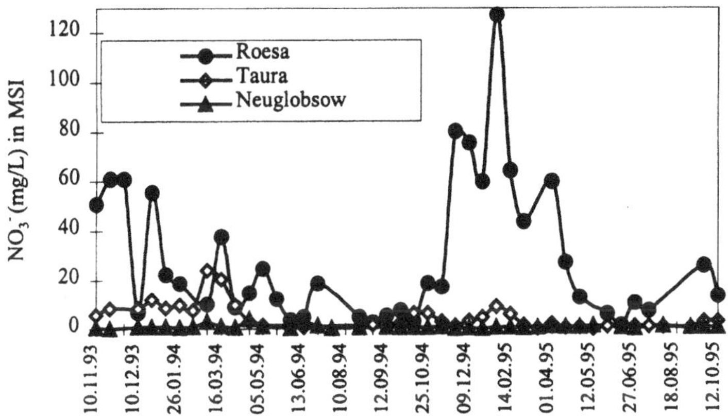

Figure 6. Time courses of the NO_3^- concentrations in the mineral soil input (MSI) at the three sites

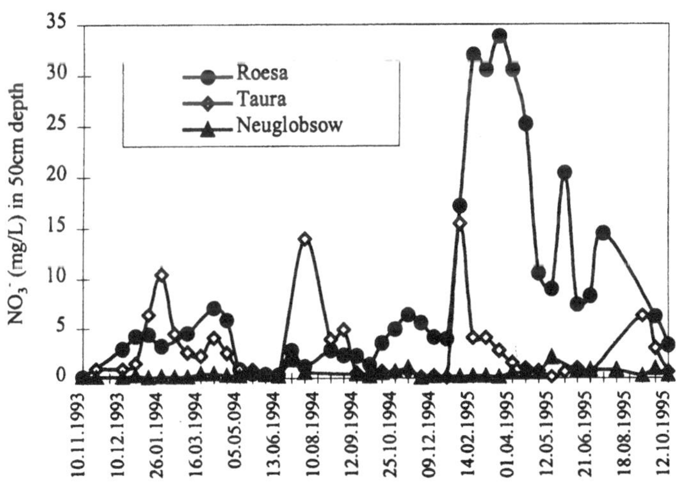

Figure 7. Time courses of the NO_3^- concentrations at 50 cm soil depth at the three sites

Al^{n+} fluxes in the mineral soil are relatively low. In P2, not only lower mineralization, but also lower seepage water fluxes (Table 1) are responsible for the lower element output rates a soil depth, of 100 cm.

At Taura, all element fluxes except for N increase strongly along the passage of the humus layer, especially for Ca^{2+}, Al^{n+}, SO_4^{2-}-S , and DOC (Table 17). Also higher H^+ amounts are leached into the mineral soil compared to the situation at Rösa. Chemical conditions of the humus layers obviously reduce nitrification rates and thus prevent elevated NO_3^--N fluxes from the organic

Table 16. Element fluxes at Rösa in P1 (1.11.1993–31.10.1994) and P2 (1.11.1994–31.10.1995)

	H^+	Ca^{2+}	Mg^{2+}	K^+	Na^+	Fe^{3+}	Mn^{2+}	Al^{n+}	NH_4^+ -N	SO_4^{2-} -S	NO_3^- -N	Cl^-	DOC
						kg/ha*yr							
P1													
BP	0.42	13.49	1.61	3.12	4.21	0.47	0.06	1.52	7.52	13.88	5.86	11.72	29.88
TF	0.64	24.57	3.65	11.61	6.57	1.16	0.51	2.12	11.22	27.07	8.85	14.64	85.85
MSI	0.07	147.02	14.57	10.31	13.01	0.90	0.30	4.87	1.76	76.59	17.44	16.24	174.4
20cm	0.11	109.82	5.87	7.37	7.30	0.63	0.28	5.37	0.61	74.46	5.02	14.27	86.06
50cm	0.14	133.60	5.31	3.72	6.91	0.23	0.34	7.24	0.39	108.81	2.64	10.83	42.08
100cm	0.14	168.09	7.18	6.07	6.95	0.20	0.27	5.51	0.52	111.91	4.41	13.90	45.21
P2													
BP	0.16	8.12	0.94	4.49	9.00	0.20	0.10	0.35	7.83	15.81	6.93	12.43	29.87
TF	0.19	16.44	2.79	14.07	11.95	0.57	0.53	0.98	9.42	25.40	9.01	15.45	88.86
MSI	0.05	76.59	5.26	4.65	8.49	1.21	0.34	2.52	0.80	37.42	24.10	12.95	102.3
20cm	0.03	56.82	4.02	3.14	6.77	0.42	0.10	2.05	0.33	31.02	14.45	13.39	55.18
50cm	0.05	38.11	1.60	1.20	3.65	0.18	0.15	2.13	0.14	29.25	5.24	8.30	21.38
100cm	0.05	62.56	2.54	2.66	4.22	0.18	0.29	2.71	0.15	54.16	3.62	7.35	20.41

BP: bulk precipitation
TF: throughfall
MSI: mineral soil input

surface layer. In P1 fluxes of Ca^{2+}, Al^n, and SO_4^{2-}-S strongly increase again below the Bw horizon. Ca^{2+} fluxes increase from 58 kg $ha^{-1} yr^{-1}$ in the MSI to 79 kg ha^{-1} yr^{-1}, Al^{n+} fluxes from 17 kg ha^{-1} yr^{-1} to 48 kg ha^{-1} yr^{-1}, and SO_4^{2-}-S from 56 kg ha^{-1} yr^{-1} to 144 kg ha^{-1} yr^{-1}. Both high mineralization rates in the organic surface layers and the source function of the Bw horizon contribute to elevated element outputs at a soil depth of 100 cm. As at Rösa, high DOC amounts are leached from the mineral soil, but NO_3^--N output is distinctly lower at Taura. Also at this site we calculate clearly lower element fluxes in P2 (Table 17).

Except for Al^{n+} and DOC element fluxes from the humus layer at Neuglobsow are significantly lower in both periods compared to the other two sites (Table 18). In P1, only 13 kg Ca^{2+} ha^{-1} yr^{-1} (P2: 16 kg ha^{-1} yr^{-1}), 9.0 kg Al^{n+} ha^{-1} yr^{-1} (P2: 7.4 kg ha^{-1} yr^{-1}), and 13 kg SO_4^{2-}-S ha^{-1} yr^{-1} (P2: 13 kg SO_4^{2-}-S ha^{-1} yr^{-1}) are leached from the Oa horizon. The high mobilization rates of DOC point to extensive mineralization. Both NO_3^--N and NH_4^+-N fluxes strongly decrease after passage of the humus layer and consequently very low amounts of inorganic N are leached from the mineral soil at a depth of 100 cm (0.4–0.5 kg N ha^{-1} yr^{-1}). Annual outputs amount to 24–26 kg Ca^{2+} ha^{-1} yr^{-1}, and to 36–37 kg SO_4^{2-}-S ha^{-1} yr^{-1}. Clearly, higher amounts of Mn^{2+} are leached from the mineral soil were compared to Rösa and Taura.

Table 17. Element fluxes at Taura in P1 (1.11.1993–31.10.1994) and P2 (1.11.1994–31.10.1995)

	H^+	Ca^{2+}	Mg^{2+}	K^+	Na^+	Fe^{3+}	Mn^{2+}	Al^{n+}	NH_4^+ -N	SO_4^{2-} -S	NO_3^- -N	Cl^-	DOC
						kg/ha*yr							
P1													
BP	0.57	13.65	1.67	2.61	4.98	0.59	0.06	1.07	8.93	14.57	7.10	12.98	31.56
TF	0.69	21.04	3.13	9.08	6.66	1.27	0.46	1.78	11.62	26.10	9.38	15.95	80.23
MSI	0.29	58.42	8.90	13.71	12.81	1.95	0.47	16.90	2.50	55.67	7.89	12.36	202.4
20cm	0.27	52.13	5.40	10.48	6.77	1.22	0.35	18.55	0.74	58.81	4.52	12.86	111.8
50cm	0.39	79.24	5.87	4.94	8.81	1.22	0.20	47.89	0.51	143.79	2.61	11.87	57.03
100cm	0.32	65.22	4.84	5.88	5.92	0.23	0.17	31.47	0.39	92.17	2.80	12.29	38.89
P2													
BP	0.28	7.98	1.02	7.16	10.00	0.38	0.08	0.52	7.87	18.64	7.18	12.55	35.68
TF	0.28	11.72	1.91	13.20	11.25	0.47	0.31	0.91	9.78	25.29	9.22	13.47	77.83
MSI	0.20	31.41	3.88	3.46	11.33	1.21	0.31	10.53	1.38	34.26	2.44	8.16	163.8
20cm	0.14	24.75	3.28	3.12	8.53	0.79	0.27	9.14	0.38	29.76	3.87	13.44	84.20
50cm	0.17	18.02	1.62	2.80	4.89	0.16	0.10	14.12	0.25	44.23	1.85	7.67	30.11
100cm	0.16	19.16	1.74	3.19	4.44	0.16	0.12	13.21	0.19	42.01	2.56	7.66	21.36

BP: bulk precipitation
TF: throughfall
MSI: mineral soil input

Table 18. Element fluxes at Neuglobsow in P1 (1.11.1993–31.10.1994) and P2 (1.11.1994–31.10.1995)

	H^+	Ca^{2+}	Mg^{2+}	K^+	Na^+	Fe^{3+}	Mn^{2+}	Al^{n+}	NH_4^+ -N	SO_4^{2-} -S	NO_3^- -N	Cl^-	DOC
						kg/ha*yr							
P1													
BP	0.57	13.65	1.67	2.61	4.98	0.59	0.06	1.07	8.93	14.57	7.10	12.98	31.56
TF	0.69	21.04	3.13	9.08	6.66	1.27	0.46	1.78	11.62	26.10	9.38	15.95	80.23
MSI	0.29	58.42	8.90	13.71	12.81	1.95	0.47	16.90	2.50	55.67	7.89	12.36	202.42
20cm	0.27	52.13	5.40	10.48	6.77	1.22	0.35	18.55	0.74	58.81	4.52	12.86	111.81
50cm	0.39	79.24	5.87	4.94	8.81	1.22	0.20	47.89	0.51	143.79	2.61	11.87	57.03
100cm	0.32	65.22	4.84	5.88	5.92	0.23	0.17	31.47	0.39	92.17	2.80	12.29	38.89
P2													
BP	0.28	7.98	1.02	7.16	10.00	0.38	0.08	0.52	7.87	18.64	7.18	12.55	35.68
TF	0.28	11.72	1.91	13.20	11.25	0.47	0.31	0.91	9.78	25.29	9.22	13.47	77.83
MSI	0.20	31.41	3.88	3.46	11.33	1.21	0.31	10.53	1.38	34.26	2.44	8.16	163.71
20cm	0.14	24.75	3.28	3.12	8.53	0.79	0.27	9.14	0.38	29.76	3.87	13.44	84.20
50cm	0.17	18.02	1.62	2.80	4.89	0.16	0.10	14.12	0.25	44.23	1.85	7.67	30.11
100cm	0.16	19.16	1.74	3.19	4.44	0.16	0.12	13.21	0.19	42.01	2.56	7.66	21.36

BP: bulk precipitation
TF: throughfall
MSI: mineral soil input

4. Discussion

The results of our study reveal distinct differences between the three sites with respect to the general soil chemical properties, the soil solution composition, and element fluxes. Likewise, the differences between the two investigation periods show a substantial influence of the seasonal pattern of intensity, distribution, and amount of rain as well as of deposition loads on seepage water composition.

4.1. Element budgets and proton loads of the humus layers

The distinct differences in acidity, ionic strength, and ionic composition of the soil solution of the MSI are of great importance for adsorption/desorption and dissolution/precipitation processes in the mineral soil. To obtain information about important processes in the humus layers, we calculated the element budgets as well as the internal proton consumption/production reactions for the two observation periods.

At Rösa, the element budgets show considerably negative values for Ca^{2+}, Mg^{2+}, SO_4^{2-}-S, and NO_3^--N in P1, whereas the budgets are positive for NH_4^+-N and H^+ due to high nitrification rates and extensive buffering (Table 19). In P2, distinctly lower amounts of Ca^{2+} and SO_4^{2-}-S but elevated NO_3^--N amounts are mobilized and leached from the humus layer. Higher nitrification rates in P2 are probably due to moderate temperatures from mid October to December 1994 when the concentrations of NO_3^- strongly increase (mean daily temperature of the period 10.93–12.93: 1.7°C; 10.94–12.94: 5.7°C). The negative NO_3^--N budgets of the humus layer are in accordance with the results of Bergmann *et al.* (1997, this volume), who find high nitrification rates only at Rösa.

Table 19. Element budgets of the humus layer at Rösa in P1 and P2 (values in kmol(c) ha^{-1} yr^{-1})

	H^+	Ca^{2+}	Mg^{2+}	K^+	Na^+	Fe^{3+}	Mn^{2+}	Al^{n+}	NH_4^+-N	SO_4^{2-}-S	NO_3^--N	Cl^-
P1												
TF	0.64	1.23	0.30	0.30	0.29	0.04	0.02	0.24	0.80	1.69	0.63	0.41
MSI	0.07	7.34	1.20	0.26	0.57	0.03	0.01	0.54	0.13	4.79	1.25	0.46
Budget	0.57	-6.11	–0.90	0.03	–0.28	0.01	0.01	–0.31	0.68	-3.10	–0.61	–0.05
P2												
TF	0.19	0.82	0.23	0.36	0.52	0.02	0.02	0.11	0.67	1.59	0.64	0.35
MSI	0.05	3.82	0.43	0.12	0.37	0.04	0.01	0.28	0.06	2.34	1.72	0.44
Budget	0.15	-3.00	–0.20	0.24	0.15	–0.02	0.01	–0.17	0.62	–0.75	-1.08	–0.09

TF: throughfall / MSI: mineral soil input

These results point to completely different conditions in P1 compared to P2. P1 was characterized by an accumulation phase during dry autumn 1993, an extremely cold November with high S inputs, and a humid December and January with moderate temperatures. The second investigation period shows clearly more humid conditions in autumn 1994 and a moderate temperature in winter with lower S and H^+ deposition. Although the high SO_4^{2-}-S fluxes from the humus layer in P1 might be partly due to the previous accumulation phase, our results point to the importance of the mineralization of organically bound S in the humus layer. Despite clearly lower SO_4^{2-}-S fluxes from the humus layer in P2, the SO_4^{2-}-S budgets were distinctly negative also in this period due to mineralization and showed a substantial influence of the high S-storage caused by the S inputs in the past. For Rösa, our conclusion is in accordance with David *et al.* (1991), who underline the importance of the humus layer in controlling leaching chemistry, but contradict the conclusions of Alewell and Matzner (1993), who found no obvious influence of organically bound S on soil solution chemistry.

The element budgets at Taura (Table 20) show similar results with regard to the differences between P1 and P2, but point to quite different (bio-)chemical processes in the humus layer. We calculated negativ budgets not only for Ca^{2+} and SO_4^{2-}-S but also distinctly higher negative budgets for Al compared to Rösa. As at Rösa, we found lower fluxes of Ca^{2+} and SO_4^{2-}-S compared to P1 mainly due to the seasonal patterns described above. In contrast to Rösa, the budgets for NO_3^--N are positive due to lower nitrification.

Table 20. Element budgets of the humus layer at Taura in P1 and P2 (values in kmol(c) ha^{-1} yr^{-1})

	H^+	Ca^{2+}	Mg^{2+}	K^+	Na^+	Fe^{3+}	Mn^{2+}	Al^{n+}	NH_4^+-N	SO_4^{2-}-S	NO_3^--N	Cl^-
P1												
TF	0.69	1.05	0.26	0.23	0.29	0.05	0.02	0.20	0.83	1.63	0.67	0.45
MSI	0.29	2.92	0.73	0.35	0.56	0.07	0.02	1.88	0.18	3.48	0.56	0.35
Budget	0.40	-1.87	-0.48	-0.12	-0.27	-0.02	0.00	-1.68	0.65	-1.85	0.11	0.10
P2												
TF	0.28	0.58	0.16	0.34	0.49	0.02	0.01	0.10	0.70	1.58	0.66	0.38
MSI	0.20	1.57	0.32	0.09	0.49	0.04	0.01	1.17	0.10	2.14	0.17	0.23
Budget	0.08	-0.98	-0.16	0.25	0.00	-0.03	0.00	-1.07	0.60	-0.56	0.48	0.15

TF: throughfall / MSI: mineral soil input

At Neuglobsow both periods are characterized by a obviously lower release of elements from the humus layer compared to Rösa and Taura. The budgets are distinctly negative only for Al^{n+} (Table 21). There is no considerable mobilization and release of SO_4^{2-}-S at Neuglobsow corresponding to the distincly lower S content of the humus layer. As at Taura, N budgets of the humus layer are positive. Unfavourable chemical conditions obviously prevent nitrification of mineralized N. Since NH_4^+-N fluxes are also very low we assume

Table 21. Element budgets of the humus layer at Neuglobsow in P1 and P2 (values in kmol(c) ha^{-1} yr^{-1})

	H^+	Ca^{2+}	Mg^{2+}	K^+	Na^+	Fe^{3+}	Mn^{2+}	Al^{n+}	NH_4^+ -N	SO_4^{2-} -S	NO_3^- -N	Cl^-
P1												
TF	0.49	0.51	0.19	0.25	0.32	0.02	0.04	0.09	0.27	0.78	0.38	0.51
MSI	0.24	0.64	0.31	0.18	0.37	0.04	0.06	1.01	0.05	0.79	0.07	0.46
Budget	0.25	–0.12	–0.12	0.07	–0.05	–0.02	–0.03	–0.92	0.22	–0.01	0.31	0.05
P2												
TF	0.24	0.52	0.18	0.32	0.64	0.01	0.04	0.06	0.37	0.99	0.46	0.60
MSI	0.20	0.81	0.28	0.07	0.50	0.05	0.07	0.83	0.03	0.78	0.03	0.46
Budget	0.04	–0.29	–0.10	0.25	0.14	–0.03	–0.03	–0.76	0.34	0.21	0.44	0.13

TF: throughfall / MSI: mineral soil input

N storage in the humus layer and N uptake/storage by plants. It is possible that N is partly released into the mineral soil as dissolved organic N (DON) (Bergmann *et al.*, 1997, this volume). Despite of high DOC fluxes from the humus layer at Neuglobsow in P1, the other element fluxes are distinctly lower compared to the other sites.

Besides deposition input and mineralization, buffering processes can be an important source of element release from the humus layer. Therefore we calculated the proton load of the organic surface layers (Tables 22–24). Especially mineralization of N and S may result in the production of considerable amounts of H^+. The release of 1 kmol(c) SO_4^{2-}-S $ha^{-1}yr^{-1}$ leads to an equivalent proton load of 1 kmol(c) H^+ $ha^{-1}yr^{-1}$ (Matzner, 1988). N-mineralization and resulting proton loads can be calculated according to equation 3 (Matzner, 1988).

$$H^+(N) = NH^+_{4\,In} + NO^-_{3\,Out} - NH^+_{4\,Out} - NO^-_{3\,In} \qquad \text{(Eq. 3)}$$

At Rösa, S mineralization results in a release of 2.54 kmol(c) SO_4^{2-}-S $ha^{-1}yr^{-1}$ from the humus layer in P1 and consequently to the corresponding production of protons (Table 22). We probably overestimated the H^+ loads due to S mineralization in P1 because the high SO_4^{2-}-S fluxes are partly the result of the mobilization of SO_4^{2-} accumulated during the dry autumn of 1993. In P2, clearly lower S mineralization leads to lower H^+ loads. Proton production due to nitrification amount to 1.29 kmol(c) H^+ ha^{-1} yr^{-1} in P1 and 1.69 kmol(c) H^+ ha^{-1} yr^{-1}in P2. Including the H^+ input via BP as well as from canopy processes, we calculated total proton loads of 4.80 and 2.96 kmol(c) H^+ $ha^{-1}yr^{-1}$ in P1 and in P2, respectively. These proton loads can be buffered completely by exchange reactions with Ca^{2+}. Consequently, a weak acidic solution (mean pH: 4.72) of high ionic strength (mean I: $3.39*10^{-3}$ M) infiltrates into the mineral soil. The ionic strength (I) was calculated with equation 4 (Adams, 1971).

$$I = \sum 0.5\, Ci{*}Zi^2 \qquad \text{(Eq. 4)}$$

[C: molar concentration of ion (i) in soil solution, Z: valence]

Table 22. External and internal proton loads of the humus layer at Rösa (values in kmol(c) ha^{-1} yr^{-1})

H^+-input/-production			H^+-output/-consumption	
		11/93–10/94		
Input (TF)	0.64		Ca^{2+}	6.11
H^+–leaching	0.33		Mg^{2+}	0.90
S–mineralization	3.10		Al^{3+}	0.31
N–transformation	1.29		Output (MSI)	0.07
Sum	**5.36**		**Sum**	**7.39**
		11/94–10/95		
Input (TF)	0.19		Ca^{2+}	3.00
H^+–leaching	0.33		Mg^{2+}	0.20
S–mineralization	0.75		Al^{3+}	0.17
N–transformation	1.69		Output (MSI)	0.05
Sum	**2.96**		**Sum**	**3.42**

MSI: mineral soil input

Table 23. External and internal proton loads of the humus layer at Taura (values in kmol(c) ha^{-1} yr^{-1})

H^+-input/-production			H^+-output/-consumption	
		11/93–10/94		
Input (TF)	0.69		Ca^{2+}	
H^+–leaching	0.39		Mg^{2+}	1.82
S–mineralization	1.85		Al^{3+}	1.68
N–transformation	0.54		Output (MSI)	0.29
Sum	**3.47**		**Sum**	**4.27**
		11/94–10/95		
Input (TF)	0.28		Ca^{2+}	0.98
H^+–leaching	0.39		Mg^{2+}	0.16
S–mineralization	0.56		Al^{3+}	1.07
N–transformation	0.11		Output (MSI)	0.20
Sum	**1.34**		**Sum**	**2.41**

MSI: mineral soil input

Lower S contents and distinctly lower nitrification result in a lower internal H^+ production in the humus layer at Taura. (Table 23). As at Rösa, the proton load is higher in P1 (3.47 kmol(c) H^+ $ha^{-1}yr^{-1}$) than in P2 (1.34 kmol(c) H^+ ha^{-1} yr^{-1}) partly due to higer S and H^+ inputs in P1. These internally produced protons are also buffered but, in contrast to Rösa, mainly by exchange of Al^{3+}. Therefore, clearly higher amounts of Al^{n+} are mobilized from the humus layer. At Taura, a more acidic solution (mean pH: 4.27) of high ionic strength (mean I: $2.08*10^{-3}$M) is percolating into the mineral soil. The calculation of the

Table 24. External and internal proton loads of the humus layer at Neuglobsow (values in kmol(c) ha^{-1} yr^{-1})

H^+-input/-production			H^+-output/-consumption	
		11/93–10/94		
Input (TF)	0.49		Ca^{2+}	0.12
H^+–leaching	0.07		Mg^{2+}	0.12
S–mineralization	0.01		Al^{3+}	0.92
			Output (MSI)	0.24
			N–transformation	0.09
Sum	**0.57**		**Sum**	**1.49**
		11/94–10/95		
Input (TF)	0.24		Ca^{2+}	0.29
H^+–leaching	0.07		Mg^{2+}	0.10
S–mineralization			Al^{3+}	0.76
			Output (MSI)	0.20
			N–transformation	0.09
Sum	**0.31**		**Sum**	**1.44**

MSI: mineral soil input

proton load may be incomplete, especially at Taura, due to the disregarding of the NH_4^+-N uptake by plants being disregarded.

At Neuglobsow we calculated considerably lower amounts of H^+ originating from S and N mineralization compared to the levels at Rösa and Taura (Table 24). Also due to a low proton input via deposition, the total proton load is only 0.57 and 0.31 kmol(c) H^+ ha^{-1} yr^{-1} in P1 and in P2, respectively. As at the other sites, protons can be buffered completely through exchange reactions especially with Al^{3+}. Due to the low proton load, base cation release from the humus layer is mainly controlled by mineralization. Thus, an acidic solution (mean pH: 4.28) of low ionic strength (mean I: $1.05*10^{-3}$ M) is leached from the organic surface layer. In contrast to Rösa and Taura, we find no substanial differences between P1 and P2 due to well balanced intensities and distribution of rainfall as well as to the generally lower S and H^+ inputs.

4.2. Sulfate dynamics and element budgets of the Bw horizons

Many intensive studies point to the importance of S retention or release processes in the mineral soil, since they may affect base cation storage or release and may produce negative charge thus increasing the ECEC (McBride, 1994). Furthermore, desorption of previously adsorbed SO_4^{2-} may delay the de-acidification after reduced SO_4^{2-} inputs (Matzner, 1989; Alewell and Matzner, 1993). Thus, the dynamics of SO_4^{2-} adsorption and desorption need to be considered when attempting to qualify both the consequences for the development of soil chemical properties and the response of the ecosystem as a whole to changes in pollutant deposition (MacDonald *et al.*, 1994).

As presented in 3.2, we determined high amounts of inorganically bound SO_4^{2-} in the mineral soils at all sites, but especially in the Bw horizons. These results confirm the strong increases of the SO_4^{2-} fluxes at a soil depth of 50 cm, in particular at Rösa and Taura in P1. Because the SO_4^{2-} fluxes strongly increase below Bw horizons, we focus intensively on SO_4^{2-} adsorption capacity as well as on the binding forms in this soil compartment.

In the lower mineral soil, SO_4^{2-} is predominantly adsorbed at positively charged surfaces of Al and Fe hydrous oxides. Nevertheless, the relationship between SO_4^{2-} and Al_o/Fe_o and Al_d/Fe_d is quite variable and depends on various soil chemical properties (Singh, 1984; Xue and Harrison, 1991; Fuller *et al.*, 1985, MacDonald and Hart Jr., 1990). A positive correlation between KH_2PO_4-extractable, specifically and nonspecifically bound SO_4^{2-} and Al_o/Fe_o was found for Bw, Bhs and AB horizons at Taura and Neuglobsow (Figure 8). In the Bw horizon of Rösa we determined similar Alo/Feo amounts, but distinctly lower KH_2PO_4-extractable SO_4^{2-} amounts. These results indicate less positively charged surfaces in the Bw at Rösa caused by alkalinization due to the specific deposition regime in the past.

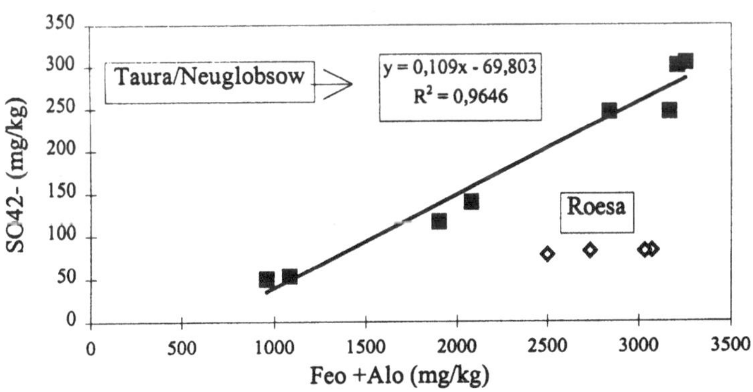

Figure 8. Relation between specifically and nonspecifically adsorbed SO_4^{2-} and contents of Al/Fe hxdrous oxides in the Bw horizons at the three sites

The results of the $MgCl_2$ extraction contradict our expectation that at Taura high amounts of SO_4^{2-} in the Bw horizons are nonspecifically bound at protonated surfaces of Al/Fe hydrous oxides. Our findings point to non-specifically adsorbed SO_4^{2-} in considerable amounts only at Neuglobsow (Figure 9). Here, we measured only very small differences of <0.1 pH units between $pH(H_2O)$ and $pH(CaCl_2)$ in the Bw horizon in 1995 (Table 25), indicating that the soil is close to the point of zero charge (PZC) (Skyllberg, 1995). Under these conditions SO_4^{2-} is hydrated and retained as non-specifically bound outer-sphere complexes (Fuller *et al.*, 1985) (Eq. 5).

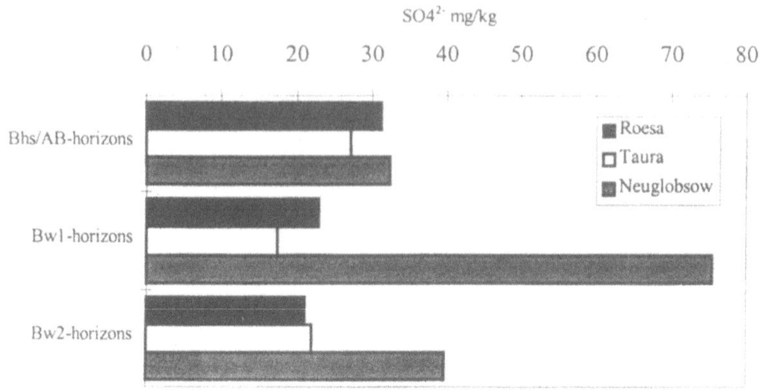

Figure 9. Amounts of nonspecifically adsorbed sulfate ($MgCl_2$-extractable) in the B horizons at the three sites

Table 25. pH values of the B horizons in 1995

	pH H_2O	pH $CaCl_2$		pH H_2O	pH $CaCl_2$		pH H_2O	pH $CaCl_2$
Horizon	Rösa		Horizon	Taura		Horizon	Neuglobsow	
Bhs	4.9	4.3	Bhs	4.3	3.8	AB	4.5	4.2
Bw1	4.7	4.3	Bw1	4.4	4.2	Bw1	4.4	4.4
Bw2	4.7	4.3	Bw2	4.3	4.1	Bw2	4.5	4.5

The general binding mechanisms are given in equations (5)–(8) (McBride 1994. Kaiser 1992).

$$Fe/Al\text{-}OH + H^+ + A^{n-} \rightarrow Fe/Al\text{-}OH_2^+ ....A^{(n-1)-} \quad \text{(Eq. 5.)}$$
$$Fe/Al\text{-}OH_2^+......A^{(n-1)-} \rightarrow Fe/Al\text{-}A]^{(n-1)-} + H_2O \quad \text{(Eq. 6.)}$$
$$Fe/Al\text{-}OH] + A^{n-} \rightarrow Fe/Al\text{-}A]^{(n-1)-} + OH^- \quad \text{(Eq. 7.)}$$
$$Al(OH)_3 + H_2SO_4 \rightarrow AlOHSO_4 + 2H_2O \quad \text{(Eq. 8.)}$$

At Rösa and Taura, the high ionic strength of the soil solutions may lead to a stronger compression of the double layer and thus, to reduced anion exclusion (Marsh *et al.*, 1987). Despite high pH values at Rösa, the closer approach of SO_4^{2-} to the mineral surfaces may enable ligand exchange reactions whereby SO_4^{2-} enters into coordination with metal oxides (Al/Fe hydrous oxides) through displacement of a coordinated hydroxyl ion (Eq. 7). This conclusion is in accordance with the results of Bolan (1993), who found elevated SO_4^{2-} retention capacities in Ca dominated solutions. At Taura, both a two step reaction including protonation of Al/Fe hydrous oxides and following H_2O release due to closer approach of SO_4^{2-} to mineral surfaces (Zhang and Sparks, 1990) and/or the direct ligand exchange reaction may be responsible for the very high SO_4^{2-} retention capacity (Eq. 6 and 7). This explains that we found lower Cl extractable amounts of SO_4^{2-} at Taura than at Neuglobsow. In contrast, low ionic strength together with relatively low pH values may be the explaination for the considerable amounts of nonspecifically adsorbed SO_4^{2-} at Neuglobsow.

A further sulfur retention mechanism may be the precipitation of minerals like jurbanite and alunite (Eq. 8) (Reuss and Walthall, 1989). Due to the very low Al^{n+} concentrations, high DOC concentrations, and in particular the low NaOH extractable SO_4^{2-} amounts (Table 9), the precipitation of Al-hydroxy-sulfates as a pathway for S retention seems to be unlikely at Rösa and Neuglobsow. Furthermore, the high pH values (>4.3) at 20/50 cm soil depth at these sites (Table 12) make the dissolution of possibly existing Al-hydroxy-sulfates as a mechanism of sulfur release improbable, since this process occurs only at pH <4.0–4.2 (Ulrich, 1991). We find somewhat higher NaOH extractable SO_4^{2-} amounts in the Bw horizons at Taura pointing to the possible occurrence of $AlOHSO_4$ minerals, but these amounts present only 10% of total extractable inorganic SO_4^{2-}. Despite high Al^{n+} and SO_4^{2-} concentrations in the soil solution at Taura, the high DOC concentrations in the lower mineral soil (16–17 mg/L) result in high complexation of Al and thus, may prevent or reduce precipitation of Al-hydroxy-sulfates. Since 90% of the inorganically stored SO_4^{2-} are water soluble (20%) or (non-)specifically adsorbed (70%) in the mineral soil at Taura, the dissolution of Al-hydroxy-sulfates as a considerable mechanism for S and Al release also seems to be unlikely at this site.

As for the humus layers, we calculated completely different budgets in the Bw horizons for the two investigation periods. At Rösa, 1.19 kmol(c) Ca^{2+} and 2.15 kmol(c) SO_4^{2-}-S $ha^{-1}yr^{-1}$ were released from the Bw horizon in P1, whereas in P2 the budgets were slightly positive (Table 26). We found similar results at Taura, but the SO_4^{2-} and the Al^{n+} budgets were also negative in P2 (Table 27). These results can be partly explained by wash-off processes of SO_4^{2-} from the soil accumulated during the dry autumn of 1993. The clearly higher tensions and corresponding lower water fluxes at a soil depth of 20 cm in autumn 1993/1994 compared to the same period in P2 point to an accumulation phase at Rösa and Taura, whereas at Neuglobsow the time courses of the tensions were similar (Figures 10–12). Thus, at Neuglobsow, P1 and P2 reveal similar element

Table 26. Element budgets of the Bw horizon at Rösa in P1 and P2 (values in kmol(c) ha^{-1} yr^{-1})

	H^+	Ca^{2+}	Mg^{2+}	K^+	Na^+	Fe^{3+}	Mn^{2+}	Al^{n+}	NH_4^+-N	SO_4^{2-}-S	NO_3^--N	Cl^-
P1												
20 cm	0.11	5.48	0.48	0.19	0.32	0.02	0.01	0.60	0.04	4.65	0.36	0.40
50 cm	0.14	6.67	0.44	0.10	0.30	0.01	0.01	0.80	0.03	6.80	0.19	0.31
Budget	−0.03	-1.19	0.05	0.09	0.02	0.01	0.00	−0.21	0.02	-2.15	0.17	0.10
P2												
20 cm	0.03	2.84	0.33	0.08	0.29	0.02	0.00	0.23	0.02	1.94	1.03	0.38
50 cm	0.05	1.90	0.13	0.03	0.16	0.01	0.01	0.24	0.01	1.83	0.37	0.23
Budget	−0.02	0.93	0.20	0.05	0.14	0.01	0.00	−0.01	0.01	0.11	0.66	0.14

Table 27. Element budgets of the Bw horizon at Taura in P1 and P2 (values in kmol(c) ha^{-1} yr^{-1})

	H^+	Ca^{2+}	Mg^{2+}	K^+	Na^+	Fe^{3+}	Mn^{2+}	Al^{n+}	NH_4^+ -N	SO_4^{2-} -S	NO_3^- -N	Cl^-
P1												
20 cm	0.27	2.60	0.44	0.27	0.29	0.04	0.01	2.06	0.05	3.68	0.32	0.36
50 cm	0.39	3.95	0.48	0.13	0.38	0.04	0.01	5.33	0.04	8.99	0.19	0.33
Budget	–0.12	–1.35	–0.04	0.14	–0.09	0.00	0.01	–3.26	0.02	–5.31	0.14	0.03
P2												
20 cm	0.14	1.24	0.27	0.08	0.37	0.03	0.01	1.02	0.03	1.86	0.28	0.38
50 cm	0.17	0.90	0.13	0.07	0.21	0.01	0.00	1.57	0.02	2.76	0.13	0.22
Budget	–0.04	0.34	0.14	0.01	0.16	0.02	0.01	–0.55	0.01	–0.90	0.14	0.16

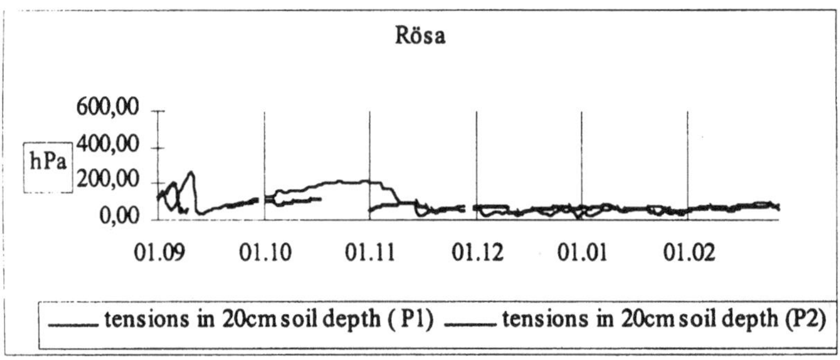

Figure 10. Time courses of the tensions at 20 cm soil depth at Roesa (September–February 1993/1994 (P1) and 1994/95 (P2), respectively)

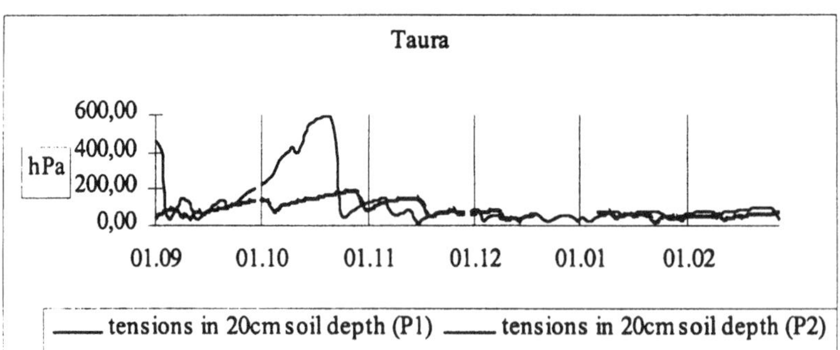

Figure 11. Time courses of the tensions at 20 cm soil depth at Taura (September–February 1993/1994 (P1) and 1994/95 (P2), respectively)

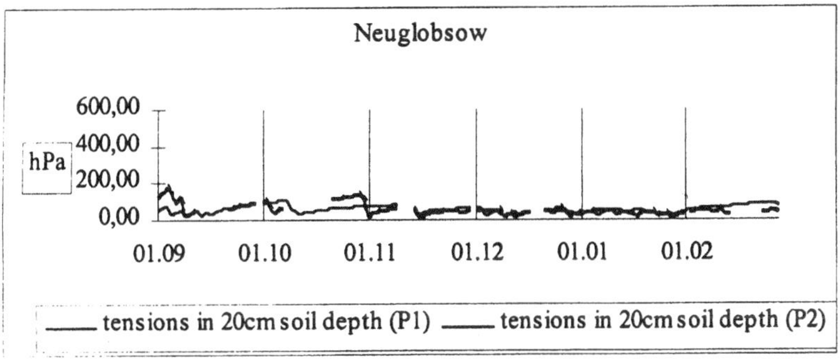

Figure 12. Time courses of the tensions at 20 cm soil depth at Neuglobsow (September–February 1993/1994 (P1) and 1994/95 (P2), respectively)

budgets. The more humid autumn and lower SO_2 depositions in 1993/94 prevented seasonal elevation of SO_4^{2-} concentrations at the background site (Table 28).

Temporal variations in SO_2 emissions and their influence on leaching losses from the mineral soil are well documented (MacDonald *et al.*, 1990; Khanna *et al.*, 1987). The results presented in Table 29a,b show substantially elevated SO_4^{2-} fluxes in particular at a soil depth of 50 cm in the period 11/93–1/94. Furthermore, the data show clearly lower pH as well as distinctly higher ionic strength of the soil solution infiltrating into the Bw horizon at Rösa and Taura for this period compared to the same period in winter 1994/95.

Table 28. Element budgets of the Bw horizon at Neuglobsow in P1 and P2 (values in kmol(c) ha^{-1} yr^{-1})

	H^+	Ca^{2+}	Mg^{2+}	K^+	Na^+	Fe^{3+}	Mn^{2+}	Al^{n+}	NH_4^+ -N	SO_4^{2-} -S	NO_3^- -N	Cl^-
P1												
20 cm	0.17	0.67	0.22	0.07	0.31	0.01	0.10	0.73	0.03	1.23	0.02	0.37
50 cm	0.19	0.66	0.21	0.10	0.29	0.01	0.10	0.69	0.02	1.73	0.01	0.33
Budget	–0.02	0.01	0.01	–0.03	0.03	0.00	0.00	0.05	0.00	–0.50	0.01	0.04
P2												
20 cm	0.08	0.57	0.19	0.04	0.35	0.02	0.08	0.48	0.02	0.95	0.02	0.43
50 cm	0.11	0.58	0.19	0.07	0.30	0.01	0.09	0.42	0.01	1.55	0.02	0.39
Budget	–0.03	–0.01	0.00	–0.04	0.05	0.00	–0.02	0.05	0.01	–0.60	0.00	0.04

At Taura (Table 29b), about 60% of the total SO_4^{2-} flux in P1 are released from the Bw horizon during these three months. Here, not only the elevated SO_4^{2-} and proton inputs, the mobilization of SO_4^{2-} accumulated during dry autumn months, and the extensive S mineralization due to high S stores in the humus layer, but the strong release of SO_4^{2-} from the Bw horizons lead to the very high SO_4^{2-} fluxes below 50cm soil depth. Especially at Taura, the leaching losses from the Bw horizons can be explained by decreasing pH of the soil solution due to higher proton inputs and by elevated ionic strength of the soil solution (Singh, 1984). Xue and Harrison (1991) found maximal SO_4^{2-} adsorption in Spodosols at pH 4,0. Likewise, at that pH level, they found the beginning of a strong release of Al^{n+}. Thus, we conclude that at Taura, the dissolution of Al hydrous oxides in the course of buffer reactions lead to a strong release of Al^{n+} and correspondingly of (non-)specifically adsorbed SO_4^{2-} (Table 29b). The clearly higher ionic strength in P1 compared to P2 at Taura enhances these processes by additionally decreasing the pH in P1.

Table 29a-c. Element fluxes in winter 1993/1994 and in winter 1994/1995

11/93–1/94	Element fluxes in kmol(c)/ha for Nov.–Jan. Throughfall input	Mineral soil input	20 cm	50 cm	Mean ionic strength (I). Mean pH (x), andminimum pH value (min) of the seepage water in 20 cm soil depth (Nov.–Jan.)	Total fluxes in P1 in 50 cm soil depth (kmol(c)/ha)
a. Rösa						
SO_4^{2-}-S	0.57	2.03	2.43	3.44		7.36
Al^{3+}	0.07	0.19	0.27	0.28	I=5.23*10^{-3} M	0.80
Ca^{2+}	0.34	3.22	2.80	3.74	pH(x)=4.70	6.67
NO_3^--N	0.10	0.54	0.11	0.06	pH(min)=4.30	0.19
H^+	0.19	0.01	0.03	0.03		0.14
11/94 –1/95	Througfall input	Mineral soil input	20 cm	50 cm	I; pH(x);pH(min) of the seepage water in 20 cm soil depth (Nov.–Jan.)	P2
SO_4^{2-}-S	0.45	0.73	0.68	0.81		1.83
Al^{3+}	0.03	0.08	0.09	0.10	I=3.49*10^{-3}M	0.24
Ca^{2+}	0.23	1.48	1.01	0.75	pH(x)=4.82	1.90
NO_3^--N	0.10	0.87	0.33	0.08	pH(min)=4.70	0.37
H^+	0.05	0.01	0.01	0.01		0.05

Table 29a–c. Element fluxes in winter 1993/1994 and in winter 1994/1995

	Element fluxes in kmol(c)/ha for Nov.–Jan.				Mean ionic strength (I). Mean pH (x), andminimum pH value (min) of the seepage water in 20 cm	Total fluxes in P1 in 50 cm soil depth
11/93–1/94	Throughfall input	Mineral soil input	20 cm	50 cm	soil depth (Nov.–Jan.)	(kmol(c)/ha)
b. Taura						
SO_4^{2-}-S	0.54	1.86	2.15	5.30		8.99
Al^{3+}	0.07	1.06	1.28	3.48	$I=4.03*10^{-3}M$	5.33
Ca^{2+}	0.26	1.15	1.27	2.40	pH(x)=4.21	3.95
NO_3^--N	0.13	0.21	0.16	0.08	pH(min)=3.97	0.19
H^+	0.20	0.08	0.09	0.10		0.39
11/94–1/95	Througfall input	Mineral soil input	20 cm	50 cm	I; pH(x);pH(min) of the seepage water in 20 cm soil depth (Nov.–Jan.)	P2
SO_4^{2-}-S	0.40	0.68	0.53	0.90		2.76
Al^{3+}	0.04	0.26	0.28	0.51	$I=1.62*10^{-3}M$	1.57
Ca^{2+}	0.16	0.55	0.32	0.29	pH(x)=4.28	0.9
NO_3^--N	0.14	0.08	0.11	0.08	pH(min)=4.24	0.13
H^+	0.06	0.04	0.04	0.05		0.17
c. Neuglobsow						
SO_4^{2-}-S	0.28	0.44	0.60	0.79		1.73
Al^{3+}	0.02	0.39	0.34	0.34	$I=1.18*10^{-3}M$	0.69
Ca^{2+}	0.12	0.23	0.28	0.30	pH(x)=4.41	0.66
NO_3^--N	0.08	0.02	<0.01	<0.01	pH(min)=4.34	0.01
H^+	0.20	0.07	0.05	0.06		0.19
11/94–1/95	Througfall input	Mineral soil input	20 cm	50 cm	I; pH(x);pH(min) of the seepage water in 20 cm soil depth (Nov.–Jan.)	P2
SO_4^{2-}-S	0.33	0.32	0.38	0.68		1.55
Al^{3+}	0.02	0.34	0.19	0.19	$I=0.97*10^{-3}M$	0.42
Ca^{2+}	0.17	0.28	0.25	0.29	pH(x)=4.59	0.58
NO_3^--N	0.09	0.01	0.01	<0.01	pH(min)=4.39	0.02
H^+	0.05	0.01	0.01	0.01		0.11

At Rösa, short term high proton loads especially in November 1993 and subsequent buffering may also result in elevated Al^{n+} and SO_4^{2-} fluxes. We found an elevated Al^{n+} release at Rösa in this period, but in amounts that cannot explain the high SO_4^{2-} fluxes (Table 29a). Thus, dissolution of Al hydrous oxides cannot be the main process that caused elevated SO_4^{2-} fluxes at Rösa. The low fluxes in P2 at Rösa rather point to the mobilization of Ca^{2+} and

SO_4^{2-} accumulated during autumn 1993 as the main impulse for elevated SO_4^{2-} release in P1 beside high S-inputs.

At Neuglobsow, high pH values (>4.3) and low ionic strength of the soil solution in 20 cm soil depth in P1 and P2 prevent accelerated SO_4^{2-} release from the Bw horizon (Table 29c). The relatively high pH value of the seepage water in 20 cm soil depth is mainly the result of the high buffer capacity of the AB horizon.

5. Conclusions

The site specific historical deposition regimes influence strongly the present soil chemical status and the element budgets of three Scots pine ecosystems. At Rösa, the effects of former dust and sulfur inputs are manifested in high stores of S and Ca^{2+} in the humus layer, elevated pH and BS throughout the whole profile, favourable mineralization and nitrification conditions, and high Ca^{2+} and SO_4^{2-} concentrations in the soil solution. Due to S mineralization and nitrification we find very high internal proton production in the humus layer, but protons are buffered completely by exchange reactions especially with Ca^{2+}. Thus, at present, the humus layer at Rösa, especially the Oe horizon seems to possess sufficient buffer capacity to prevent accelerated re-acidification. Comparably high sulfur but lower alkaline dust inputs in the past at Taura are reflected in the clearly lower pH and BS, in lower nitrification, and in high SO_4^{2-} and Al concentrations in the seepage water. Due to the lack of elevated dust depositions, the soil at Taura shows the most severe symptoms of soil acidification. However, the calculated molar ratios are well above the critical values of 1 for the Ca/Al ratio and exceed the critical value of 0.2 for the Mg/Al ratio (Ulrich, 1989). Thus, Al toxicity or nutrient imbalances are not to be expected. At Neuglobsow, the lack of alkaline dust inputs and lower sulfur depositions are reflected in clearly lower Ca^{2+} and S contents of the humus layer and a low BS throughout the profile. Due to the lack of nitrification and low S mineralization the internal proton load of the humus layer is low. Therefore, cation mobilization from the humus layer in the course of buffer reactions is also very low and the cation release is mainly controlled by mineralization.

The investigation of the inorganic SO_4^{2-} retention mechanisms reveal a high SO_4^{2-} retention capacity in the mineral soils of all three sites, but differences in the main binding forms. These differences are mainly the result of the former deposition regimes. Our results show that the inorganic SO_4^{2-} retention capacity of the three mineral soils depends on the amount of Al/Fe hydrous oxides, the pH values, the ionic strength, and the soil solution composition. Despite high pH values, we find relatively high amounts of specifically adsorbed SO_4^{2-} in the mineral soil at Rösa mainly due to ligand exchange reactions with OH groups on surfaces of Fe/Al hydrous oxides. Decreasing pH values in future may increase the amount of positively charged surfaces and thus temporarily increase the SO_4^{2-} retention capacity of the mineral soil at Rösa. Due to very low pH values and high ionic strength of the soil solution as well as to high contents of Al/Fe hydrous oxides, we found the highest amounts of specifically adsorbed SO_4^{2-} in the Bw horizon at Taura. Here, the strong SO_4^{2-} and Al^{n+} release from the Bw horizon is mainly due to buffer processes. At Taura, high precipitation amounts after dry periods, especially in winter, when high proton amounts are deposited in combination with elevated SO_4^{2-} inputs, will intensify both the dissolution of Al hydrous oxides and the mobilization of accumulated Ca^{2+}, Al^{n+}, and SO_4^{2-}. Although we find very high amounts of SO_4^{2-} stored in the Bw horizon at Neuglobsow, the low ionic strength as well as the relatively high pH values of the soil solution that infiltrates into the Bw horizon, prevent enhanced leaching losses of SO_4^{2-} and Al^{n+} in the course of buffer reactions. But decreasing buffer capacity of the AB horizon in the future could strongly increase the SO_4^{2-} and Al^{n+} release from the Bw horizon.

Acknowledgements

This project was financed by the Federal Ministry of Education and Research (BMBF, Bonn/Germany) as part of the SANA (Redevelopment of the Atmosphere above the New Federal States) research program.

References

Adams F. 1971. Ionic concentrations and activities in soil solutions: Soil Sci Soc Am Proc. 35, 420–426.

Ajwa HA, Tabatabai MA. 1995. Metal-induced sulfate adsorption by soils: Effect of pH and ionic strength. Soil Sci. 159(1), 32–42.

Alewell C, Matzner E. 1993 Reversibility of soil solution acidity in acid forest soils. Water Air Soil Pollut. 71(1–2), 155–166.

Alewell C, Matzner E. 1995. Water, $NaHCO_3$-, NaH_2PO_4- and NaCl-extractable SO_4^{2-} in acid forest soils. Z Pflanzenernähr Bodenk. 159, 235–24.

Beier C, Rasmussen L, de Visser P *et al.*, 1993. Effects of changing the atmospheric input to forest ecosystems – Results of the 'EXMAN' project. In: Experimental Manipulation of Biota and Biogeochemical Cycling in Ecosystems. CEC-Ecosytsems Research Report 4. Eds. Rasmussen L, Brydges T, Mathy P. 138–154.

Bergmann C, Fischer T, Hüttl RF. 1997. Decomposition of needle-, herb-, and root-litter and humus in three Scots pine (*Pinus sylvestris L.*) stands in NE-Germany (this volume).

Blume HP, Schwertmann U. 1969. Genetic evaluation of profie distribution of aluminum, iron, and manganese oxides. Soil Sci Soc Am J. 33, 438–444.

Bolan NS, Syers JK, Summer ME. 1993. Calcium induced sulfate adsorption by soils. Soil Sci Soc Am J. 57, 691–696.

Bredemeier M, Dohrenbusch, a Murach D. 1995. Response of soil water chemistry and fine roots to clean rain in a spruce forest ecosystem at Solling, FRG. Water Air Soil Pollut. 85, 1605–1611.

Courchesne F, Hendershot WH. 1990. The role of basic aluminum sulfate minerals in controlling sulfate retention in the mineral horizons of two spodosols. Soil Sci. 150(3), 571–578.

David MB, Vance GF, Fasth WJ. 1991. Forest soil response to acid and salt additions of sulfate: II. Aluminium and base catios. Soil Sci. 151(3), 208–218.

Fuller RD, David MB, Driscoll CT. 1985. Sulfate adsorption relationships in forested Spodosols of the Northeastern USA. Soil Sci Soc Am J. 49, 1034–1040.
Jansson PE. 1991. Simulation model for soil water and heat conditions. Swedish University of Agricultural Sciences Uppsala, Report 165.
Johnson DW, Lindberg SE. Eds. 1991. Atmospheric deposition and forest nutrient cycling: A synthesis of the integrated forest study. Ecological Series 91. Springer Verlag, New York. 707pp.
Kaiser K. 1992. Salz- und Säureeffekte auf die Zusammensetzung der Bodenlösung und die Sorptionseigenschaften saurer Waldböden. Bayreuther Bodenkundliche Berichte Band. 29, 1-128.
Khanna PK, Prenzel J, Meiwes KJ, Ulrich B, Matzner E. 1987. Dynamics of sulfate retention by acid forest soils in an acidic deposition environment. Soil Sci Soc Am J. 51, 446–452.
Koopmans CJ, Lubrecht WC, Tietema A. 1995. Nitrogen transformations in two nitrogen saturated forest ecosystems subjected to an experimental decrease in nitrogen deposition. Plant Soil. 175(2), 205–218.
Lüttschwager D, Wulf M, Rust S, Forkert J, Hüttl RF. 1997. Tree canopy and forest floor transpiration in three Scots pine (*Pinus sylvestris L.*) stands (this volume).
MacDonald NW, Burton AJ, Witter JA, Richter DD. 1994. Sulfate adsorption in forest soils of the Greate Lakes Region. Soil Sci Soc Am J. 58, 1546–1555.
MacDonald NW, Hart, JB Jr. 1990. Relating sulfate adsorption to soil properties in Michigan forest soils. Soil Sci Soc Am J. 54, 238–245.
Marschner B. 1990. Elementumsätze in einem Kiefernökosystem auf Rostbraunerde unter dem Einfluß einer Kalkung/Düngung. ber. d. Forschungszentrums Waldökosysteme. Universität Göttingen, Reihe A, Bd. 60.
Marsh KB, Tillman RW, Syers JK. 1987. Charge relationships of sulfate sorption by soils. Soil Sci Soc Am J. 51, 318-323.
Matzner E. 1989. Acidic precipitation: Case study Solling. In: Acidic Precipitation. Volume I: Case Studies. Eds. Adriano DC, Havas M. Springer Verlag.
Matzner E. 1988. Der Soffumsatz zweier Waldökosysteme im Solling. Ber. d. Forschungszentrums Waldökosysteme. A40, S. 1–217.
Matzner E, Murach D. 1995. Soil changes induced by air pollutant deposition and their implications for forests in Central Europe. Water Air Soil Pollut. 85, 63–76.
McBride MB. 1993. Enviromental chemistry of soils. Oxford University Press. New York – Oxford.
Meiwes KJ, König ,. Khanna PK, Prenzel J, Ulrich B. 1984. Chemische Untersuchungsverfahren für Mineralboden, Auflagehumus und Wurzeln. Ber d Forschungszentrums Waldökosysteme. Universität Göttingen, Bd. 7, S. 1–67.
Moldan F, Hultberg H, Andersson I. 1995. Covered catchment experiment at Gardsjön. Changes in runoff chemistry after four years of experimentally reduced acid deposition. Water Air Soil Pollut. 85, 1599–1604.
Nodvin SC, Driscoll CT, Likens GE. 1986. The effect of pH on sulfate adsorption by a forest soil. Soil Sci. 142(2), 69–75.
Reuss JO, Walthall PM. 1989. Soil reaction and acidic deposition. In: Acidic Precipitaion, Volume 4: Soils, Aquatic Processes and Lake Acidification. Eds. Adriano DC, Havas M.
Schaaf W. 1992. Elementbilanz eines stark geschädigten Fichtenökosystems und deren Beeinflussung durch neuartige basische Magnesiumdünger. Bayreuther Bodenkundl Ber. 23, 1–169.
Schlichting E, Blume H-P, Stahr K. 1995. Bodenkundliches Praktikum. 2. neu-bearbeitete Auflage. Pareys Studientexte 81.
Singh BR. 1984. Sulfate sorption by acid forest soils: 3. Desorption of sulfate from adsorbed surfaces as a function of time, desorbing ion, pH, and amount of adsorption. Soil Sci. 4(5), 346–353.
Skyllberg U. 1995. Solution/soil ratio and release of cations and acidity from Spodosol horizon. Soil Sci Soc Am J. 59, 786–795.
Sposito G. 1989. The chemistry of soils. Oxford University Press, New York – Oxford.
Trüby P, Aldinger E. 1986. Eine Extraktionsmethode zur Bestimmung der austauschbaren Kationen im Boden. Z Pflanzenernähr Bodenk. 152, 301–306.

Ulrich B. 1991. An ecosystem approach to soil acidification. In: Soil Acidity. Eds. Ulrich B, Summer ME. New York (Springer), 28–79.

Ulrich B. 1989. Forest decline in ecosystem perspective. In: Internationaler Kongress Waldschadensforschung: Wissensstand und Perspektiven, Friedrichshafen, 2.-6.10.1989, Vorträge Band 1. Ed. Ulrich B. 21–41.

Ulrich B. 1981. Theoretische Betrachtungen des Ionenkreislaufes in Waldökosystemen. Z Pflanzenernähr Bodenk. 144, 647–659.

de Visser PHB, van Breemen N. 1995. Effects of water and nutrient applications in a Scot pine stand to tree growth and nutrient cycling. Plant Soil. 173(2), 299–310.

Weisdorfer M, Schaaf W, Hüttl RF. 1995. Auswirkungen sich zeitlich ändernder Schadstoffdepositionen auf Stofftransport und -umsetzung im Boden. In: Atmosphärensanierung und Waldökosysteme. Reihe Umweltwissenschaften der BTU Cottbus, Bd 4. Eds. Hüttl RF, Bellmann K, Seiler W. 56–74.

Xue D, Harrison B. 1991. Sulfate, aluminum, iron, and pH relationships in four Pacific-Northwest forest subsoil horizons. Soil Sci Soc Am J. 55, 837–840.

Zhang PC, Sparks DL. 1990. Kinetics and mechanisms of sulfate adsorption/desorption on goethite using pressure-jump relaxation. Soil Sci Soc Am J. 54, 1266–1273.

13
Radial increment of Scots pine stands

U. NEUMANN and G. WENK

1. Introduction

The radial growth of forest trees is an integral, easily measurable quantity, which reflects the effects of the entire positive and negative environmental impacts on forest growth.

The objective of this investigation consists of verifying and quantifying the effects of pollutant inputs on radial growth of pine with special emphasis on the possible effects of the reduction in air pollution and deposition which has taken place since about 1990.

Since the annual fluctuations in increment are mainly caused by weather and climate conditions, the question of the effect of the reduction in pollutants may not be correctly answered, unless the effects of the climate are eliminated. On the assumption that increment variations due to the weather are only a short-term occurrence, and thus compensated on a medium-term basis, they might be filtered out, without being explicitly determined. This, however would presuppose long-term observational periods prior to and subsequent to the changed pollutant situation. Hence, such a procedure is not possible because of the short period since the pollutant reduction. Thus, the direct registration of the impact of the weather on radial growth is an important precondition for answering the above-raised question. Besides this, however, it is of interest per se.

The incremental investigations consider both the annual radial increment over the entire life span and the intra-annual incremental course over the observational period to register the weather-dependent intra-annual fluctuations.

In addition to their usefulness in these evaluations, the measured data mainly serve as initialization and verification for the ecosystem model FORSANA (Grote *et al.*, this volume).

2. Materials and methods

The efforts undertaken focus on radial increment investigations. They comprise boring core (in brief: core) and stem disk analyses to investigate the long-term development of radial growth on the one hand, and microdendrometer

R.F. Hüttl and K. Bellmann, Changes of Atmospheric Chemistry and Effects on Forest Ecosystems, 227–250.

measurements and girth measurements using plastic tape measures to register the intra-annual course of radial growth on the other. Height growth and volume increment have been investigated on a supplementary scale.

2.1. Long-term investigations of the radial increment using cores and stem disks

The first core extraction took place in Taura and Rösa in the spring of 1992 as well as in Neuglobsow in the spring of 1993.

On two trial plots of each investigation site (Rösa 3 and 4, Taura 2 and 3, Neuglobsow 1 and 6), one core was extracted from each of 76 trees throughout the diameter range. These trees may be regarded as representative of the stands. However, many problems arose during the crossdating of the ring-width time series of these core series concerning the sequences of very narrow tree-ring arrangement. Often, weak trees appeared to have rather atypical incremental courses, which was due to the competition situation. In order to investigate more closely the influence of the weather on ring width, a second core extraction took place in October 1994. Here, 25 sample trees per trial plot, with medium- and upper-diameter classes, were selected.

The cores were measured, using the 'Göttingen measuring system', at the Institute of Forest Growth and Forest Computer Science, Tharandt. For this purpose the cores that had been glued onto a slotted strip were subjected to high precision grinding, so that the individual ligneous cells became readily visible under a microscope at 80 fold magnification. Automatic synchronization was renounced in favor of high accuracy and exact control. In August/September 1995, 5 sample trees were felled in one plot of each of the three investigation sites: Rösa 3, 65-year-old; Taura 4a, 48-year-old; Neuglobsow 1, 61-year-old. The sample trees were selected according to the diameter classes, i.e.

1st class: 30–35 cm	3rd class: 20–25 cm	5th class: 10–15 cm
2nd class: 25–30 cm	4th class: 15–20 cm	

Stem disk measurements were conducted on the sample trees that were felled. Along eight radial lines the stem disks were similarly measured, using the 'Göttingen measuring system'.

In order to search for trend modifications possibly caused environmentally, and as a basis for the investigations into the impact of the weather and climate, the age trend was determined and eliminated. To this end the age trend A_t, $t = t_0, \ldots, t_E$ was estimated ($t_0, \ldots, t_E$: the respective observation period) and the respective ring-width indices I_t, $t = t_0, \ldots, t_E$ deduced, i.e. the relative values of tree-ring widths Z_t, $t = t_0, \ldots, t_E$ with respect to the assumed trend function A_t:

$$I_t = Z_t/A_t \, , \; t = t_0, \ldots, t_E \, . \qquad (1)$$

When selecting the type of age-trend function, it is important to consider what effects the trend function should involve. If indeed only the age influence is to be determined, a relatively rigid trend function will be appropriate. If other influences such as thinning, fertilization and infestation with pests are to be surveyed, more flexible trend functions are needed, e.g. in terms of spline functions (Kublin and Gantert, 1993). However, such flexible trend functions imply the danger in involving increment deviations which are caused by impacts which themselves are the research subject, i.e. the impact of the weather. Deviations of this kind would then be lost in index formation.

Since not enough is known about other explanatory variables, the trend function is related here exclusively to the age trend. Because tree-ring analyses in our case do not include the juvenile phase of tree growth, functions like

$$A_t = a * e^{b*t},\ t = t_0, \ldots, t_E \qquad (2)$$

have been taken as a basis for the age-trend estimation. Trend estimations and the respective index formation were conducted by single trees. Subsequently, the ring-width indices have been averaged by plots.

2.2. Determination of the intra-annual radial increment by means of microdendrometer measurements and girth plastic tape measures

In two trial plots per investigation site (the same plots as for the core extraction) 76 sample trees, representative of the stand, were selected for increment precision measurements, using microdendrometers.The microdendrometer measurements were conducted in Rösa and Taura from 1992 to 1995, and in Neuglobsow from 1993 to 1995.

To apply a microdendrometer, i.e. a dial gauge fixed onto an angular frame, every sample tree was provided with three screws, the heads of which formed a firm plain running in parallel to a tangential plain of the tree. The radial increment causes the distance from the measuring point on the bark surface to the reference plane to decrease, cf. Figure 1.

The measurement accuracy is about ± 20 µm. The microdendrometer measurements are to indicate the intra-annual course of radial growth. Hence, the measuring period has to comprise the time points at the beginning and the end of growth. Because of the reversible changes in diameter it is not possible to determine these points of time exactly. The measurements were made from about the beginning of April until mid-October in one- to two-week intervals.

Elimination of the reversible diameter fluctuations

One problem of the increment precision measurements is, that the measured data do not reflect the actual radial growth by the actually grown ligneous substance, but they include the reversible changes in thickness by swelling and shrinkage, respectively, resulting from the weather conditions and from the internal water storage. The target quantity of the intra-annual incremental

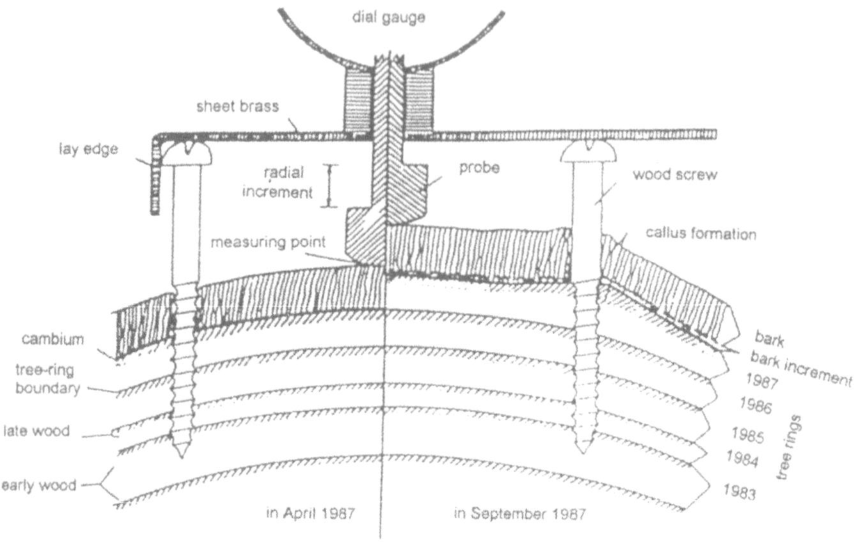

Figure 1. Tree-ring formation and microdendrometer measurement at the beginning and in the end of a growing season (Wenk *et al.*, 1988)

measurements is the actual increment. Thus, the measurement values have to be broken down into these two components, and it is impossible to measure one of the components alone. A simple, plausible method was developed for this purpose, which is based upon the comparison of incremental series of trees with low and with high annual increment.

At first the measurement results have been summarized to curves of stand means, and subsequently evaluated by diameter classes.

Also, the girth measurements using the plastic tapes served the same purpose – the registration of the intra-annual course of radial increment.

To this end 15 sample trees were selected on 5 trial plots each in the three investigation sites, i.e. 75 sample trees of each investigation site. These trees were likewise subjected to determinations of biomass, needle weight and nutritional state (Gluch *et al.*, this volume), as well as photosynthetic capacity (Dudel *et al.*, this volume). The total period and the points of time for the measurements were equivalent to those of microdendrometer measurements.

The girth measurements are advantageous compared to microdendrometer measurements in that the sample trees are not injured, the internal variation of radial increment being largely eliminated, and an intermediate readjustment as with the screws for the microdendrometer measurements, is unnecessary. The drawbacks are the low measuring accuracy, some slack in response, and a long adjustment phase of the tape measures.

2.3. Investigations of the dependence of radial increment on weather and climate conditions

An assessment of the results obtained from the radial increment measurements concerning the response to the impact of pollutants as well as to a reduction in pollutants, respectively, is not possible without taking into account the simultaneously involved climate effects, because they are the main reason for the short- and medium-term incremental fluctuations. Therefore, our task consisted of quantitatively assessing the influence of the climate, both with regard to the long-term ring-width time series and to the intra-annual incremental courses.

Linear regression models were used to investigate the dependence of the tree-ring widths on the climate factors (Wenk *et al.*, 1994; Neumann, 1996a). As independent, i.e. explanatory variables, the monthly means of temperature and the monthly sums of precipitation and, alternatively, various transformations of these basic variables, were used on the one hand, and the ring-width data of the previous years on the other. The ring-width indices, i.e. the increment data from which the age trend had been eliminated have been used as dependent variables, i.e. the variables to be explained.

The usage of monthly climate data in the investigation of the climate impact on radial increment is a very rough method. Regression models implying tree-ring width, however, do not allow the use of daily weather data as independent variables. Only through the registration of the radial-increment course within the growing season in small time intervals, is a direct comparison with the weather data measured daily possible.

To evaluate the influence of the weather on the course of radial growth within the growing season an empirical-statistical model (Neumann, 1996b) has been developed, which refers to daily data of temperature and precipitation (Weisdorfer *et al.*, this volume).

The model connects basic considerations about the growing conditions, applying statistical methods; modeling of physiological processes is, however, not included. The basic relations of the model are schematically outlined in Figure 2.

Supposedly, the temperature (T) has an immediate and via the radiation a mediate positive influence on growth, it may, however, through the negative influence of the water supply, also have a growth-inhibiting effect. The impact of precipitation (N) is expressed by the effect of soil moisture (B) or suction tension in soil in connection with the temperature. All the other impacts are summarized in a trend function (p_t, $t=t_0, \ldots, t_E$), which corresponds to the possible growth under optimal weather conditions. This trend function includes the time points of the beginning (t_0) and the culmination of growth. The annual radial increment to be expected without knowing the weather during the growing season (Z_J) has been estimated based upon previous climate conditions using a regression model as mentioned above. Through multiplicative linkage of the trend function with the impact functions as for

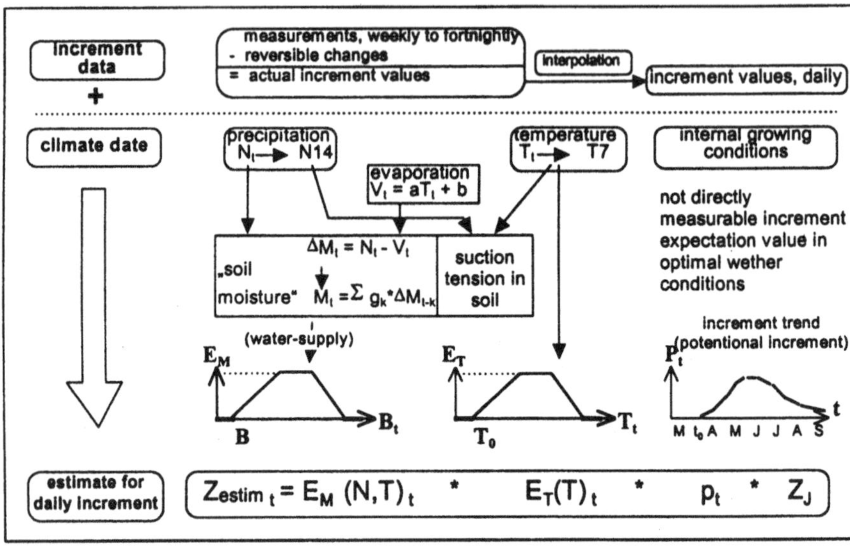

Figure 2. Outline of model for intra-annual radial increment

temperature (E_T) and moisture (E_M) the potential increment data are reduced corresponding to the actual weather conditions.

The starting point for the calculations are the daily precipitation (N_t) and temperature (T_t) data. Only the interpolation of data measured either weekly or every two weeks are available in terms of comparative values for the estimated radial increment. The model equations are used as non-linear regression models. The coefficients have been separately estimated for the trend function (p), the dependence of soil moisture or suction tension of soil on temperature and precipitation, as well as for the impact functions ($E_M(N,T)$ and $E_T(T)$). Although the goodness of fit could be improved by a simultaneous estimation of all parameters, the great number of parameters would then, however, cause numerical problems, and the interpretability of the parameters would be affected.

3. Results

3.1. *Past development of radial increment*

Tree-ring analyses on cores

In general, the results obtained from the core analyses refer to the 1994 core series. The 1992/93 core series have been used for comparison only. Although they better correspond to the stand means because of the more stand-representative selection of sample trees as compared with those obtained from the 1994 series, they supply, however, less information on climate impact because of the lower proportion of dominant trees.

Figure 3 shows the 1994 mean ring-width courses of the 1994 core series as for the six trial plots.

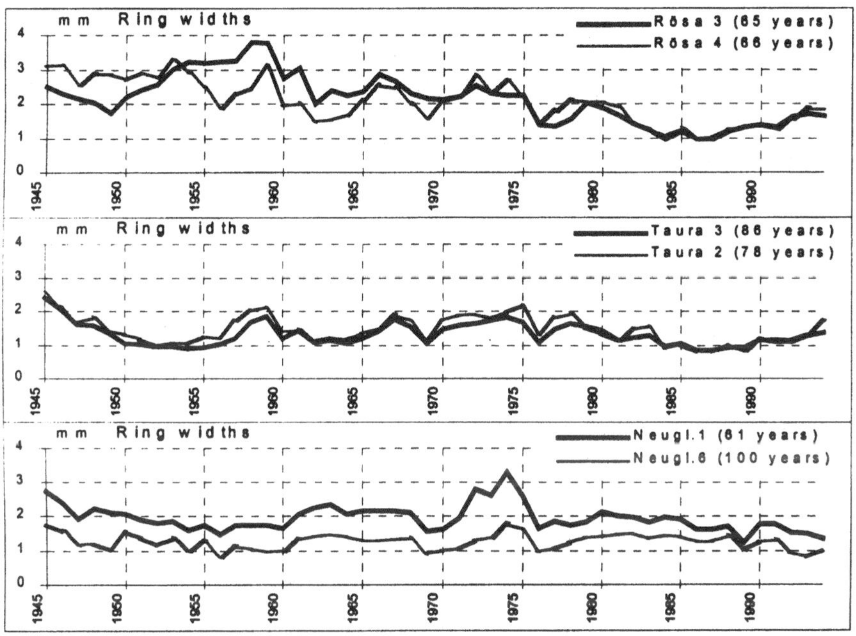

Figure 3. Ring widths of the six trial plots according to cores of 1994

The ring-width series of the plots Rösa 3 and 4 have remained rather similar since 1983, and the same applies to Taura 2 and 3 throughout the observation period. The initial deviations between Rösa 3 and 4 are much more pronounced than the standard errors, emphasizing a significant difference between these plots in the initial phase. The lower growth at Neuglobsow 6 compared with Neuglobsow 1 is substantiated by the age differences. The increment courses of the plots at Rösa and Taura are very similar, whereas those of Neuglobsow reflect a special own development, except for the period of time between the joint indicator years of 1969 and 1976. The different incremental trends obvious in the individual investigation sites become even more distinct after the elimination of the age-related trends, cf. Figure 5.

Figure 4 demonstrates the courses of the averaged ring widths of the two core series by investigation sites. In this connection the plot Neuglobsow 6 was excluded because it is older (cf. Figure 4).

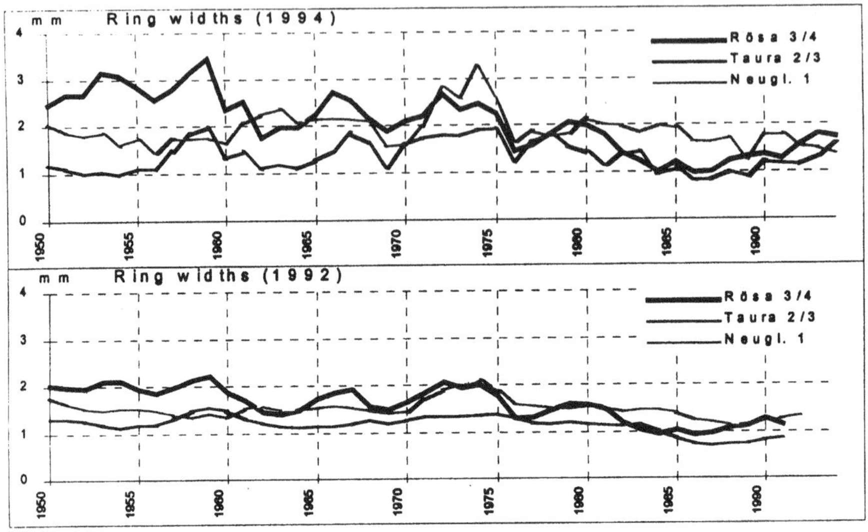

Figure 4. Ring-width series for Rösa 3/4, Taura 2/3, and Neuglobsow 1 according the cores of 1992/93 and 1994

Table 1. Mean ring width (cores of October 1994) for the six trial plots

	Rösa 3	Rösa 4	Taura 2	Taura 3	Neugl. 1	Neugl. 6
Mean ring width 1950–94	2.13	1.98	1.41	1.26	1.90	1.24
Mean ring width 1980–90	1.27	1.35	1.09	1.06	1.80	1.35
Mean ring width 1990–94	1.53	1.56	1.32	1.21	1.58	1.06

The comparisons of the 1980–1990 and 1990–1994 mean increments with respect to the three sites appear to point to the typical differences regarding the increment trend: As for Rösa and Taura an unambiguous increment increase could be verified, which in Rösa is slightly stronger than in Taura, whereas the increment slightly decreases in Neuglobsow according to age trend.

Figure 5 shows the ring-width indices, i.e. the relative increment values as related to the general, exponentially estimated age trend, of the three investigation sites on the basis of the cores as of 1994 again without Neuglobsow 6. The values of 1995 were complemented from microdendrometer measurements.

The ring-width index curves show clearly the results of increment change; they do not, however, provide information on the absolute increment level:

- Rösa — sharp increase since the mid-80s, starting from a very low relative level,
- Taura — sharp increase since the late 1980s, starting from a very low relative level

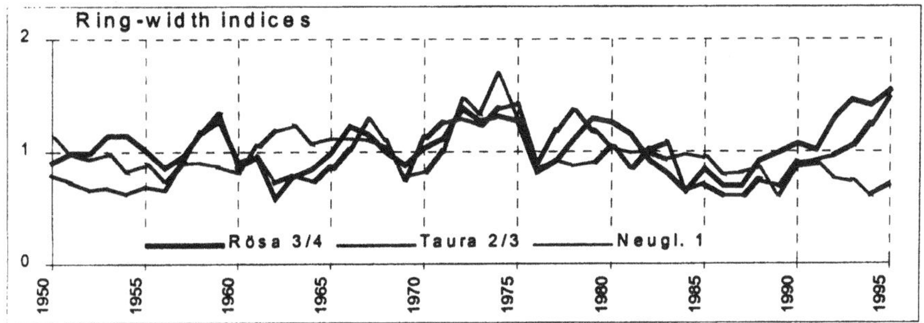

Figure 5. Ring-width indices for Rösa 3/4, Taura 2/3 and Neuglobsow 1

- Neuglobsow slight decrease starting from the highest relative increment level in the late 1980s, as indicated by comparison of sites.

3.2. Relationships between ring width and diameter at chest height

A linear relationship, which may vary in the statistical significance, exists between the annual diameter increment at chest height and the chest height diameter (dbh) itself.

The higher the coefficient of determination of this relationship the more uniform are the relative increments, thus the more pronounced some homogeneity of increment within the stand. In contrast to this, a small coefficient of determination points to unknown, non-uniform impacts on the radial increment because of the correspondingly high residual variance. These impacts which might be a consequence of thinning operations, fertilization or infestation with pests, play an important role in stand growth assessment.

Table 2 shows the coefficients of determination as for the relationships between the increment and chest height diameter derived from the core analyses and the microdendrometer measurements for all 6 plots and various periods of time. In this connection the following becomes obvious:

- The coefficients of determination are very small on average. Concerning the 1994-cores this is partly due to the missing lower diameter range.

- Distinct differences exist between the investigation sites: a decrease of the coefficients of determination from Neuglobsow to Taura to Rösa. A connection with the various extents of pollution load, which as an influential factor causes additional variation, may be supposed in this respect.

- However, there are also strong differences between the plots of the same investigation site, e.g. between Rösa 3 and Rösa 4.

Table 2. Coefficients of determination for the dependence of the mean annual radial increment on the diameter at breast height (bdh)

		Rösa 3	Rösa 4	Taura 3	Taura 2	Neugl.1	Neugl.6	Mean
Microdendro-meter measurements (75 data each)	1994	0.041	0.470	0.168	0.325	0.278	0.265	0.258
	1995	0.083	0.487	0.051	0.333	0.248	0.438	0.273
Core 1994 (20 data each)	1992	0.014	0.000	0.034	0.025	0.352	0.002	0.071
	1993	0.006	0.000	0.161	0.022	0.132	0.010	0.055
	1994	0.058	0.011	0.246	0.058	0.191	0.008	0.095
	1985–94	0.006	0.083	0.361	0.005	0.414	0.109	0.163
	1951–60		0.093	0.598	0.229	0.364	0.603	0.340
Core 1992 (75 data each)	1982–91	0.019	0.095	0.109	0.367	0.408	0.180	0.196
Plot means		0.056	0.194	0.216	0.171	0.298	0.202	
Site means		Rösa: 0.125		Taura: 0.193		Neugl.: 0.250		0.190

- There are differences for time periods: The coefficient of determination related to the 10-year mean was found to exceed distinctly that of the years 1951–60 (Rösa 3: 1965–74) compared to the past 10 years, however with the exception of Neuglobsow 1.

- There are also differences occurring quite incidentally, which, supposedly, result from estimation errors due to the small sample sizes. The magnitude of the actual coefficient of determination, however, is independent of the sample size.

Diameter classification

Due to the small coefficients of determination regarding the relationships between the annual diameter increment and chest height diameter of the SANA trial plots, the evaluation by the 5 dbh classes (1st: 30–35 cm, 2nd: 25–30 cm, 3rd: 20–25 cm, 4th: 15–20 cm, 5th: 10–15 cm) is supposed to become problematic.

The cores as of 1994 belong to classes 1 to 3. Figure 6 shows the class mean-value curves as for Rösa 3: Since 1960, hardly any difference is obvious any more between classes 2 and 3.

Here a certain regularity becomes apparent: The large-diameter trees lose their dominance that resulted from the rapid juvenile development. The growth in diameter of the previous years is nearly independent of the chest height diameter.

In this context the plot Rösa 3 is taken as an example only. A subdivision into classes as for the ring-width time series based upon the cores as in 1994, is not allowed from the statistical point of view.

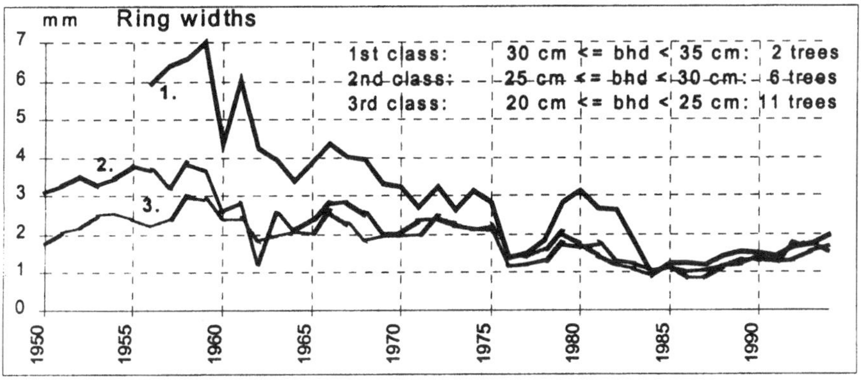

Figure 6. Radial increment Rösa 3, class mean-value curves from core analyses of 1994

Class-related mean-value curves may be derived from the total mean-value curves by means of the increment-diameter relationship.

Stem disk analyses

Figure 7 illustrates along with the respective mean-value curves the ring-width development of the five trees each of Rösa 3, Taura 4a, and Neuglobsow 1, which belong to the 5 dbh classes.

The increment–diameter straight lines as related both to the 1981–1990 and to the 1993–1995 mean dbh increment, which refers to these five trees each, are outlined in Figure 8. Here, owing to the specific selection of the sample trees and the 8 radial lines per stem disk the coefficients of determination (R^2) drastically exceed those obtained from the core analyses:

- Rösa 3: $R^2_{R3} = 0.82$
- Taura 4a: $R^2_{T4a} = 0.66$
- Neuglobsow 1: $R^2_{N1} = 0.77$

Despite the high coefficients of determination, six cases may be stated, referring to the period from 1993 to 1995, in which the average annual radial increment of some tree exceeds that of a higher-class tree on the same plot.

Thus, the increment data of the trees that were selected in the individual classes, are not very typical of those classes.

This fact is yet underlined by the numerous crossings of the individual tree-ring curves in Figure 8, which theoretically cannot exist in the actual mean-class curves.

Statistically significant differences between the plots cannot be deduced from the increment lines. Nevertheless, here as in the core analyses it becomes once more obvious, that the mean relative increment in Neuglobsow was leading

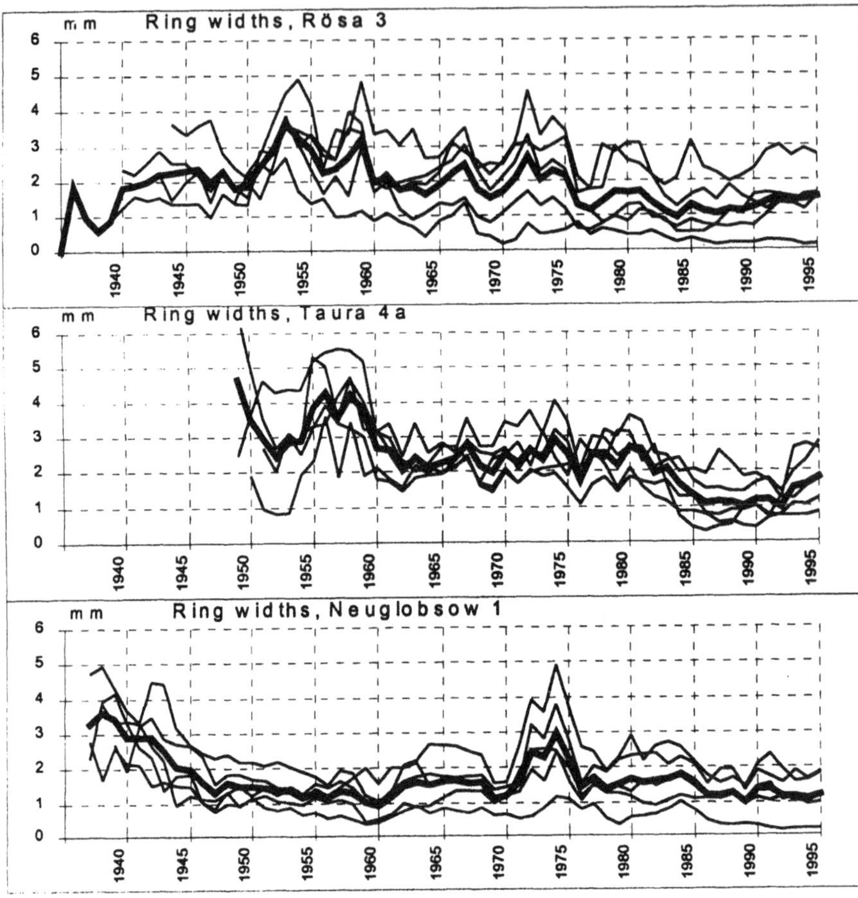

Figure 7. Ring-width series according to the stem disks of the 5 sample trees of each investigation site, with the mean-value curves

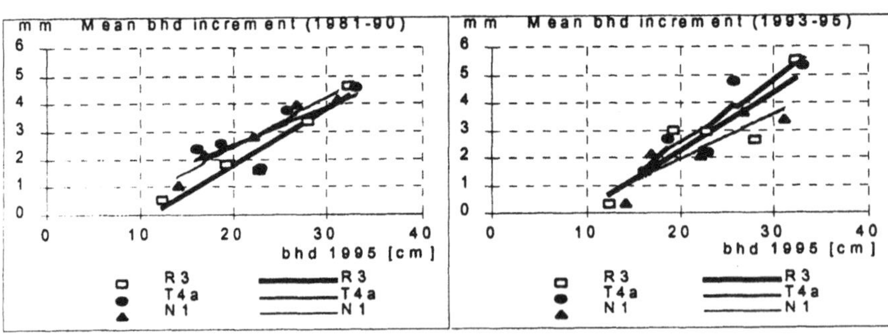

Figure 8. Increment lines for the time periods 1981–90 and 1993–95 for the three trial plots Rösa 3, Taura 4a, and Neuglobsow 1 based upon stem disk analyses of the 5 sample trees each of 1995

over the past decade up to 1990, however, it was below that of Rösa during the past three years. The fact that the increment in Taura exceeded that of Rösa, may be attributed to the newer, younger trial plot of Taura 4a.

The mean tree-ring series obtained from stem disk analyses (Figure 9) are linked with a very high risk regarding reliability, and, concerning characterization for stand increment, they are inferior to the core tree-ring series. However, they approximately correspond to the increment series obtained from the core measurements, thus corroborating the results of the core analyses concerning the increment trends over the past few years: Rise in increment in Rösa and Taura, no rise in Neuglobsow. As for Taura it should be noted that the stem disks were obtained from the younger plot Taura 4a in contrast to the cores (Taura 2 and 3).

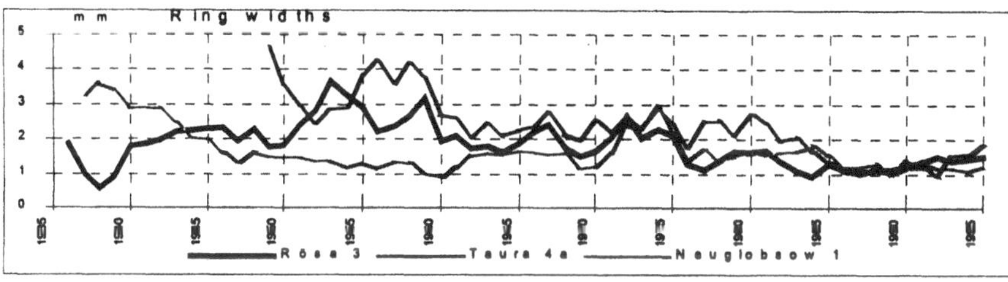

Figure 9. Mean ring-width curves developed from stem disk measurements of the five sample trees each, averaged by plots

3.3. *The course of the intra-annual radial increment*

Microdendrometer measurements

Families of curves describing single trees

The curve families denoting the increment courses of single trees show a uniform pattern: The individual curves run largely in 'parallel', i.e. the sequence of trees relative to increment level hardly changes over the growing season. Furthermore, differences between high- and low-increment trees in connection with the beginning and the end of increment could not be proven. These general characteristics of the single-tree curve families derived from the microdendrometer measurements are illustrated at the example of Rösa 3, 1995 in Figure 10.

In these mircrodendrometer measurement data, however, the actual radial increment is still superimposed by the reversible fluctuations caused by swelling and shrinkage.

The proportion of the reversible changes and its elimination are illustrated by the same example in Figure 11.

Figure 11 gives the mean-value curve of all measurement series, and that of the 10 lowest-increment trees, the estimated reversible fluctuations, and the corresponding actual increment data of the stand.

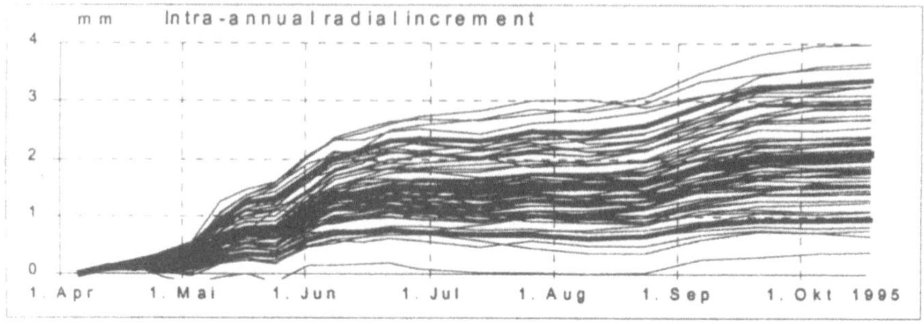

Figure 10. Single-tree curve families of the intra-annual radial increment obtained from the microdendrometer measurements, Rösa 3, 1995, including mean value curves for all trees, for the 10 trees with the highest and the 10 trees with the lowest increment

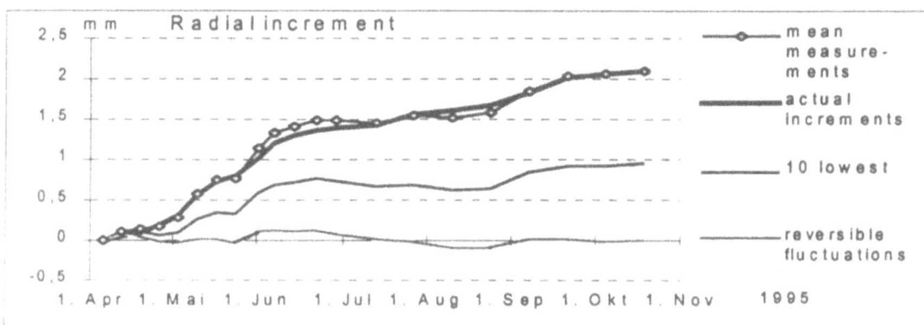

Figure 11. Microdendrometer measured data comprising reversible fluctuations and actual radial increment, Rösa 3, 1995

Figure 12 shows the intra-annual increment courses of Rösa 3, Taura 3 and Neuglobsow 1, in which the reversible fluctuations were eliminated. In this connection the systematic differences between the increment courses of the three plots on the one hand and the differences caused by weather between the four years on the other hand become obvious.

Increment–diameter relationships

The coefficients of determination for the increment-diameter relationships based upon the microdendrometer measurements and calculated for the years 1994 and 1995 exceed on average those derived from the core measurements. Here too, there are distinct differences between the plots, e.g. between Rösa 3 and 4, and between Taura 2 and 3, with each being located not far apart from one another, these differences are not supposed to be attributable to environmental impact. A relation with stand density is likewise not apparent.

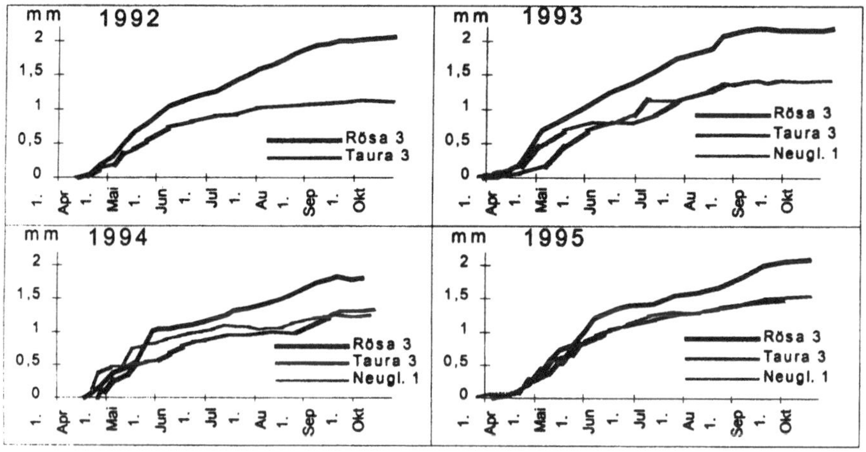

Figure 12. Intra-annual radial increment of Rösa 3, Taura 3, and Neuglobsow 1 from 1992 to 1995

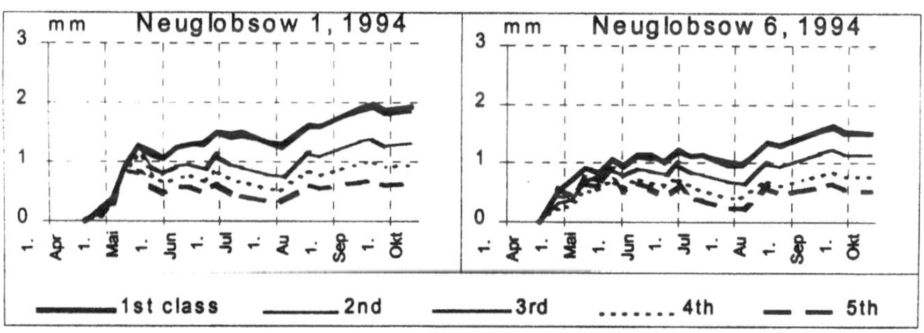

Figure 13. Intra-annual radial increment of 1994 for Rösa, Taura, and Neuglobsow, classified by diameter

Diameter classification

The diameter classification given in Hüttl and Bellmann (this volume) also applies here. The needed class-related sample size depends on the coefficient of determination of the increment-diameter relationship. Hence, the following holds true for 1994:

The class differences in Rösa 3 are not significant. Rösa 4 is distinguished by a higher coefficient of determination compared with Rösa 3. The five dbh classes have the following tree numbers: 4, 10, 25, 21, and 14. Figure 13, which represents the intra-annual radial increment courses as of 1994 from all the 6 trial stands, shows for Rösa 4 for the classes 2 to 5 correspondingly distinct differences. In contrast to this, the curves of classes 1 and 2 are found to run close to one another, from the beginning of June they even exhibit the 'wrong' sequence. There are significant differences between the values of final increment referring to classes 2 to 5, but not to the classes 1 and 3 nor 1 and 2.

The example of Rösa 1994 reveals, that in the increment precision measurements a classification by chest height diameter can only be allowed under certain aspects, depending upon the coefficient of determination and the sample size with regard to the proportions of the individual classes.

In the given sample size the use of increment lines to determine the increment-diameter relationship seems to be more appropriate.

Girth measurements

Proper experiences for the tree species spruce have been made in girth measurements using plastic tape measures (Spelsberg, 1992; Weihe, 1976). Until recently, own experiences with pine were missing. The time required for adjustment of the plastic tape measures on pine proved to be longer than in spruce. The outer bark should have been removed to a greater extent before the tapes were attached. Probably it was gradually compressed under the pressure of the tapes. Plausible values were only registered starting from the second year in Rösa (1993); in Taura, however, from the third year (1994), and in Neuglobsow even not yet in the third year (1995). The measured diameter increment of Rösa from 1992 to 1995 is represented cumulatively in Figure 14. In this context it should be taken into account that the initial values measured in spring always clearly exceeded those measured in the previous autumn. Almost no actual growth has taken place in the intermediate intervals.

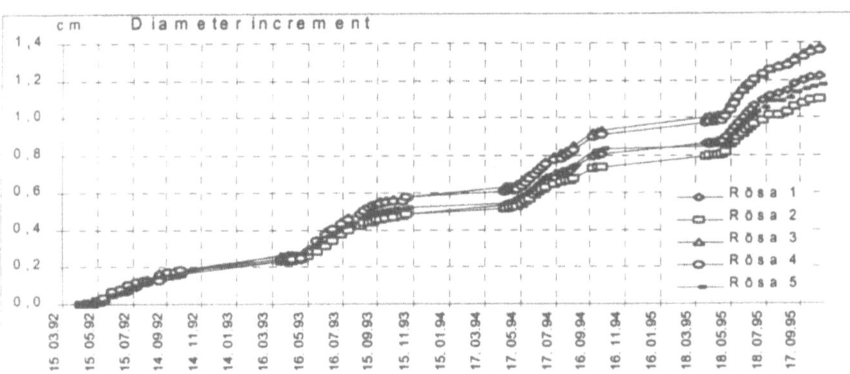

Figure 14. Diameter increment subsequent to the girth measurements for Rösa, cumulative from 1992 to 1995

However, swellings occurred due to the various degrees of water storage which are also included in the annual growth data as well. Apparently, the wood had a higher moisture content during the observational period in April compared to October.

Compared with the microdendrometer measurements, the girth measurements appeared to be less suitable for the relatively short observational period of 3 or 4 years, respectively, because of the inaccuracies described. Therefore, further investigations on the intra-annual radial increment are exclusively based on the microdendrometer measurements.

3.4. Comparison of measurement techniques

A detailed comparison of measurement results obtained from the various techniques of radial increment measurements is not given here.

When comparing core and stem disk measurements it should considered, that the variance of core measurements consists of the internal variance of single trees and the variance between trees. In stem disk measurements the internal variance is largely eliminated due to the computation of means from 8 radial lines. The standard error of the mean estimations obtained from the ring widths of the cores is reduced by the higher number of cores. In this case the estimations of the means derived from the core measurements are more precise than those obtained from the respective five stem disks.

With regard to Taura, it should be noticed that the five sample trees from which stem disks had been taken belonged to the 48-year-old plot Taura 4a, whereas the core and microdendrometer measurements belonged to the much older plots of Taura 2 (78-year-old) and Taura 3 (86-year-old).

In general, the girth tape measure data were too small.

The comparison of the annual increment values by the various measuring techniques relative to the stand means is given in Figure 15.

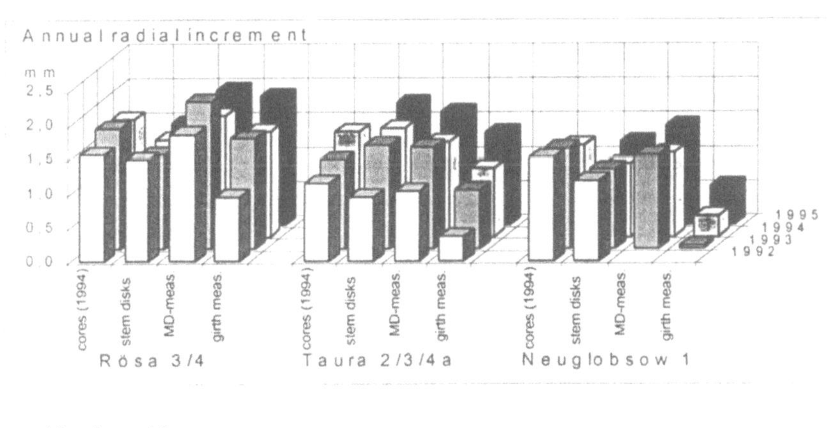

Figure 15. Annual increment means by the various measuring techniques

Microdendrometer measurements were conducted on 8 of the 15 sample trees that had been cut in August/September 1995. Hence, the respective annual values may be directly compared with the tree-ring widths at the microdendrometer measuring points of the stem disks. The microdendrometer data including the bark increment have been found to exceed the corresponding tree-ring widths of the stem disks by approximately 20%.

3.5. Climate–increment models

Linear regression models

By the extensive investigations, among others, the following results could be achieved (Neumann 1996a):

1. The results obtained from the regressions are very similar to those for the three investigation sites, hence common increment-climate relations could be formulated for the three investigation sites.

2. Concerning the data of the previous years, only the last two have a significant influence on the ring-width index I_t of the year t, i.e.

$$I_t = a_0 + a_1 I_{t-1} + a_2 I_{t-2} + e_t \tag{3a}$$

with the conditions

$$a_{t-1} > 0,\ a_{t-2} < 0 \text{ and } a_{t-2} < a_{t-1}$$

being valid throughout. Thus, from (3a) it follows

$$I_t = a_0 + c_1 I_{t-1} + c_2 (I_{t-1} - I_{t-2}) + e_t \text{ with } c_1, c_2 > 0, \tag{3b}$$

i.e. the current year index is both positively affected by the preceding increment level, and by the latest 'trend'.

3. Among the climate conditions of the period from January to April, i.e. prior to the growing season, only the temperature has a significant effect. The February and/or the March temperatures have a strong effect, always being associated with a positive sign, i.e. cold late winter periods may be expected to result in a reduction of growth.

   If the effect of the temperature in January is significant, it will always be linked with a negative coefficient, i.e. high temperatures will have a negative impact on growth – a phenomenon which could be attributed to a decrease of frost hardiness. In most cases, however, the significance level is not very high. Hence, there is always a number of years without any negative consequences of high temperatures in January, e.g. in a phase when there is no severe frost in late winter.

The impact of the temperature in April is significant only in rare cases, if so, however, with negative sign again. This might be explained by the analogy with the January temperature.

4. A distinctly positive impact on radial increment during the growing season could in general only be shown to exist in connection with the precipitation in June and/or in August. In May, if at all, only the temperatures are crucial, while in July, partly temperatures and precipitation.

   To sum up, the regression analyses regarding the increment-climate relations result in the following:

$$I_t = a_1 I_{t-1} + a_2 I_{t-2} + b_1 T1 + b_2 T2 + b_3 T3 + c_6 Ne6 + c_8 Ne8 + d + E_t \qquad (4)$$

with T1, T2, and T3 being the temperature means for January, February, and March; and Ne6 and Ne8 denoting transformations of the total precipitation of June and August, which correspond to exponential saturation functions. E_t is the random error.

Hence, because of the independence of the explaining variables

$$D^2(I_t) = D^2(a_1 I_{t-1} + a_2 I_{t-2}) + D^2(b_1 T1 + b_2 T2 + b_3 T3 + c_6 Ne6 + c_8 Ne8) + D^2(E_t) \qquad (5)$$

Figure 16a displays these three parts of variance of (5) to assess the explanatory value of the climate and of the previous two years, the height of bars corresponds to the respective part of variance, while the width corresponding to the T-test statistics for the significance of the dependence.

Figure 16b represents the actual climate part of variance broken down into the individual monthly variables according to

$$D^2(b_1 T1 + b_2 T2 + b_3 T3 + c_6 Ne6 + c_8 Ne8)$$
$$D^2(b_1 T1) + D^2(b_2 T2) + D^2(b_3 T3) + D^2(c_6 Ne6) + D^2(c_8 Ne8) \qquad (6)$$

The ring-width indices, for which the climate-induced effect has been eliminated based on the results obtained from the regressions, are given in a slightly smoothed representation by moving averages over three years in Figure 17.

Thus, the results from the core analyses according to Ende and Hüttl (this volume) are confirmed and stated more precisely:

- In Rösa, where the relative radial increment reached its lowest point in the late 1980s, a rise in radial increment has been identified since 1987, from which the age trend was eliminated and which is not attributable to climate impacts. Here, the mean relative radial increment is presently higher than ever before.

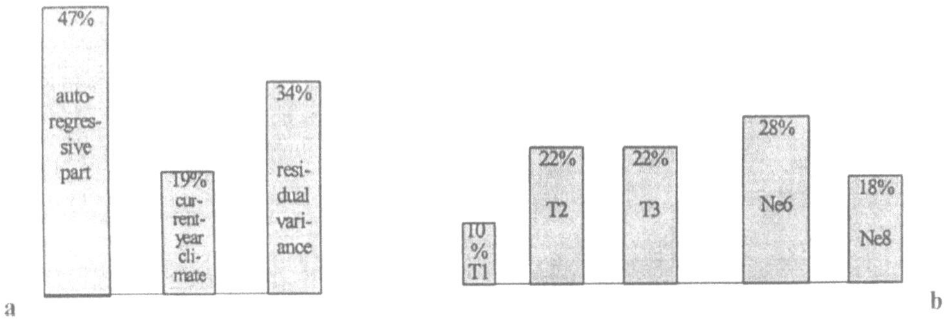

Figure. 16. (a) Parts of variance according to (5). (b) Parts of variance according (6)

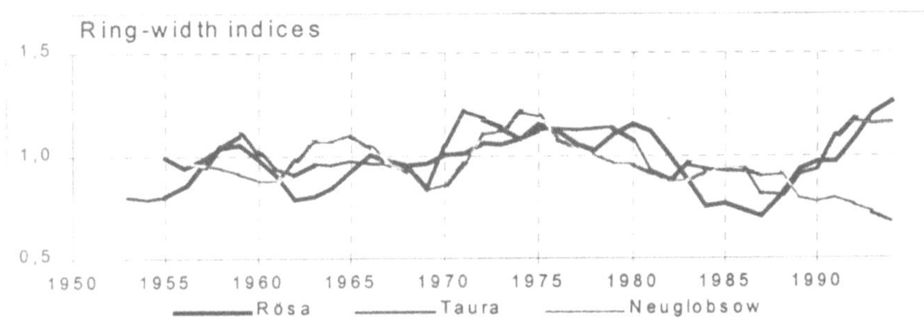

Figure 17. Ring-width index series from which the climate impacts were eliminated

- In Taura there has also been a rise in increment independent of the climate, which began at a relatively low level in the late 1980s and which, however, was not so strong as in Rösa.

- This trend could not be evidenced at all in Neuglobsow; on the contrary, here the relative radial increment was decreasing following its maximum in the mid 1970s until 1994 – the year in which it achieved its lowest point so far. It might be assumed that this decline was caused by the age, i.e. the age trend had not been taken into account adequately. However, the plot of Neuglobsow 6 being 39 years older reveals a similar incremental decline over the same period.

A complete consideration of all climate impacts on radial growth is, of course, impossible, the more so, if based upon monthly data. Moreover, exact data of thinning, infestation with pests, fertilization, fructification etc. are missing.

Model for the intra-annual weather-dependent radial increment
By the model concept according to Figure 2, (Hüttl and Bellmann, this volume) proper adaptations of estimated values may be obtained by regression analysis. In this context, the function typ

$$p(t) = a\,t^{b}\,e^{-c\,t} \qquad (7)$$

proved to be suitable for intra-annual trend function (p).

Another intermediate step in this context was the estimation of soil suction tension based on temperature and precipitation. Figure 18 gives the result of the regressions made for Rösa as of 1994, using the logarithmic-linear model

$$\ln S50_t = a\,N14_t + b\,T7_t + c + E_t\,, \qquad (8)$$

S50 is the suction tension at a depth of 50 cm (Weisdorfer *et al.*, this volume). The temperature and precipitation diurnal values (Weisdorfer *et al.*, this volume) have been smoothed of 7 (T7) or 14 (N14) days, respectively.

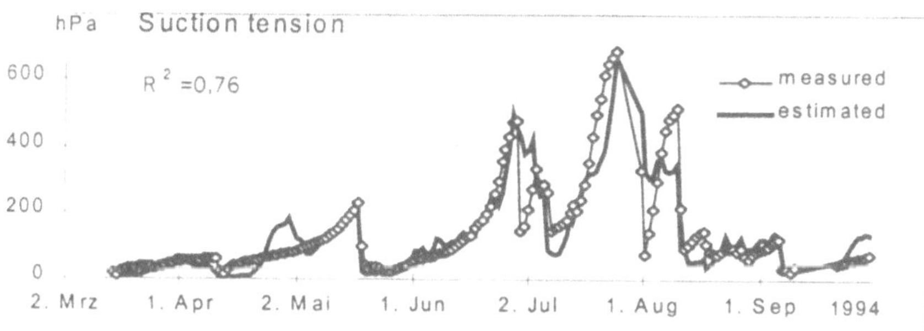

Figure 18. Suction tension in soil as a function of temperature and precipitation, Rösa 1994

In Figure 19 the interpolated daily increment values of Rösa 3, 1994, are compared to the model values. This example shows, that by this concept the incremental course can be estimated, in a flexible response to weather changes.

Finally, Figure 20 shows the measured and the estimated incremental courses as accumulative for the three investigation sites in the years 1994 and 1995, with the year 1994 serving the calibration, i.e. the adjustment of the annual total value.

Unfortunately, a sufficient validation of the results is not possible because of the short observational period. Complete data sets are available only for the years 1994 and 1995, the great variety of possible weather conditions has not been covered by far. For that reason the specific results of modelling cannot be introduced here. Therefore, the concept can only be regarded as a possibility to

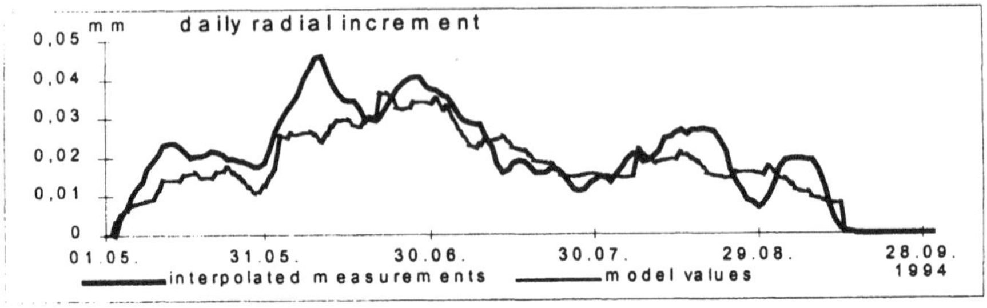

Figure 19. Intra-annual radial increment as daily values, Rösa 3, 1994, comparison of interpolated measured and model values

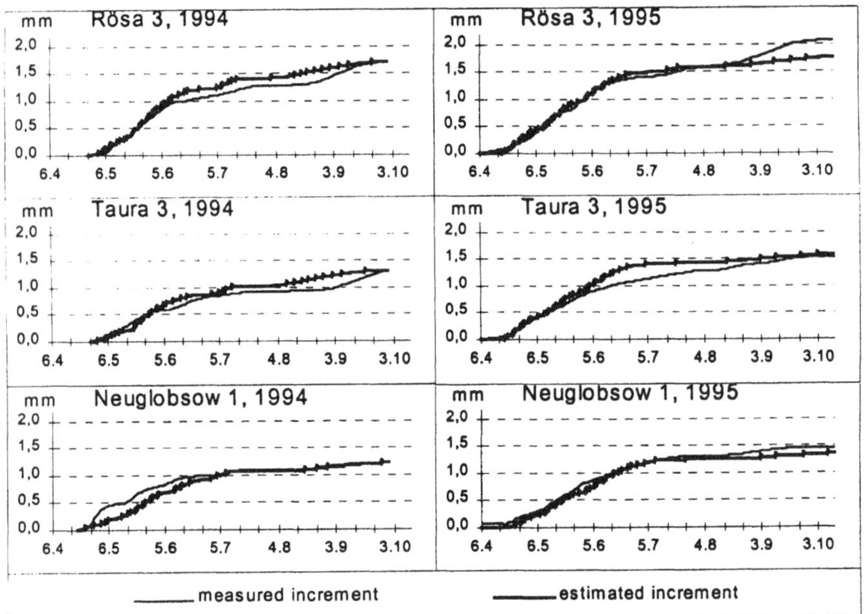

Figure 20. Intra-annual radial increment of Rösa, Taura, and Neuglobsow in 1994 and 1995, comparison of measured and model values

quantitatively determine the increment-weather relationships by optimally utilizing the described initial data. An assessment of the results cannot be given on the basis of the data available.

4. Discussion

Two methods were used to check the radial increment for a possible recovery of the pine stands after reduced air pollution: (1) a comparison between three differently polluted sites and (2) elimination of age trend and weather influence for the direct assessment of the tree-ring series.

The results obtained from the radial increment analysis are highly consistent. They substantiate the assumption that the increased increment is an effect of recovery. However, this conclusion is not yet binding on the basis of incremental investigations in consideration of climate and weather influence alone. The complex interactions of many influential factors has to be taken into account when interpreting the incremental changes. The question e.g. of the increment increase beginning before the clear reduction of air pollution remains open.

Based upon monthly values, the regression analyses of the weather impact show uniform results in the compared sites. In addition to the weather-neutral increment series, the investigations of the impact of the weather yield interesting independent statements. For instance, it can be shown to what proportion the increment is determined by the levels of the previous years, i.e. based upon after-effects, and to what proportion it is directly influenced by the winter temperatures and the summer precipitation. However, the direct influence of the weather has only an explanation value of approximately 20%. The actual influence being certainly higher. The monthly values are, however, too coarse to be registered e.g. extreme frost events or periods of drought. Thus, the 'weather-neutral' increment series are not free from errors as far as the unexplained weather-conditioned proportion of the changes in increment is concerned.

On stand average, the results for the incremental trend and the intra-annual growing course are reliable. Regarding the various methods of radial increment determination, the possibility to differentiate the results by diameter classes has to be critically examined.

The empirical-statistical model for the intra-annual course of increment is adapted to the data base and gives good preliminary results. However, the observation period was too short to cover the multitude of possible weather conditions and so no generalization or final assessment can be reached. The intra-annual increment investigations provide important information about the course of increment during the growing season, but do not, however, render any direct contribution to the question of trend changes.

Acknowledgements

We are glad to have had the opportunity to make a contribution to the environmental research in the scope of the program 'Environment research and environmental technology' supported by the Bundesministerium für

Bildung, Wissenschaft, Forschung und Technologie (BMBF) within the themes of SANA project.

Therefore, our thanks are due to the representatives responsible in SANA project. Likewise, we are indebted to the forest district supervisors for their valuable assistance. and all those who have made available to us useful data from their data bases.

References

Abetz P. 1966. Sind kurzfristige Zuwachsreaktionen an jungen Fichten nach einer Durchforstung meßbar? Allg Forst- und Jagdzeitung. Frankfurt a.M. 137, 41–49.

Cook ER, Kairukstis LA. 1990. Methods of Dendrology. Kluwer Academic Publishers, Dordredcht, The Netherlands. 394 p.

Dixon RK *et al.* 1990. Process Modeling of Forest Growth Responses to Environmental Stress. Timber Press. Portland, Oregon.

Fiedler F. 1990. Fichtenzuwachs 1989: Ergebnisse von Zuwachsfeinmessungen an Fichten im Jahre 1989 in ausgewählten Probeflächen des Erzgebirges. Der Wald Berlin. 40, 11, 335–336.

Fritts HC. 1976. Tree Rings and Climate. Academic Press, London.

Kahle H-P. 1993. Witterungssignale im Zuwachs von Fichten (Picea abies [L.] Karst.) aus montanen Lagen des Schwarzwaldes. In Tagungsberichte der AG Ökologie, Heft 4.

Kublin E, Gantert C. 1993. Trendanalysen von Zuwachsreihen mit Hilfe von Spline-Regression, Zustandsraum- und nichtparametrischen Regressionsmodellen. In: Tagungsberichte der Arbeitsgruppe Ökologie, Heft 4.

Neumann U. 1996a. Methodische Untersuchungen zur Radialzuwachsanalyse und Auswertungsergebnisse für drei Kiefernbestände aus dem ostdeutschen Tiefland. In: Tagungsberichte der Arbeitsgruppe Ökologie der Internationalen Biometrischen Gesellschaft, Deutsche Region, 8. Herbstkolloquium in Göttingen, Heft 7.

Neumann U. 1996b. Modell für den innerjährlichen Radialzuwachs. In: Tagungsberichte der Sektion Forstl. Biometrie und Informatik, 8. Tagung in Grillenburg/Sachsen 1995. Ed. G Hempel.

Spelsberg G. 1992. Erfahrungsbericht über fünf Jahre Zuwachsmessung mit dem Dauer-umfangsmeßband. Forstarchiv 63, 112–116.

Riemer Th. 1994. Über die Varianz von Jahrringbreiten. In: Bericht des Forschungszentrums Waldökosysteme. Reihe A, Bd. 121.

Wätzig H. 1996. Das innerjährliche Radialwachstum von Kiefern in Abhängigkeit von Temperatur und Niederschlag. Wiss. Zeitschrift der TU Dresden, 45 Heft 2.

Weihe J. 1976. Erfahrungen mit einem Kuststoffmeßband. In: Jahrest. Deut. V. Forstl. Forschungsanstalt. Sektion Ertragskunde, Paderborn. pp 70–71.

Wenk G, Antanaitis V, Smelko St. 1990. Waldertragslehre. Dt. Landwirschaftsverlag, Berlin.

Wenk G, Wätzig H, Neumann U. 1995. Untersuchungen über den Einfluß wesentlicher Umweltfaktoren auf den Radialzuwachs von Kiefern in den SANA-Testflächen. In: Atmosphärensanierung und Waldökosysteme. Ed. RF Hüttel, K Bellmann, W Seiler. pp 129–143, Blottner, Taunusstein.

Wenk G, Vogel M, Antálek B. 1994. Quantifizierung der Jahrringbreitenschwankungen durch heuristische Modelle und multiple Regressionen. In: Tagungsberichte der Arbeitsgruppe Biometrie.

Wenk G *et al.* 1988. Lehrbrief Holzmeßkunde 2. Tharandt, TU Dresden.

Zollfrank M. 1990. Ergebnisse von Zuwachsfeinmessungen in immissionsgeschädigten Fichtenbeständen im StFB Eibenstock mit Mikrodendrometern. Tharandt, TU Dresden, Sektion Forstwirtschaft. Diplomarbeit.

14
Modelling of carbon-, nitrogen-, and water balances in Scots pine stands

R. GROTE, F. SUCKOW and K. BELLMANN

1. Introduction

The aim of the project is to describe the development of pine ecosystems as well as the carbon, nitrogen and water balances at the selected sites as dependent on climate, air pollution and deposition. Because these driving forces develop independently from each other and not homogeneously within the area, measurement results from a limited number of sites cannot easily be transferred to other sites. Thus, a simple statistical approach or the simulation of growth from integrated environmental influences was not adequate to fulfil the tasks of the project. Consequently, a process-based approach has been developed to consider every impact by means of the changes in the basic physiological process involved, and which accounts explicitly for linkages and feedback mechanisms on different time scales.

The new model FORSANA addresses particularly the following tasks:

1. Long-term simulations (several decades) of pine forest development under changing environmental conditions, including climate, air pollution, deposition and management.

2. Full representation of carbon, nitrogen and water balance, considering the linkages and feedback reactions between those cycles.

3. Initialisation of all necessary state variables from forest inventory data and coarse soil mappings in order to simulate even aged stands in a region of approximately 400 km^2 (see Erhard and Flechsig, this volume).

Concerning long-term applicability, process based models have previously been used in case studies to investigate impacts of environmental changes (e.g. Cropper and Gholz, 1993; Eckersten, 1994; Heij *et al.*, 1991). However, air pollution effects were included in only a few of these investigations (e.g. Chen, 1993; Meng and Arp, 1994; Mohren *et al.*, 1992), and the results are difficult to generalise because parameterization and validation were mostly restricted to a single site. Furthermore, the validation period was only short in most cases. Exceptions can only be found in investigations which focus on specific

R.F. Hüttl and K. Bellmann, Changes of Atmospheric Chemistry and Effects on Forest Ecosystems, 251–281.

questions and which are based on the assumption that no change in boundary conditions occurs (e.g. Sheriff *et al.*, 1996). In contrast, the long-term investigation of carbon, nitrogen and water fluxes under changing air pollution and deposition as well as climate are needed to account also for developing stand properties (e.g. leaf area), which are known to change under these conditions (Gluch, 1988). Thus, short-term (daily) and medium-term (annual) adjustment reactions of growth and mortality have been included in the new modelling approach (Grote, 1998).

Furthermore, a detailed simulation of soil processes is also required to account for long-term changes in nutrient pools and potential water uptake affecting plant growth via their impact on nutrition and transpiration. However, in most of the current process-based stand models the interrelation between soil and vegetation processes are not adequately represented. For example, the amount of soil water is generally assumed to restrict transpiration, but the role of fine root density, root depth and distribution in different soil layers is widely neglected (Sheriff *et al.*, 1996; Waring and Running, 1976; Whitehead and Kelliher, 1991), but see also Schlichter (1983). Also, nitrogen availability is mostly accounted for as a soil specific parameter (e.g. Aber *et al.*, 1996; Hoffmann, 1995; Running and Gower, 1991), without considering the linkages to mineralisation and/or water cycle. To account for the feedback reactions between (1) fine root distribution and water uptake, (2) water uptake and nitrogen uptake as well as (3) nitrogen supply and fine root mass, root dynamics were modelled explicitly as dependent on soil conditions in each layer as well as on the actual supply rate of water and nitrogen.

The second limitation of process-based models is their restriction to local case studies, because of their high requirements on proper initialisation. Because FORSANA has been developed to be used in a region of approximately 400 km^2 with more than 5000 different pine stands, the model initialisation was restricted to the use of the available forest inventory data and simple soil characteristics.

In this paper, simulations focus on balances of matter at the three investigation sites. Verification of some state variables and fluxes has been accomplished by direct comparison with measurements. The model has then been applied to the stands in order to simulate the retrospective development of the last 27 years, taking into account the specific climatic and air pollution conditions as well as nitrogen deposition, N-fertilisation and thinning.

2. Model description

2.1. General information and research approach

Firstly, a system analysis has been applied, from which the structure of the mechanistic model and the necessary processes have been derived. The same analysis served for the development of the experimental design at the three test sites, which was then executed by other groups within the research program. Thus, the onset of the measurements according to the modelling requirements assured the availability of necessary data for model parameterization and verification. Figure 1 shows an overview of the co-operation structure within the impact research program.

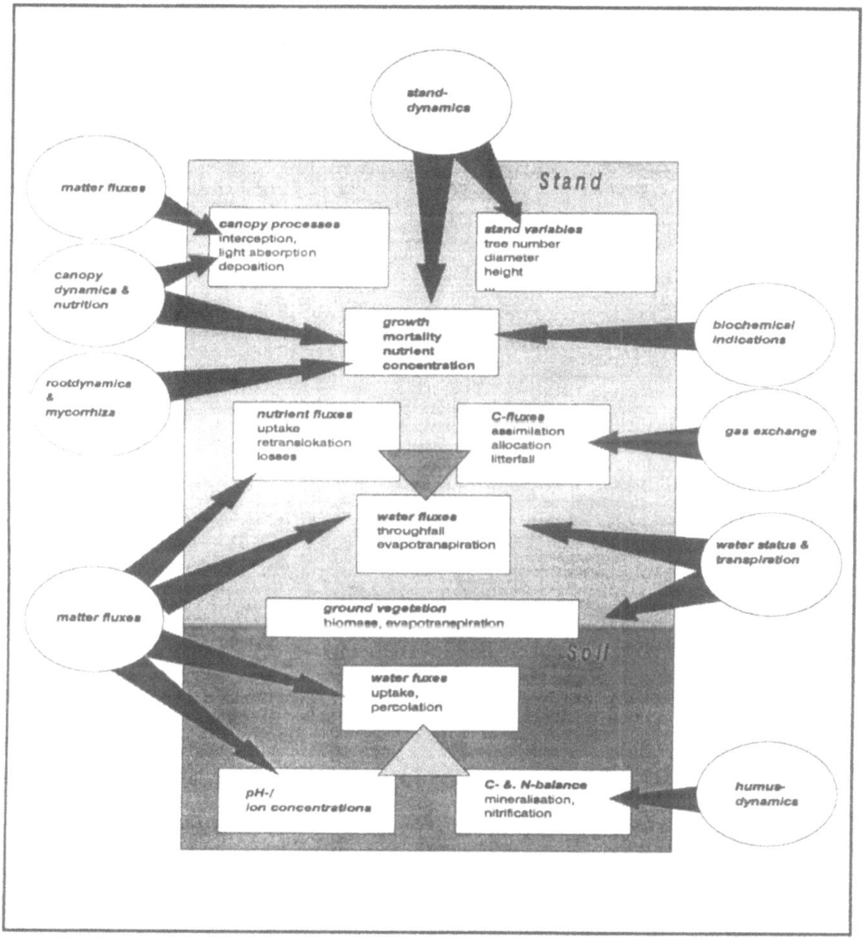

Figure 1. Overview about the research network, showing field investigation groups (in circles) and model features, which were supplied with information (in rectangles)

Consequently, FORSANA uses a wide range of ecological relations, reaching from basic physiological processes with small time steps to stand disturbances, which occur only once in several years. The coupling of these processes in space and time is done by a nested system of higher resolution models embedded in lower resolution models (Kittel and Coughenor, 1988; Luan *et al.*, 1996). For calculations of light absorption, photosynthesis, respiration and SO_2 effects on assimilation, the respective FORGRO submodel is used (Mohren, 1987, Grote, and Suckow, 1998). The soil module is derived from an agricultural model (Kartschall *et al.*, 1990) and adapted to forest stands. A simplified overview of the whole model is given in Figure 2, showing the main interactions within the submodels operating on a daily time

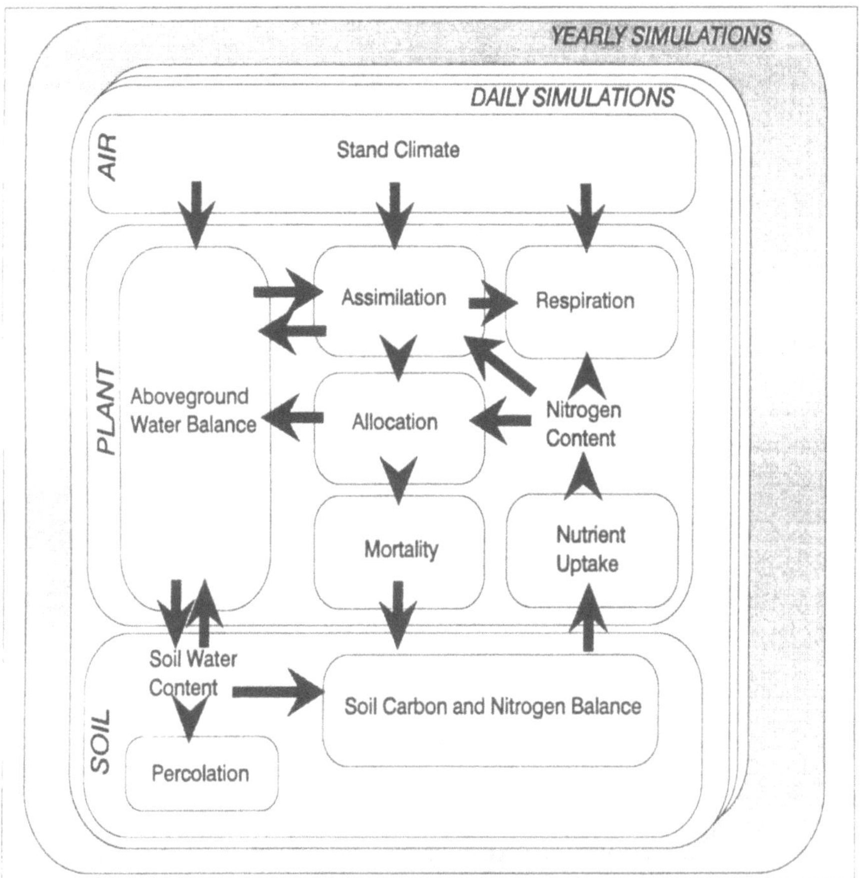

Figure 2. Model overview of interactions between major processes

step. The influence of biomass increment on the stand development and the reciprocal influence of stand properties on allocation and thus on environmental conditions in the canopy and at the ground provide the linkage between the daily and annual time steps in the model (Figure 3).

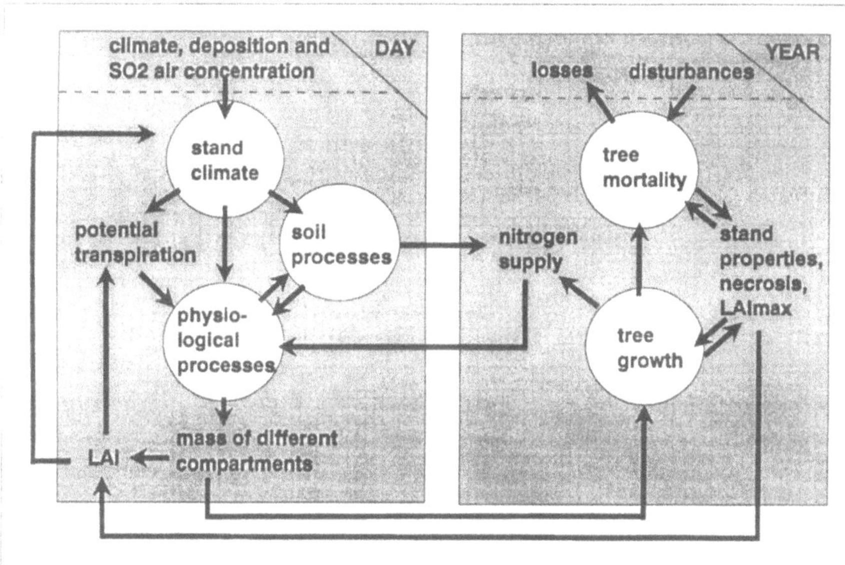

Figure 3. Scheme of the interaction between daily and yearly submodules

The current version of FORSANA can be applied only to even aged coniferous forests, because the forest stand is assumed to be horizontally homogeneous and all information is scaled up from average stems to the stand level. The canopy and the rooted soil are divided into horizontal layers with different amounts of foliage and fine roots, respectively. The thickness of the soil layers is derived from the soil profile at the site. Canopy layers are of equal thickness, whose absolute value depends on canopy thickness and initial number of layers. The above-ground part of the herbaceous vegetation is modelled as a separate layer of foliage beneath the canopy and the underground part shares the soil layers down to an initialised depth with the tree roots. Besides leaves and fine roots, the model assumes that the stand consists of woody compartments (sapwood, heartwood, branches and coarse roots) and a pool of reserve carbohydrates, which provides assimilates for flushing and from which the release of reproductive tissue is supplied.

Because only coarse site description data could be obtained at a regional level, the model was constructed to require only broadly available forest inventory data (average diameter, stand height, standing stemwood volume) as initial stand variables. Concerning soil data, the required information (number of layers, hydraulic parameters, humus content, rooting depth) can be obtained from soil mappings or by taking standard soil properties for specific types of soil profiles. However, if more detailed information about tree and soil properties are available (e.g. crown dimension, root distribution, mineral content) they could be used to improve parameters and thus simulation for a specific site.

Output data of state variables (e.g. biomass of all stand compartments and ground vegetation, C and N content of each compartment and the soil, water amount in the soil-, sapwood- and interception storage) and fluxes (e.g. photosynthesis of canopy and ground vegetation, respiration, growth of all compartments, transpiration and evaporation, soil water fluxes) are provided with a daily resolution. In addition, stand properties (e.g. stem number, height, diameter, canopy closure, standing stemwood volume and export of stemwood) as well as the cumulative values of the water, carbon and nitrogen balances are calculated on an annual basis.

In the following, only an overview of the physical and physiological processes is given, but more detailed information can be obtained from Grote and Suckow (1998) and Grote (1998). All model statements are written in FORTRAN notation suitable for PC as well as UNIX systems. For description of model variables see Table 1; indices are given in Table 2.

2.2. *Physiological processes*

The calculation of the stand water balance (equation 2.2.1) is not only needed for information about ground water recharge (PE(z,t) = percolation from last rooted soil layer), but also because insufficient water supply has a major influence on assimilation, allocation and mortality of the vegetation.

$$\begin{aligned} PRC(t) = {} & INT_{T,GV}(t) + TRA_{T,GV}(t) + EVS(t) - \\ & \Delta WC_{INT_{T,GV} SAP,SOI}(t) - PE(z,t) \end{aligned} \qquad (2.2.1)$$

Intercepted water in the canopy of the trees (INT_T) and the ground vegetation cover (INT_{GV}), water storage at the leaves (WC_{INT}), in the sapwood (WC_{SAP}) and the soil in each layer within rooting depth (WC_{SOI}) are considered as state variables. Interception loss is determined according to the potential interception storage and evaporation from the canopy (Liu, 1997). Transpiration is obtained from potential evaporation demand at the dry surface, calculated from the Penman-Monteith equation (Hancock *et al.*, 1983; Monteith, 1965), and the potential water supply. The latter includes the potential soil water uptake (UPT_{pot}, see equation 2.3.8) as well as a pool of sapwood water storage. Soil evaporation (EVS) is estimated from the saturation deficit of the air beneath the ground cover layer and the soil resistance according to water content of the uppermost soil layer.

The amount of nitrogen, which is incorporated in the vegetation, is formed by the uptake from the soil ($UPTN_{SOI}$) and the uptake through the canopy ($UPTN_{NDL}$). Losses occur only with litterfall in every compartment (LIT), taking into account that a certain share of nitrogen is retranslocated in other tissues for further plant use.

Table 1. Variable names of vegetation and soil processes and description

Abreviation	Description	Type	Unit
p	Number pi	Constant	–
k	Conductivity	Initial	
A	Gross assimilation	Variable	kgCH_2O ha^{-1}
C	Carbon	Variable	kgDW ha^{-1}
CN	Carbon:nitrogen ratio	Variable	–
CRAI	Crown area index	Variable	–
D	Carbon demand	Variable	–
DBH	Diameter at breast height	Variable	m
DENS	Wood density	Parameter	kg dm^{-3}
EVS	Soil evaporation	Variable	mm
FC	Field capacity	Initial (derived from pore volume data)	mm
FGR	Growth efficiency factor	Parameter	–
FRET	Fraction of retranslocated nitrogen	Variable	–
FTOT	Fraction of dying trees	Variable	–
G	growth	Variable	kgDW ha^{-1}
GDBH	Diameter increment	Variable	m
GTOP	Height growht intervall	Variable	m
HD	Height:diameter ratio	Variable	–
HFC	Fraction of available water	Variable	–
INT	Interception	Variable	mm
LIT	Litter	Variable	kgDW ha^{-1}
M	Compartment mortality	Variable	kgDW ha^{-1}
N	Total nitrogen	Variable	kgN ha^{-1}
NH4	NH_4 nitrogen	Variable	gN m^{-2}
NO3	NO_3 nitrogen	Variable	gN m^{-2}
NTRE	Number of trees	Variable	–
PE	Percolation	Variable	mm
POOL	Total available carbon for growth	Variable	kgDW ha^{-1}
PRC	Precipitation	Driving variable	mm
QCD	Crown diameter:stem diameter ratio	Parameter	–
R	Maintanance respiration	Variable	kgCH_2O ha^{-1}
SLO	Slope of the tree mortality function	Parameter	–
SWC	Water content at soil water saturation	Initial (derived from pore volume data)	mm
TRA	Transpiration	Variable	mm
TS	Temperature	Variable	°C
UPT	Water uptake	Variable	mm
UPTN	Nitrogen uptake	Variable	kgCH_2O ha^{-1}
VIT	Vitality index	Variable	–
W	Compartment weight (biomass)	Variable	kgDW ha^{-1}
WC	Actual water content	Variable	mm
WP	Water content at wilting point	initial (derived from pore volume data)	mm

Table 2. Indices of vegetation and soil variables/parameters and description

Abreviation	Description
110	110 percent of the original value
90	90 percent of the original value
AOS	Active organic matter (humus)
bas	Basic value
BRA	Branch wood
CRT	Coarse roots
EXS	Exsudation
FRT	Fine root
GV	Ground vegetation
H	Heat
max	Maximum
min	Minimum
NDL	Needles
nit	Nitrogen
OPS	Primary organic substance (litter)
opt	Optimum
pot	Potential
REN	Regenerative tissue
SAP	Sapwood
SOI	Soil
STE	Stem
syn	Synthesis
t	Time
T	Tree
W	Water
z	Depth of soil layer

$$\Delta N_{T,GV}(t) = UPTN_{SOI_{T,GV}}(t) + UPTN_{NDL}(t) - \sum_i \left\{ LIT_i(t) \cdot CN_i(t) \cdot \left(1 - FRET_{T,GV}(t)\right)_{T,GV} \right\} \quad (2.2.2)$$

To account for the importance of canopy uptake (Grosch, 1990; Helmisaari and Mälkönen, 1989; Lovett and Lindberg, 1984; Skeffington and Wilson, 1988), in particular in polluted areas (McLeod *et al.*, 1990), canopy uptake is estimated as a function of NH_3 and NO_X deposition, leaf area and stomatal restriction due to water stress. The formulation was constructed empirically to meet the field measurements at the study sites, which also showed that the atmospheric uptake contributes considerably to total nitrogen input into the ecosystem, especially in stands with sub-optimal nutrition (Weisdorfer *et al.*, this volume). The distribution of nitrogen is done regardless of the nitrogen source according to the size and the optimum nitrogen concentration in each compartment.

The daily pool of carbohydrates available for growth (G) is determined from daily assimilation gains (A), losses from maintenance respiration (R), and the share of carbon, which is lost due to the conversion into dry matter (FGR):

$$G(t) = \left(A(t) - R(t)\right) \cdot FGR \qquad (2.2.3)$$

Net assimilation dependency on light is represented with a negative exponential equation (Goudriaan, 1982). The temperature influence is described with an optimum function (Bossel, 1994), and the response to foliage nitrogen concentration is linear over a wide range (Aber *et al.*, 1996). The effect of water shortage was considered by reducing the carbon gain of the whole canopy, determined without stomatal limitations, by a drought stress factor, which is calculated from the relationship between transpiration demand and potential water uptake. Maintenance respiration in each tree compartment depends on compartment mass, temperature and nutrient content (Penning de Vries *et al.*, 1989).

The distribution of carbohydrates to the reserve pool, fine root and sapwood compartments are simulated by assuming an organ-specific growth demand (D), which is calculated from optimum and actual biomass values (W_{opt}, W).

$$D_i = \max\left(0, \frac{W_{opt_i} - W_i}{W_{opt_i}}\right) \qquad (2.2.4)$$

Optimum weights are determined according to their relation to foliage mass. This relationship is allowed to shift either annually with stem dimensions (woody biomass) or daily according to the ratio between the supply and the demand of water and nitrogen (fine roots). The seasonal dynamic of allocation results from the temperature induced flushing of foliage and the different rates of mortality in each compartment.

The total amount of mortality in each compartment (M) is calculated from a base biomass (W_{bas}) and a compartment-specific mortality function (F_{mort}). This function considers base longevity parameters, which are modified under environmental stress.

$$M_i(t) = W_{bas_i} \cdot F_{mort_i}(t) \qquad (2.2.5)$$

With respect to foliage mortality, the total amount is calculated in advance on an annual basis, dependent on stand density and SO_2 concentration. The daily rate of needle fall, however, is determined from the ratio between gross photosynthesis and total maintenance respiration. If this ratio decreases, the mortality is increased. Thus, an increasing litter fall during the end of the vegetation period is simulated without any direct dependency on the day of the year. W_{bas} for fine roots is the actual fine root biomass and the mortality is enhanced with decreasing soil water content relative to maximum soil water content.

Total sapwood mortality is determined annually. Because very little is known about the underlying mechanisms, it is based on the following assumptions: Firstly, the sapwood that dies is not required for foliage supply as calculated at one specific time in a year (with the additional assumption that xylem conductivity is a constant) and secondly, the distribution of sapwood mortality within the year follows the same pattern as that of foliage. Dead stem-sapwood is converted into heartwood, whereas the coarse roots and branches are shed into the litter pool of the soil.

2.3. Soil processes

Physical and chemical soil processes are simulated for any desired soil layer with the humus layer being viewed as the first layer. Water content, soil temperature, pH value and element content of each soil layer are estimated as functions of the basic soil parameters, air temperature, precipitation beneath the ground-cover vegetation and deposition (see Figure 4).

The essential interfaces between soil and vegetation are litterfall, which supplies the soil with organic material, and uptake of water and nitrogen by the vegetation. Litter of each compartment (needles, wood, ground vegetation, fine roots) is divided into carbon and nitrogen content. Fine root litter is mineralised in the soil layer in which the respective living roots were located, whereas other litter forms a part of the humus layer.

Mineralisation is influenced by soil water content, soil temperature and pH value (Franko, 1990; Kartschall *et al.*, 1990). The dominant process is carbon mineralisation, which provides the energy for the whole turnover of the organic matter with a distinction being made here between primary organic matter (C_{OPS} = litter, dead roots) and active organic matter (C_{AOS} = humus) (Franko, 1990).

The basic turnover is described as a reaction of the first order with a reduction function F_x = f (water, temperature and pH):

$$\frac{\partial}{\partial t} C_{OPS}(z,t) = -k_{OPS} \cdot F_{OPS}(z,t) \cdot C_{OPS}(z,t) \tag{2.3.1}$$

The transformation of primary- to active organic matter is controlled by a synthesis coefficient (k^*_{syn}) specific to the substratum. The turnover of carbon in the active organic matter is made up of the synthesised portion and the carbon used in the mineralisation process:

$$\begin{aligned}\frac{\partial}{\partial t} C_{AOS}(z,t) = k^*_{syn} \cdot k_{OPS} \cdot F_{AOS}(z,t) \cdot \frac{\partial}{\partial t} C_{AOS}(z,t) \\ -k_{AOS} \cdot F_{AOS}(z,t) \cdot C_{AOS}(z,t)\end{aligned} \tag{2.3.2}$$

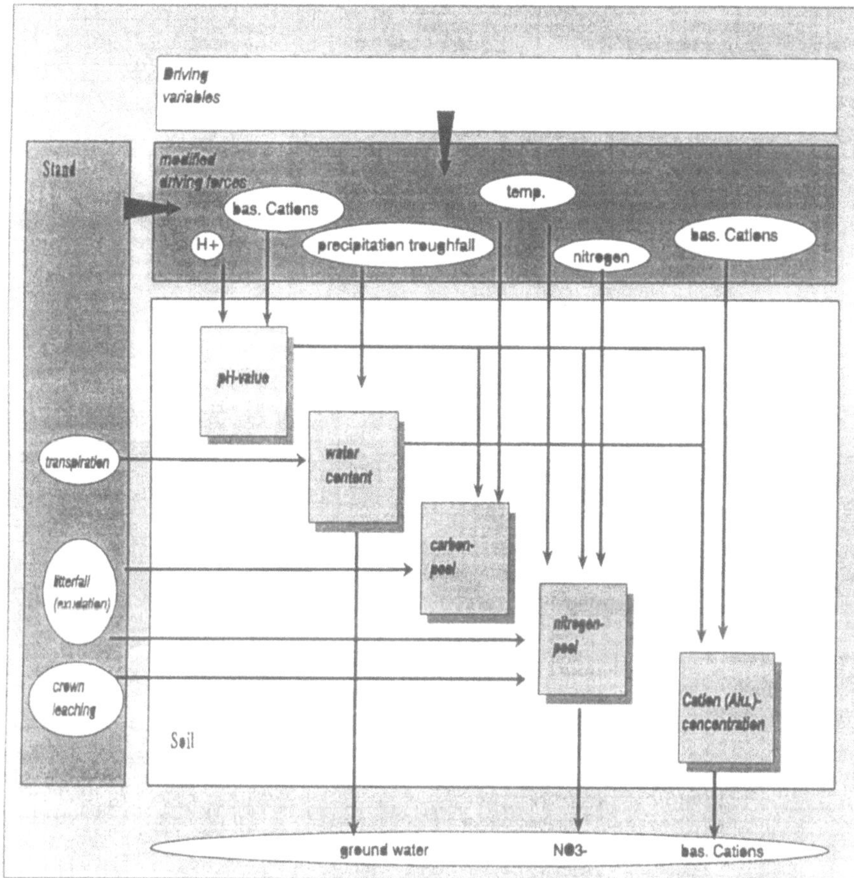

Figure 4. Essential processes of the soil model and their mechanistic links

The amount of nitrogen bounded to the active organic matter (N_{AOS}) and the mineralised proportion depends on the C/N ratio of both organic fractions (OPS, AOS) and on the carbon used in the synthesis of the active organic matter. The nitrogen turnover in the primary and active organic matter is similarly calculated to the turnover of carbon, whereas the C/N ratios of both organic fractions modify the synthesis coefficient k^*_{syn} into k_{syn}. In addition, the pools of ammonia (NH_4) and nitrate (NO_3) are considered:

$$\frac{\partial}{\partial t} N_{OPS}(z,t) = -k_{OPS} \cdot F_{OPS}(z,t) \cdot N_{OPS}(z,t) \qquad (2.3.3)$$

$$\frac{\partial}{\partial t} N_{AOS}(z,t) = k_{syn} \cdot k_{OPS} \cdot F_{OPS}(z,t) \cdot N_{OPS}(z,t) - k_{AOS} \cdot F_{AOS}(z,t) \cdot N_{AOS}(z,t) \tag{2.3.4}$$

$$\frac{\partial}{\partial t} NH4(z,t) = (1 - k_{syn}) \cdot k_{OPS} \cdot F_{OPS}(z,t) \cdot N_{OPS}(z,t) + k_{AOS} \cdot F_{AOS}(z,t) \cdot N_{AOS}(z,t) - k_{nit} \cdot F_{nit}(z,t) \cdot NH4(z,t) \tag{2.3.5}$$

$$\frac{\partial}{\partial t} NO3(z,t) = k_{nit} \cdot F_{nit}(z,t) \cdot NH4(z,t) \tag{2.3.6}$$

The set of differential equations (2.3.3) to (2.3.6), with the appropriate initial values, can be solved by means of the Laplace transformation. The parameters for these processes ($k_{OPS}, k_{AOS}, k_{nit}$) were gained in the course of investigations by Fischer *et al.* (in Anonymus, 1997).

Chemical substances are transported with the water moving in the soil and are taken up by plant roots. A transport of organic matter was not considered here, since this flux is small compared to the absolute quantities available in the soil (Weisdorfer, this volume). Also nitrogen loss in gaseous form is not considered because the process was neither investigated nor could sufficient information be found in the literature.

The quantity of percolation water (PE) is calculated according to Koitzsch (1977) with the initial condition $PE(0,t)$ = effective stand precipitation and by accounting for the loss of water through plant uptake ($UPT_{C,GV}$) and soil evaporation (EVS) with $EVS(z,t) = 0$ for $z > 40$ cm:

$$\frac{\partial}{\partial t}(WC(z,t) - FC(z)) = PE(z,t) - \left(UPT_{T,GV}(z,t) + EVS(z,t)\right) - \lambda_W \cdot (WC(z,t) - FC(z))^2 \tag{2.3.7}$$

The solution of (2.3.7) by means of a simple finite difference method provides the current water content for each layer. At temperatures below 0°C, no percolation occurs. In this case, precipitation is stored as a water equivalent in a 'pool' of snow, which is emptied as a function of temperature. The melt-water will then be added to the uppermost soil layer.

For each layer with the middle depth z^*, the amount of water available for plant uptake (UPT_{pot}) is calculated from the fine root mass of the tree and the ground vegetation ($W_{FRT_{T,GV}}$), the conductivity of the roots (which is the maximum limiting condition for equation 2.3.8) and the relationship between water content (WC), wilting point (WP), saturation water content (SWC) and field capacity (FC) (Chen, 1993). It is assumed that optimal conditions for water uptake exist when the water content does not vary by more than 10% from the field capacity and water uptake is not constrained by fine root conductivity:

$$\mathrm{UPT_{pot}(t)} = \frac{\mathrm{HFC(z^*,t)} \cdot \left(\mathrm{WC(z^*,t) - WP(z^*)}\right) \cdot \mathrm{W_{FRT_T}(z^*,t)}}{\left(\mathrm{W_{FRT_T}(z^*,t) + W_{FRT_{GV}}(z^*,t)}\right)} \tag{2.3.8}$$

$$\mathrm{HFC(z^*,t)} = \begin{cases} 1 - \frac{\mathrm{FC_{90}(z^*) - WC(z^*,t)}}{\mathrm{FC_{90}(z^*) - WP(z^*)}} & \text{for } \mathrm{WC(z^*,t) < FC_{90}(z^*)} \\ 0.3 + 0.7 \cdot \frac{\mathrm{SWC(z^*) - WC(z^*,t)}}{\mathrm{SWC(z^*) - FC_{110}(z^*)}} & \text{for } \mathrm{WC(z^*,t) > FC_{110}(z^*)} \\ 1 & \text{otherwise} \end{cases} \tag{2.3.9}$$

The dynamics of soil temperature are described by a one-dimensional heat conduction equation with the appropriate initial and boundary conditions (Suckow, 1989).

$$\mathrm{K_H(z,t)} \cdot \frac{\partial}{\partial t}\mathrm{TS(z,t)} = \frac{\partial}{\partial z}\left(\lambda_{\mathrm{H}}(\mathrm{z,t}) \cdot \frac{\partial \mathrm{TS(z,t)}}{\partial z}\right) \tag{2.3.10}$$

The solution of the heat conduction equation provides the soil temperature (TS) in each soil layer. The heat capacity (K_H) is calculated by means of a linear function from the specific heat of the solid soil components, the bulk density and the heat capacity. The thermal conductivity (λ_H) is determined by means of a non-linear function of the bulk density and the water content (Suckow, 1986).

The description of the geochemical processes used to determine the pH value essentially rests on the SMART model (De Vries *et al.*, 1989). Its basis is a soil-chemical equilibrium model combined with the principles of mass conservation. Depending on soil reaction (pH value), various buffer reactions acquire importance (Bossel *et al.*, 1985). The pH value has a reducing influence on the mineralisation and nitrification in particular ranges. As with N transport, ion transport occurs through water movement. Thus, the outflow from the last root-containing layer determines the loss from the ecosystem of water, nitrogen compounds and other solutes.

2.4. Stand dynamics

Annual total wood growth is distributed into growth of branches, coarse roots and stemwood. The fraction of coarse roots is assumed to be constant, whereas the branch wood fraction (FBRA) is assumed to decrease with increasing diameter (DBH) relative to maximum diameter (Bossel, 1994). In the next step, stemwood growth of the average tree (G_{STE}) is distributed into height growth (GTOP) and diameter growth (GDBH) by the use of an optimum height/diameter ratio (Hd_{opt}), which is calculated from the stand density (or crown area) index (CRAI, Bossel, 1994) and parameters:

$$GDBH = \frac{G_{STE}}{3 \cdot \partial \cdot HD_{opt} \cdot DBH^2} \tag{2.4.1}$$

$$GTOP = HD_{opt} \cdot GDBH \tag{2.4.2}$$

$$= \frac{DENS \cdot 0.5 \cdot \pi}{4} \tag{2.4.3}$$

$$HD_{opt} = \min(HD_{max}, HD_{min} + (HD_{max} - HD_{min}) \cdot CRAI) \tag{2.4.4}$$

$$CRAI = \frac{NTRE \cdot (DBH \cdot QCD)^2 \cdot \pi}{40000} \tag{2.4.5}$$

Stand density is a very important factor in stand development. It determines not only the partitioning between height and diameter growth, but also environmental conditions for the single tree, particularly light and water availability. Nevertheless, the effect on resource availability would be overestimated without accounting for the shading between trees in case of a non-perpendicular sun angle (Mohren, 1987).

A vitality index (VIT) is introduced to characterise tree health. It is calculated by dividing net primary production (POOL) by total annual compartment mortality (M, foliage = NDL, fine roots = FRT, coarse roots = CRT, branches = BRA, regenerative tissue = REN) including core wood formation (M_{SAP}) and exsudation losses (G_{EXS}).

$$VIT = \frac{\sum POOL}{\sum(M_{NDL} + M_{FRT} + M_{SAP} + M_{CRT} + M_{BRA} + M_{REN} + G_{EXS})} \tag{2.4.6}$$

By means of VIT natural tree mortality (FTOT) is calculated, which is used for the reduction of tree number and every single biomass compartment as well as all nitrogen pools.

$$\mathrm{FTOT} = \begin{cases} 1 - \mathrm{VIT}^{SLO} & , \text{if VIT} < 1 \\ 0 & , \text{if VIT} \geq 1 \end{cases} \tag{2.4.7}$$

Because management is one of the most important and most common disturbances, in particular in the even-aged conifer plantations in Europe, the mortality function can be replaced by prescribed cutting percentages after selected height growth intervals. It has to be considered that the reduction of stem number due to management generally does not select trees randomly. Depending on silvicultural strategy, the average size of removed trees is smaller or bigger than the average size of all trees in the stand. Thus, it is accounted for in the model that the decrease in biomass (wood, foliage, roots, reserve pool) is not necessarily proportional to the decrease of tree number. In this way, different types of management can be reflected.

3. Study sites and simulation experiments

Field investigations at three pine forest stands, grown under different conditions of air pollution and deposition, provided information about important ecological relations and served for parameterization and verification of the model (further information about sites and investigations can be obtained from other reports in this volume). Climate data, used for simulation of the ecosystem processes during the investigation period, were measured close to the stands in the open field.

For the retroperspective calculations, daily meteorological data from weather stations in the region (Neuglobsow, Wittenberg) served as driving variables. SO_2 concentration was estimated from emission inventory data (Friedrich in Anonymus, 1997), whereas it was assumed that the nitrogen deposition was constant for the entire period based on the deposition of 1994.

Initial stand data for the long-term simulation of the three investigations sites were not available. Thus, they were estimated from stand age with empirically developed functions of height, diameter and stand volume, based on the forest inventory data of a nearby representative forest district. The soil data were initialised as measured in 1994 but total nitrogen content was reduced by the amount of total fertilisation at the specific site.

4. Results and discussion

4.1. *Two-year simulations*

Water balance

In Figure 5 (A–D), as well as in Table 3, simulated and measured water fluxes at the three study sites are presented. It shows that the annual sum of canopy transpiration at Neuglobsow is about 10–15 mm higher than in Rösa and

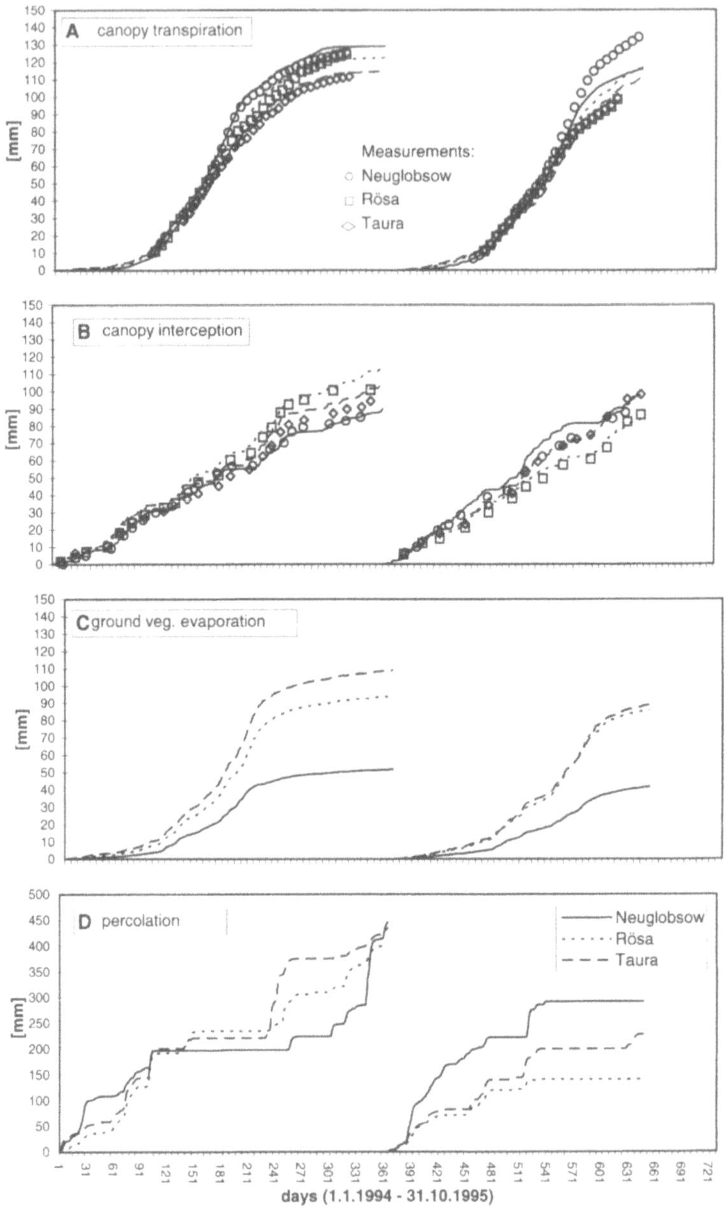

A canopy transpiration
Measurements:
Neuglobsow
Rösa
Taura
B canopy interception
C ground veg. evaporation
D percolation
Neuglobsow
Rösa
Taura
[mm]
days (1.1.1994 - 31.10.1995)

Figure 5.

Table 3. Simulated water balance for the years 1994 and 1995 at the three investigation sites (all values in mm)

	Neuglobsow		Taura		Rösa	
	1994	1995*	1994	1995*	1994	1995*
Precipitation	741.4	564.4	793.5	551.6	781.1	464.2
Interception						
– Canopy	90.0	107.0	104.9	98.5	114.0	86.4
– Ground vegetation	10.4	11.2	20.4	20.2	19.1	16.7
Transpiration						
– Canopy	129.5	118.4	114.7	116.1	122.9	120.3
– Ground vegetation	41.5	32.4	98.2	78.0	75.3	71.4
Soil evaporation	14.3	12.6	22.0	21.54	20.4	15.4
Percolation	452.8	275.1	433.3	224.4	415.0	150.6
Soil water change	2.9	7.7	0	−7.1	14.4	3.4

Taura, although it started to increase slightly later. In late summer Transpiration in Neuglobsow increases more slowly than at the other sites, which might be related to the higher concentration of roots in the upper soil horizon (see also Strubelt *et al.*, this volume), which is considered in the simulation. This makes the trees more susceptible to drought, because the small soil volume and the high rooting density of trees and ground vegetation leads to a relatively fast drying of the upper soil layers. Thus, the role of ground vegetation is perhaps most important at this site, despite the fact that the simulation based estimation of understorey transpiration is more than 30 mm lower than at Rösa and Taura. This result is a consequence of the higher leaf area index of the ground vegetation (see Lüttschwager *et al.*, this volume) and a higher potential evaporation beneath the crowns, partly because of a smaller stand density (see Neumann, this volume) at these sites.

Another reason for the differences between the sites is the site-specific rainfall distribution. In Neuglobsow, summer drought was more intense than at the other sites, especially in 1994, and precipitation as a whole was very unevenly distributed. This led to a smaller evaporation from interception in 1994 and a higher percolation rate in 1995 compared with the other sites, although e.g. at Rösa total precipitation was very similar and the measured differences in leaf area index (LAI) were only small. Differences in interception between Rösa and Taura can be explained either with a smaller LAI (1994) or with weather conditions (1995), depending on which effect is the stronger.

Although the model results and the estimations from measurements are generally in good agreement, the simulation overestimates transpiration at Rösa and Taura during drought periods, especially in 1994. A possible explanation for this result could be a different xylem conductance at these sites, what is discussed in more detail by Rust (this volume). The transpiration at Neuglobsow during the drought period of 1995 is underestimated, which corresponds with the result from fine root measurements (see Figure 7 and

Leefeld *et al.*, this volume) that indicate a larger amount of fine roots in this period than are expected from the simulation.

Regarding the percolation from the deepest rooted soil layer, the differences between the sites and also between the two observed years are remarkable. Although, in general most percolation occurred during winter and early spring, indicating high precipitation and small evaporation at these times, considerable water loss is also simulated during summer storm events.

Carbon balance

Differences in weather conditions, LAI, water and nutrient availability determine the specific net assimilation rate per day at each site. Estimates of photosynthesis, carried out during some days in 1994 at four selected branches at each site, were compared to simulations (Figure 6). The drought stress sensitivity-parameter has been estimated from this comparison by minimising the differences between measurements and simulations, although only minor drought stress deficits could be detected during most of the days on which the measurements took place. Considering the variability of carbon gain within and between trees of the same stand (see Dudel *et al.*, this volume), simulations are well in accordance with measurements, except at Neuglobsow in July. However, the measured rate seems to be unreasonably low to be representative of the whole stand, particularly because it was obtained before the drought period started (Lüttschwager *et al.*, this volume).

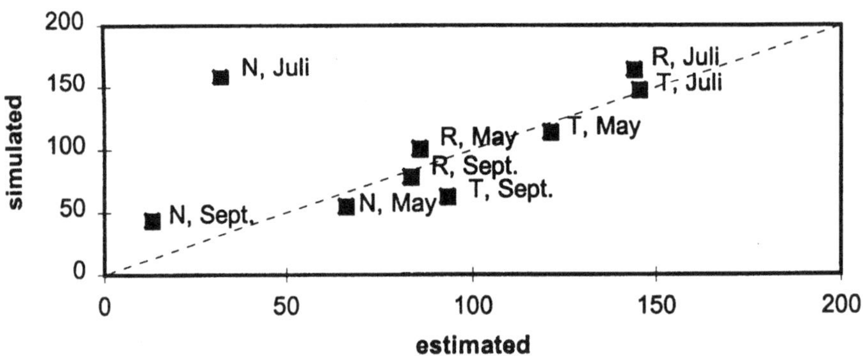

Figure 6. Simulated and independently estimated daily assimilation in kgCH_2O ha^{-1} (R = Rösa, T = Taura, N= Neuglobsow)

Allocation of net carbon gain on the one hand and compartment mortality on the other are responsible for the carbon dynamics within the tree and the stand. Only a few of the involved processes could be traced seasonally by direct measurements (e.g. litterfall of needles) or indirectly from more aggregated measurements (e.g. sapwood growth from diameter growth). These are compared with simulations in Figure 7 (A–C).

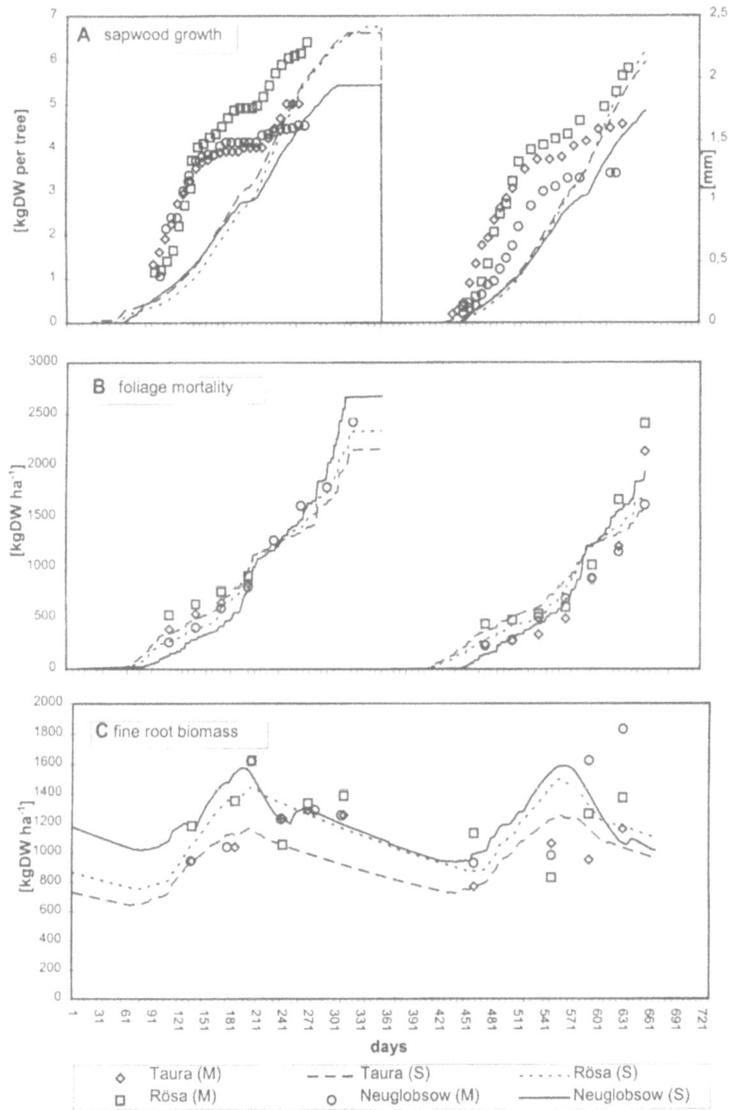

Figure 7A–C. Simulated and measured development of single elements from the carbon balance between 1.1.1994 and 31.10.1995

With respect to the seasonal sapwood growth, it can only be compared with measurements of diameter increment (Neumann *et al.*, this volume). However, it has to be considered that biomass and diameter development are not linearly related, because height and diameter growth occur over different time periods (Dougherty *et al.*, 1994). Additionally, large seasonal differences in wood density occur throughout the year. During the summer drought, which is experienced at every site, sapwood growth as well as diameter increase stops almost completely. After this period, growth increases again. This second increase is more expressed in the simulation than in the diameter measurements, which is probably due to an increase in wood density. Unfortunately, no measurements of wood density are available and possible differences between the sites cannot be judged. It seems, however, that the simulation overestimates the biomass increase in Neuglobsow during the end of the vegetation period, when the measurements indicate that almost no growth occurs anymore. With respect to the slightly smaller sapwood growth in Taura compared to Rösa, which seems to represent the results obtained from diameter measurements inadequately, it has to be considered that the trees in Rösa are smaller. This results in a larger diameter growth per unit biomass increase.

According to the simulation results, foliage mortality in Neuglobsow starts somewhat later in the year but increases more sharply in late summer, which is correlated with a larger drought stress at this site. With respect to the small deviations between simulated development and measurements, the model assumptions about the mechanisms of foliage mortality are supposed to be sufficient to describe the development of litterfall. Small deviations are possibly due to the influence of wind, which is not included in the model, and the delay in litterfall of needles, which have died in the previous year.

Simulated root dynamics show an increase of root mass, which is correlated to the increase in foliage mass. Fine root biomass decreases in summer, because the mortality is increased under dry conditions, which is particularly expressed at Neuglobsow. However, also the allocation of carbon to fine roots is increased. If fine root growth cannot compensate for fine root losses, and rainfall occurs as long as the amount of foliage is still large, a (smaller) second root flush is simulated (Neuglobsow). These findings are similar to observations from Santantonio and Santantonio (1987) who also found little root growth during the time of shoot elongation and are in accordance with general fine root dynamics in temperate regions (see Eissenstat and Van Rees, 1994; Persson, 1979).

Despite some deviations from measurements, it seems that enough information is available to estimate the development of compartment masses based on simulations and to calculate a total carbon balance of the two measurement periods (Table 4). From this, it seems obvious, that the trees as well as the soils of all three sites are acting as carbon sinks. However, the differences between the sites are only small and the ranking shifts even in the two investigated years.

Nitrogen balance

The simulated nitrogen balance based on water uptake, carbon balance and nitrogen concentrations (Ende *et al.*, this volume) is given in Table 5. Stand deposition of total nitrogen is very similar at the three stands, but canopy uptake was estimated to be different, particularly because of different nitrogen concentrations in the needles (see also Weisdorfer *et al.* and Ende *et al.*, this

Table 4. Simulated carbon balance for the years 1994 and 1995 at the three investigation sites (all values in kgC ha^{-1})

	Neuglobsow		Taura		Rösa	
	1994	1995*	1994	1995*	1994	1995*
Trees						
Gross assimilation	11 533	10 803	10 598	10 387	10 643	10 341
Maintenanance respiration	4 017	3 675	3 777	3 516	3 822	3 490
Growth respiration	939	891	853	859	853	856
Litterfall (& exsudation)	4 184	5 034	3 883	2 613	4 339	3 246
Ground vegetation						
Gross assimilation	2 162	1 896	4 682	4 785	4 061	4 428
Maintenanance respiration	400	333	1 081	1 045	842	872
Growth respiration	443	391	932	947	820	892
Litterfall	1 338	1 180	2 532	2 664	2 145	2 433
Soil respiration	4 379	2 667	3 863	3 010	4 015	3 010
Tree balance	2 393	1 203	2 085	3 399	1 629	2 749
Ground vegetation balance	–19	–8	137	128	254	231
Soil balance	1 138	3 584	2 119	1 947	1 990	2 368
Total balance	3 517	4 742	4 774	5 795	4 352	5 649

*Until 31.10., when measurements ended

Table 5. Simulated nitrogen balance for the years 1994 and 1995 at the three investigation sites (all values in kgN ha^{-1})

	Neugobsow		Taura		Rösa	
	1994	1995*	1994	1995*	1994	1995*
Stand deposition	14.2	10.9	19.6	17.1	16.2	14.6
Canopy uptake	3.7	3.7	3.1	3.1	2.9	2.6
Tree root uptake	42.4	58.9	31.1	51.1	50.2	70.3
Ground vegetation root uptake	14.1	16.5	51.0	62.3	37.8	50.9
Tree litter loss	34.9	34.2	31.9	25.9	40.7	35.3
Export	1.6	0	0.2	0	1.1	0
Ground vegetation litter loss	17.8	16.0	48.3	55.0	30.5	43.9
Percolation loss	3.0	4.2	0.7	0.5	1.6	0.6
Tree balance	9.6	28.4	2.1	28.3	11.3	37.6
Ground vegetation balance	–3.7	0.5	2.7	3.9	19.7	26.7
Soil balance	7.4	–18.5	17	–15.9	–2.2	–28
Total balance	13.3	10.4	21.8	19.7	16.4	16.6

volume). In contrast, nitrogen uptake through the roots is affected mainly by the different availability of nitrogen in the soil solution.

Total amounts of nitrogen in litter are somewhat higher than measured nitrogen litterfall (Fischer *et al.*, in Anonymus, 1997) because in the simulation, fine root litter was also included. Major differences between the sites are due to the higher nitrogen concentrations in the litter rather than the differences in litter amounts, which reflects the large differences in foliage nitrogen concentration compared to the small differences in foliage biomass (see Ende *et al.*, this volume). The differences in nitrogen turnover between sites are expressed more in ground vegetation than in trees, because not only nitrogen concentrations but also the biomass within the herb layer is different.

Finally, it can be concluded that nitrogen losses from the system with biomass export and drainage are generally smaller than the nitrogen input at all three sites, despite the difference in nitrogen storage. Nevertheless, losses clearly depend on specific annual conditions of percolation. Thus, distinct differences in the total balance from year to year may occur. Differences from year to year are also pronounced in the cycle of N within the trees and within the ground vegetation, resulting from different growth and mortality conditions for the plants. On average, tree N-uptake at Neuglobsow is estimated to be between that at Rösa and Taura but ground vegetation N-uptake in Neuglobsow was only 27% of that in Taura, with Rösa being intermediate.

4.2. Retrospective long-term simulations

Based on the model evaluation from 1994 and 1995, long-term water, carbon and nitrogen balances at the three investigation sites are simulated. As shown in Figure 8 A–C, the stands differ largely in their history, with Rösa being the most impacted site with respect to SO_2 as well as fertilisation. At this site, a higher tree mortality is simulated, leading to a relatively small crown area index compared to Taura and Neuglobsow. At Taura, however, it has to be considered that the stand is 20 years younger than at Rösa. Thus, it is likely that trees were much smaller at Taura at the beginning of the simulation period as compared to the other sites. Consequently, growth developments cannot be compared directly.

The simulated water balance at the three sites (Figure 9 A–C) reveals that precipitation at Neuglobsow is larger than at the other two sites (in average 595 to 559 mm). However, in some years, the total precipitation can also be considerably smaller (e.g. 1993). Comparing the different fractions of the water balance, the percolation turns out to be the most variable one, ranging from approximately 30 to more than 300 mm a year at every site. The most obvious differences between the sites is in the proportion of ground-vegetation transpiration that is considerably higher at Rösa, whereas the tree transpiration is smaller. This can be explained with the smaller stand density and the high ground vegetation biomass at this site.

The carbon balance (Figure 10 A–C) reflects drought years (e.g. 1975, 1976,

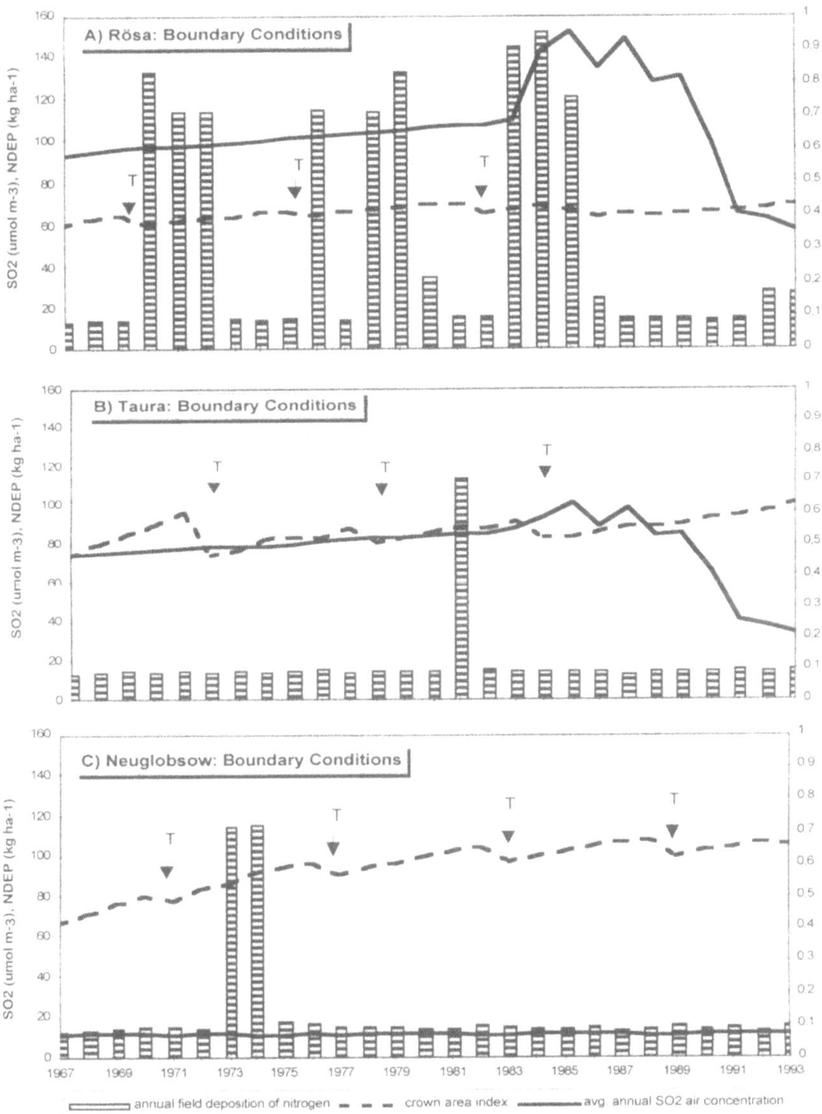

Figure 8A–C. Development of N-deposition, average SO_2 concentration and simulated crown area index during the years 1967–1993. (T = thinning events)

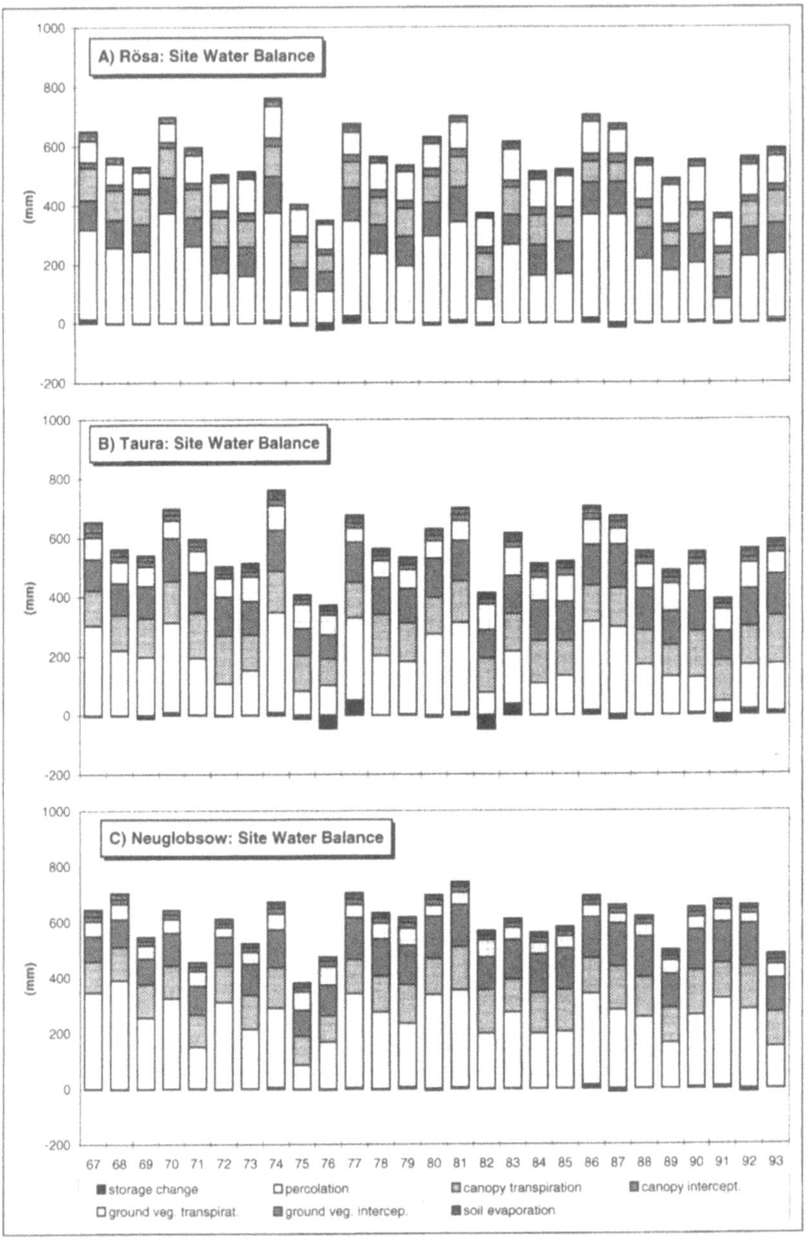

Figure 9A–C. Simulated water balances during the years 1967–1993 for the investigation sites

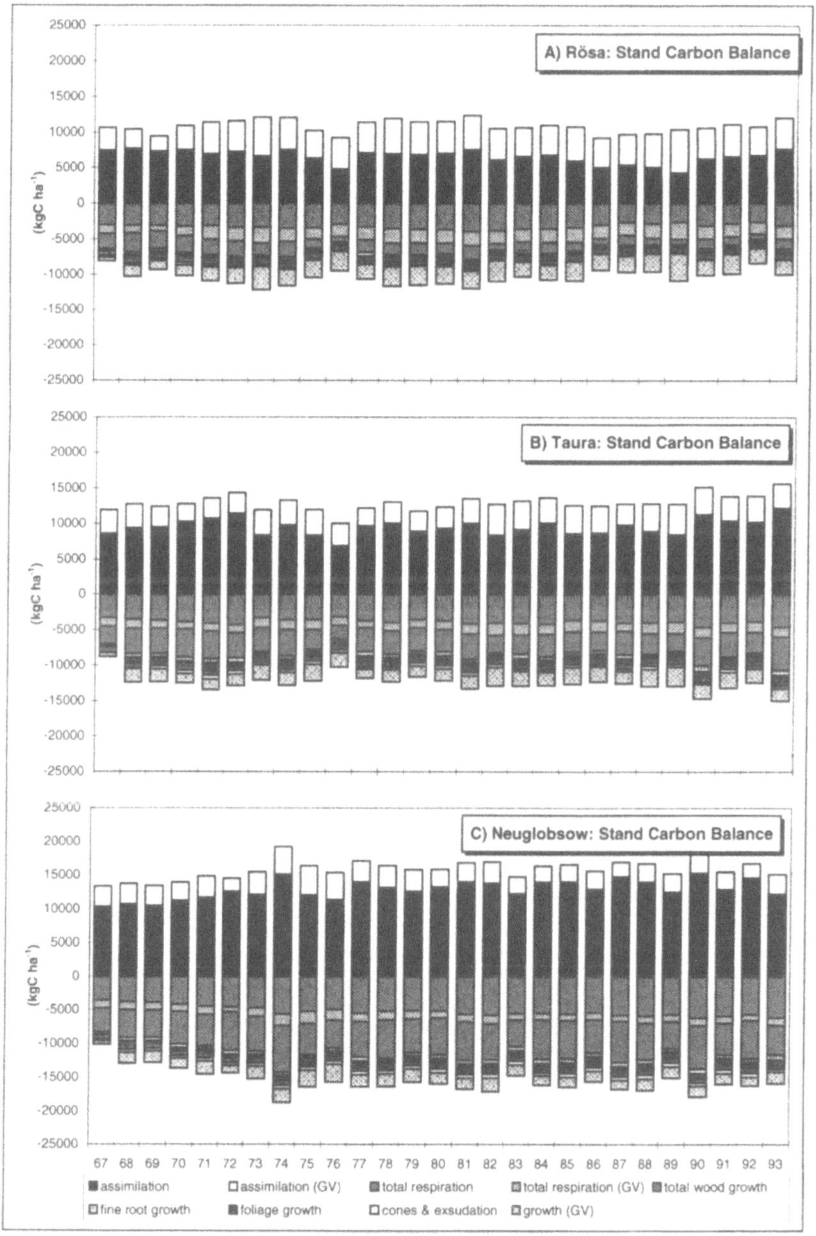

Figure 10A–C. Simulated carbon balances during the years 1967–1993 for the investigation sites

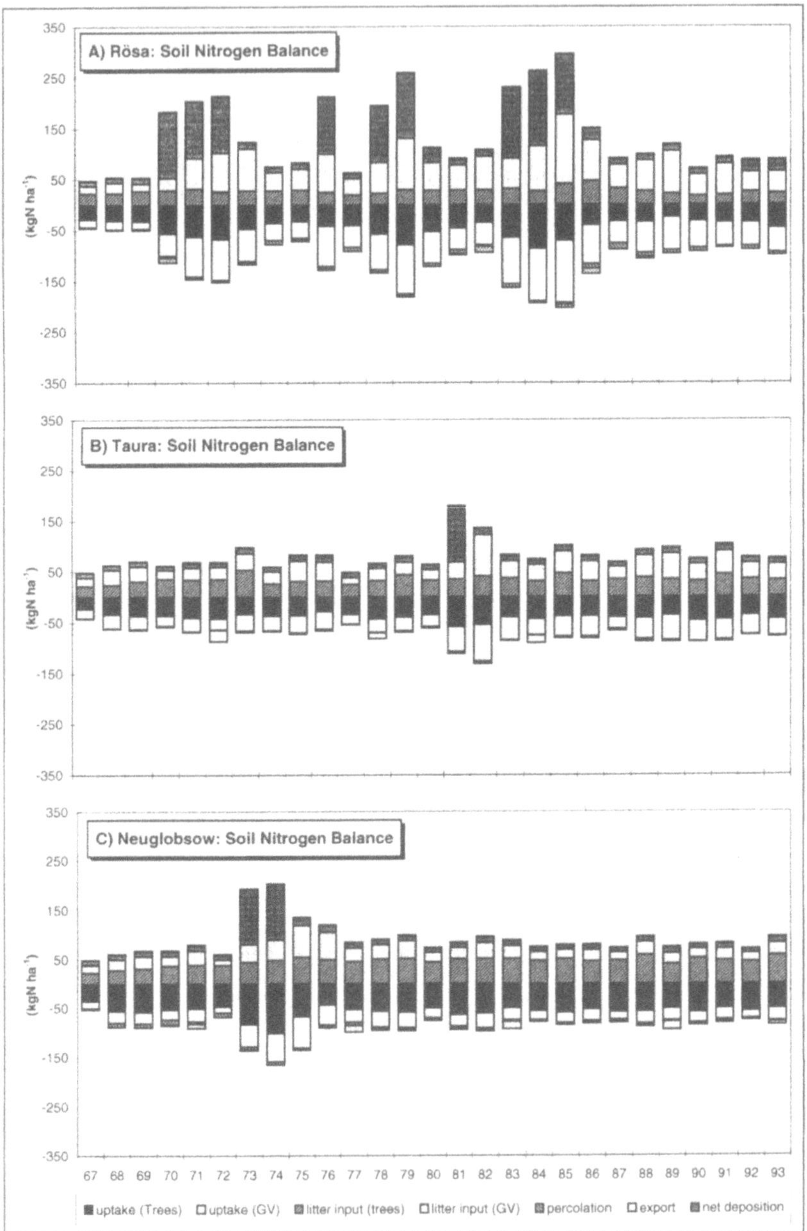

Figure 11A–C. Simulated nitrogen balances during the years 1967–1993 for the investigation sites

1982, 1991) as well as fertilisation effects. The reaction to fertilisation is often more expressed in ground vegetation than in the trees, particular if thinning has been applied in preparation of the additional nutrient supply (Rösa). The reason is partly in the decreased tree biomass and partly in the combined response to improved light and nutrient conditions in the herb layer. A distinct decrease in photosynthetic stand production during the 1980s occurs in Rösa and – much less expressed – in Taura, which is correlated with the increasing SO_2 concentration in this period. Fertilisation during this time is simulated to have hardly any effect. Since 1990, however, the SO_2 concentration has decreased and the biomass production increased again.

The effect of fertilisation on the nitrogen balance is shown in Figure 11 (A–C). The increasing supply is mostly used by trees and ground vegetation within the same year. In the following years, the nitrogen within the litter compartment is increased and after approximately 3 years, the previous input:uptake relation has been re-established. This course of events is in general accordance with literature findings, although the duration of the fertilisation effect has mostly been reported to last 1 to 2 years longer (Heinsdorf, 1967; Rodenkirchen, 1995).

Besides the effects on vegetation development, the retrospective simulation of nitrogen balance in both stands yields a net nitrogen storage in the ecosystem of the same magnitude as stand deposition. In other words, the amounts of nitrogen, which were lost by drainage and wood export (together approximately 5 kg ha^{-1} a^{-1} in average at all sites) were of the same magnitude, which was estimated to be taken up by the canopy. It is not definite whether this result is in accordance with reality, since the nitrogen pool of the soil of these particular stands in 1967 is not known. However, comparisons with regional investigations in the early 70's (see Erhard and Flechsig, this volume) led to the conclusion that a large nitrogen increase may indeed have taken place. Nevertheless, it has to be admitted that also gaseous losses (Brumme and Beese, 1992; Tietema *et al.*, 1991) or percolation together with organic compounds (Hüttl, personal communication), which are not taken into account, may play an important role in these ecosystems.

5. Summary and conclusions

The new model proved to be useful in the investigation of pine stand development under changing air pollution, nitrogen deposition (including fertilisation) and climate. By simulating the water, carbon and nitrogen balances from initial data and climate data input, measured in the field, it was possible to reproduce observations at the three investigation sites over a period of two years, and to estimate fluxes that could not be measured directly. The simulation revealed remarkable differences between the sites on the one hand and the two individual years on the other hand.

Transpiration of the canopy differed between the stands, partly because of differences in climate and LAI but also because of different conductivity responses. It is not clear, however, if these differences are linked to the ability of the rooting system to adapt to the water distribution in the soil, or if different sapwood properties are responsible for the findings. At the stand level, transpiration at the tree sites seems to be more similar than only by looking at the canopy layer, because water use by ground vegetation increases with increased understorey-LAI and decreased canopy cover. The amount of evaporated and transpired water found in this investigation was quite high, even if it is considered that ground vegetation was also found to play an important role in the water cycle of other pine stands (Wedler *et al.*, 1996).

The simulated share of ground vegetation regarding the carbon and nitrogen balance was also very high at the sites Rösa and Taura, but less important in Neuglobsow. This agrees with measurements and shows that the understorey reacts quite flexibly to changing conditions. It seems that if nitrogen is available in excess, much of it can be incorporated into the ground vegetation, leading to a storage of more than 40 kg ha^{-1}.

The differences between the simulation results of the two individual years seemed to be mainly caused by drought conditions. For example, wood growth until October has been simulated at Taura to be higher in 1995 than in 1994, whereas at Neuglobsow the opposite was the case. This pattern coincides with canopy transpiration, although not with total precipitation. From the principles of the model one might expect that nitrogen uptake is closely correlated with water uptake and that the respective pattern between the years should be the same. However, this was not always the case during the investigation period, underlining the importance of simulating nitrogen concentration of the soil solution in each layer. Nevertheless, it has to be admitted that an investigation period of two years is too short for inter-annual comparisons, because errors in initialisation of carbon and nitrogen pools in the soils as well as of foliage and litter production may have occurred, influencing the first year much more than the second. In this respect, it may be of importance that the year 1993 was a drought year at Neuglobsow but not in Rösa and Taura.

The main differences between the sites are in the nitrogen balance (and other nutrient elements, to which the model is not sensitive). Total nitrogen turnover of trees in Rösa was 17% above that of Neuglobsow, with both stands having almost the same foliage mass. The nitrogen turnover obtained at Taura is smaller than at the other two sites, because the relatively good nutrition in Taura is counteracted by its lesser foliage and fine root biomass, the steeper fine root distribution and the competition of ground vegetation in the upper soil layers. The simulations also represent the measured differences in ground vegetation between the sites. The turnover of the herb cover in Rösa and Taura reached the same magnitude as estimated for the trees, whereas in Neuglobsow it was much lower.

As stated above, the estimation of a trend in stand growth development from two years of investigation is cumbersome. However, because of the better nutrition and less drought stress a generally better tree growth in Rösa and

Taura than in Neuglobsow is reasonable. Indeed, diameter and height increase are greater at these sites, which is not only because the trees are smaller and thus the same increase in wood results in a more highly expressed dimensional change.

The dynamic linkage of element budgets with tree and stand development within a physiologically-based model has successfully served as a framework for a multitude of ecophysiological and physico-chemical measurements at specific sites. Furthermore, it has been shown that the approach is suitable if results are extrapolated in time, although the simulation of stand development suffered from a lack of knowledge about the initial stand conditions and management. It can be concluded that the model can be applied for a broad range of conditions, particular if further improvements, such as a more mechanistic description of the ground vegetation responses and the dynamic adjustment of rooting depth, could be realised.

Acknowledgement

This project was supported by the Federal Ministry of Education and Research (BMBF, Bonn, Germany) within the framework of SANA. We would like to thank all colleagues within the ecology group of SANA who have worked in the field and the laboratory, because model implementation would not have been possible without such good co-operation within the project.

References

Aber JD, Reich PB, Goulden L. 1996. Extrapolating leaf CO_2 exchange to the canopy: a generalized model of forest photosynthesis compared with measurements by eddy correlation. Oecologia. 106, 257–265.

Anonymus. 1997. SANA – Wissenschaftliches Begleitprogramm zur Sanierung der Atmosphäre über den neuen Bundesländern, BMBF, Bonn.

Bossel H. 1994. TREEDYN3 Forest Simulation Model. Forschungszentrum Waldökosysteme, B/35. Universität Göttingen, Göttingen, 118 pp.

Bossel H, Metzler W, Schäfer H. 1985. Dynamik des Waldsterbens. Mathematisches Modell und Computersimulation. Springer Verlag, Berlin Heidelberg.

Brumme R, Beese F. 1992. Effects of liming and nitrogen fertilization on emissions of CO_2 and N_2O from a temperate forest. J Geophys Res, 97, 851–858.

Chen CW. 1993. The response of plants to interacting stresses: PGSM Version 1.3 model documentation. EPRI TR-101880. Electric Power Res Inst., Palo Alto, CA.

Cropper WPJ, Gholz HL. 1993. Simulation of the carbon dynamics of a Florida slash pine plantation. Ecol Model. 66, 231–249.

De Vries W, Posch M, Kämäri J. 1989. Simulation of the long-term response to acid deposition in various buffer ranges. Water Air Soil Pollut. 48, 349–390.

Dougherty PM, Whitehead D, Vose JM. 1994. Environmental influences on the phenoloy of pine. In: Environmental Constraints on the Structure and Productivity of Pine Forest Ecosystems: A Comparative Analysis. Eds. HL Gholz, S Linder, RE McMurtrie. Ecol Bull. Copenhagen, p. 64–75.

Eckersten H. 1994. Modelling daily growth and nitrogen turnover for a short-rotation forest over several years. For Ecol Management. 69, 57–72.
Eissenstat DM, Van Rees KCJ. 1994. The growth and function of pine roots. Ecol Bull. 43, 76–91.
Franko U. 1990. C- und N-Dynamik beim Umsatz organischer Substanz im Boden. Dissertation B Thesis. Akademie der Landwirtschaftswissenschaften der DDR, Berlin.
Gluch W. 1988. Zur Benadelung von Kiefern (*Pinus silvestris* L.) in Abhängigkeit vom Immissionsdruck. Flora. 181, 395–407.
Goudriaan J. 1982. Potential production processes. In: Simulation of Plant Growth and Crop Production. Eds. FWT Penning de Vries, HH van Laar. PUDOC, Wageningen. p. 98–113.
Grosch S. 1990. Der atmosphärische Gesamteintrag auf natürlichen Oberflächen unter besonderer Berücksichtigung der trockenen Deposition in Waldgebieten. Berlin Inst Meteorologie und Geophysik Univ Frankfurt. 83, 1–123.
Grote R. 1998. Integrating dynamic morphological properties into forest growth modeling II. Allocation and mortality. For Ecol Management. 111, 193–210.
Grote R, Suckow F. 1998. Integrating dynamic morphological properties into forest growth modeling I. Effects on water balance and gas exchange. For Ecol Management. 112, 101–119.
Hancock NH, Sellers PJ, Crowther JM. 1983. Evaporation from a partially wet canopy. Geophysicae. 1, 139–146.
Heij GJ, De Vries W, Posthumus AC, Mohren GMJ. 1991. Effects of air pollution and acid deposition on forests and forest soils. In: Acidification Research in The Netherlands. Final report of the Dutch Priority Programme on Acidification. Studies in Environmental Science. Eds. GJ Heij, T Schneider. Elsevier, Amsterdam. p. 97–137.
Heinsdorf D. 1967. Untersuchungen über die Wirkung mineralischer Düngung auf das Wachstum und den Ernährungszustand von Kiefernkulturen auf Sandböden im nordostdeutschen Tiefland. IV. Trockensubstanzproduktion, Nährstoffmehraufnahme und Nährstoffspeicherung von Kiefernkulturen nach Düngung. Arch Forstwes. 16(3), 183–201.
Helmisaari H-S, Mälkönen E. 1989. Acidity and nutrient content of throughfall and soil leachate in three *Pinus Sylvestris* stands. Scand J For Res. 4, 13–28.
Hoffmann F. 1995. FAGUS, a model for growth and development of beech. Ecol Model. 83, 327–348.
Kartschall T, Döring P, Suckow, F. 1990. Simulation of nitrogen, water and temperature dynamics in soil. Syst Anal Model Simul. 7(6), 33–40.
Kittel TGF, Coughenor MB. 1988. Prediction of regional and local ecological change from global climate model results: a hierachial modelling approach. In: Monitoring Climate for the Effects of Increasing Greenhouse Gas Concentrations. Eds. RA Pilke, TGF Kittel. Fort Collins, Colorado, USA. p. 173–193.
Koitzsch R. 1977. Schätzung der Bodenfeuchte aus meteorologischen Daten, Boden- und Pflanzenparametern mit einem Mehrschichtmodell. Z Meteor. 27, 302–306.
Liu S. 1997. A new model for the prediction of rainfall interception in forest canopies. Ecol Model. 99, 151–159.
Lovett GM, Lindberg SE. 1984. Dry deposition and canopy exchange in a mixed oak forest as determined by analysis of throughfall. J Appl Ecol. 21, 1012–1027.
Luan J, Muetzfeldt RI, Grace J. 1996. Hierarchical approach to forest ecosystem simulation. Ecol Model. 86, 37–50.
McLeod AR, Holland MR, Shaw PJA, Sutherland PM, Darrall NM, Skeffington RA. 1990. Enhancement of nitrogen deposition to forest trees exposed to SO_2. Nature. 347(6290), 227–279.
Meng FR, Arp PA. 1994. Modelling photosynthetic responses of a spruce canopy to SO_2 exposure. For Ecol Management. 67, 69–85.
Mohren GMJ. 1987. Simulation of forest growth, applied to Douglas Fir stands in the Netherlands. Agricultural University, Wageningen, The Netherlands. 184 pp.
Mohren GMJ, Jorritsma ITM, Vermetten AWM, Kropf, MJ, Smeets WLM, Tiktak A. 1992. Quantifying the direct effects of SO_2 and O_3 on forest growth. For Ecol Management. 51, 137–150.

Monteith JL. 1965. Evaporation and environment. In: The State and Movement of Water in Living Organisms. Symp Soc Exp Biol. Ed. GE Fogg. Academic Press, London. p. 205–234.

Penning de Vries FWT, Jansen DM, ten Berge HFM, Bakema A. 1989. Simulation of Ecophysiological Processes of Growth in Several Annual Crops. Simulation Monographs, 29. PUDOC, Wageningen, The Netherlands.

Persson H. 1979. Fine root production, mortality and decomposition in forest ecosystems. Vegetation. 41, 101–109.

Rodenkirchen H. 1995. Nutrient pools and fluxes of the ground vegetation in coniferous forests due to fertilizing, liming and amelioration. Plant Soil. 168–169, 383–390.

Running SW, Gower ST. 1991. FOREST-BGC, A general model of forest ecosystem processes for regional applications. II. Dynamic carbon allocation and nitrogen budgets. Tree Physiol. 9, 147–160.

Santantonio D, Santantonio E. 1987. Seasonal changes in live and dead fine roots during two successive years in thinned plantation of *Pinus radiata* in New Zealand. NZ J For Sci. 17, 315–328.

Schlichter TM, Van der Ploeg RR, Ulrich B. 1983. A simulation model of the water uptake of a beech forest: testing variations in root biomass and distribution. Z. Pflanzenernährung u. Bodenkunde. 146(6), 725–735.

Sherif, DW, Mattay JP, McMurtrie RE. 1996. Modeling productivity and transpiration of *Pinus radiata*: climatic effects. Tree Physiol. 16, 183–186.

Skeffington RA, Wilson EJ. 1988. Excess nitrogen deposition: Issues for consideration. Environ Pollut. 54, 159–184.

Suckow F. 1986. Ein Modell zur Berechnung der Bodentemperatur unter Brache und unter Pflanzenbestand. Dissertation A Thesis, Akademie der Landwirtschaftswissenschaften der DDR, Berlin.

Suckow F. 1989. Ein Modell zur Berechnung der Bodentemperatur im Rahmen des Basismodells Boden (BAMO), Tag.-Ber Akad Landwirtsch-Wiss DDR, Berlin. pp. 159–164.

Tietema A, Bouten W, Wartenbergh PE. 1991. Nitrous oxide dynamics in an acid forest soil in the Netherlands. For Ecol Management. 44, 53–61.

Waring RH, Running SW. 1976. Water uptake, storage and transpiration by conifers: a physiological model. In: Water and Plant Life. Problems and Modern Approaches. Eds. OL Lange, L Kappen, ED Schulze. Berlin – Heidelberg – New York. p. 189–202.

Wedler M, Geyer R, Heindl B, Hahn S, Tenhunen JD. 1996. Leaf-level gas exchange and scaling-up of forest understory carbon fixation rates with a 'patch-scale' canopy model. Theor Appl Climatol. 53, 145–156.

Whitehead D, Kelliher FM. 1991. Modeling the water balance of a small *Pinus radiata* catchment. In: Advancing Toward Closed Models of Forest Ecosystems. Eds. MR Kaufmann, JJ Landsberg. Heron, Victoria, Canada. p. 17–33.

15
A landscape model for the investigation of atmogenic pollution effects on the dynamics of Scots pine ecosystems

M. ERHARD and M. FLECHSIG

1. Introduction

During past decades, forests in some areas of the former East Germany (GDR) have been exposed to high concentrations of air pollutants, in particular to sulphur compounds, with the consequence of severe damage. Since German reunification in 1990, the quantity and chemical composition of emissions have changed drastically (Bellmann *et al.*, this volume). By 1992, emission of SO_2 had been reduced to 40%, alkaline dusts to 20% and NO_x to 80% of the values of 1989 (see Figure 1), mainly due to changes in the economic structure and to the implementation of emission reduction technologies.

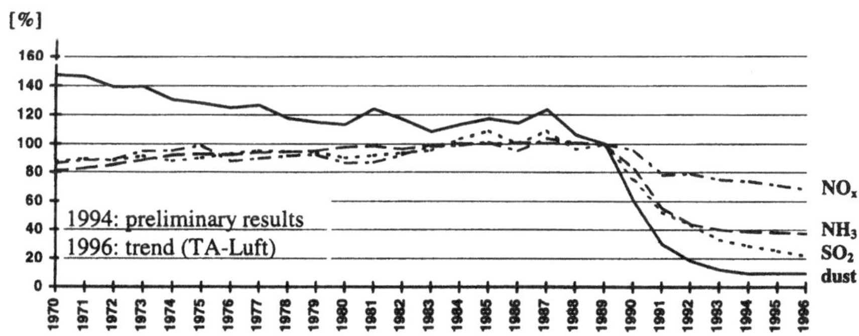

Figure 1. Total emissions between 1970 and 1992 and estimated for 1992–1998 relative to the year 1989 (= 100%); Data: IER, Stuttgart

R.F. Hüttl and K. Bellmann, Changes of Atmospheric Chemistry and Effects on Forest Ecosystems, 283–311.

In order to observe these changes in air pollution and to describe the impacts on ecosystems, the German Federal Ministry of Science and Technology (BMBF) initiated the SANA project in 1992. The goal of the subproject described in this paper was to analyse and simulate the stability and productivity of forest ecosystems under high pollution loads in the past as well as under the rapidly changing environmental conditions during the period of industrial restructuring and to estimate their further development using scenario techniques.

The spatial and temporal scales of the landscape study as well as the selection of the ecosystem type were determined by the response time of forest ecosystems to changing disposition by air-pollutants (years to decades) and the present state of the art of modelling imission/deposition and ecosystem dynamics.

The selected investigation area – the Dübener Heide – is located near to former major industrial centers. It is influenced directly by the emissions of the Halle–Leipzig–Bitterfeld region (Figure 2). With an extent of about 450 km^2, 75% of the area is covered by forests, about 50% of them even-aged stands of Scots pine (*Pinus sylvestris* L.), growing on sandy, anhydromorphic soils. The climate is characterized by relatively dry conditions (550 mm annual precipitation) and annual mean temperatures of 8.0–8.5°C. Data sets which allow soil and forest conditions since 1967 to be reconstructed and different gradients of pollution loads changing within short horizontal distances are the most interesting aspects of this area.

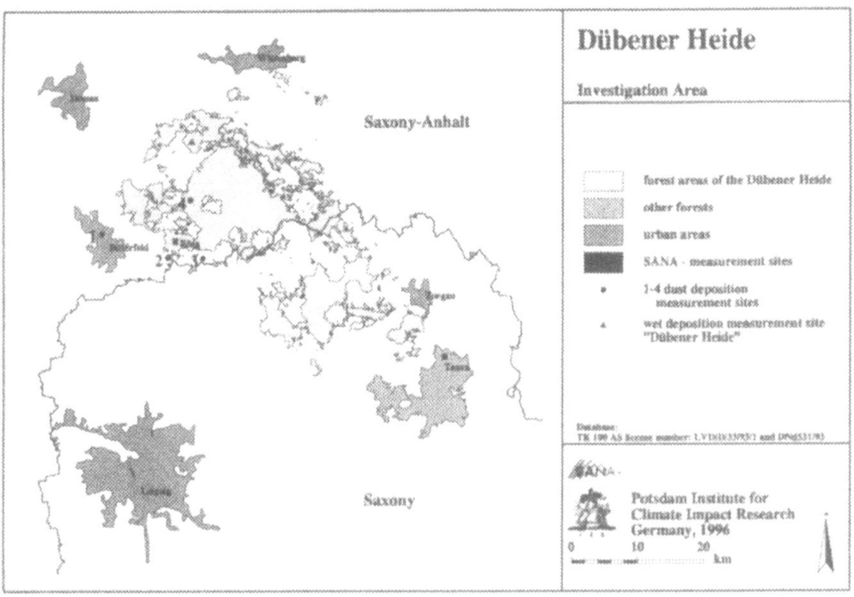

Figure 2. The investigation area Dübener Heide, the location of the SANA-measurement sites 'Rösa and Taura' and of the deposition measurement sites

Two different time intervals were considered for the model-based investigations:

- a retrospective time period between 1967 and 1989 for model validation purposes and

- a prospective, scenario-oriented simulation period from 1990 to the year 2024 to study possible developments of forest stands under changing immission/deposition.

2. Materials and methods

For modelling and mapping forest growth on a landscape scale, three elements were combined in the landscape model:

- The dynamic, process-oriented simulation model FORSANA, describing growth of Scots pine ecosystems. It is sensitive to immission/deposition and climate,

- a geo-referenced data set to manage all data for spatial calibration and validation of FORSANA and to provide model output facilities, and

- as driving forces for FORSANA, an immission/deposition and meteorological data set for the region: the spatial and temporal dynamics of the immission concentration of SO_2, NO_x, alkaline dusts, the deposition of SO_4^{2-}, NO_3^- and NH_4^+ and meteorological data (temperature, precipitation, radiation, etc.).

2.1. The forest growth model

The model FORSANA was developed for the description of the growth of coniferous forest stands within decades. It describes the reaction of physiological processes of the trees (assimilation, respiration, allocation and mortality) in relation to the impact of daily weather and air pollution. Photosynthesis and respiration processes are derived from the model FORGRO (Mohren, 1987). To model the influence of longterm environmental changes on tree growth, other physiological processes such as allocation and mortality and soil processes are implemented additionally as described in Grote (1996) and Kartschall *et al.* (1990). Forest stand characteristics (stand height, stem diameter, stem volume and number of trees per unit area) are derived from the single tree growth on annual time steps. This allows influences such as forest management activities to be simulated. A detailed description is given by Grote *et al.* (this volume) and in Bellmann *et al.* (1996).

2.2. *The regional data base*

For the calibration and validation of FORSANA, available spatial data of the area was collected and implemented in a geographic information system (GIS-ARC/INFO). Forest inventory, soil and topographic maps of the area were digitized and the attributed data sets and additional information such as soil descriptions and soil chemical analysis (see Table 1) were adjoined by importing the digital data into the relational data base system of the GIS. Additional attributes were derived using GIS – functionalities and algorithms. For example, Delaunay–Triangulation (ARC/INFO modul TIN) was used to construct a digital elevation model and for the interpolation of soil chemical data. To provide the forest growth model with a unique set of site condition parameters for each stand, so-called spatial homogenous units (SHUs) based on even aged forest patches were created. Therefore a 'supervised map overlay' was performed, assigning information from the different monothematic maps to the areas of the forest stands by different strategies (e.g. maximum area for soil type, mean value for exposure and weighted mean for slope). For the calibration of the forest growth model and for its validation as well as for the observation of changes in the environmental conditions data sets measured independently were collected for each forest stand.

The most important data sets are (Table 1):

- Forest inventory data (tree species, age, height, diameter, yield, size of the stand) for the year 1970 for the western part of the area and for the year 1992 for the whole area, which is based on the subscription of 1982 inventory data with empirical growth functions. So about 5000 different even-aged pine forest stands on an area of 20 000 ha were available for simulation runs. About 2500 of them could be used for validation purposes.

- Soil data (about 130 different soil types and their profiles, texture, field capacity down to 3 m depth) and measurements of soil chemistry at 175 different profiles (pH, nutrient contents, humus type, C/N ratio etc.) for the years 1967 (western part) and 1977 (eastern part of the area) and between 1988 and 1994. Apart from data about soil profiles which are measured along a gradient with measurements down to 2 m soil depth, most of the monitoring data for analysing the spatial distribution of pollution loads are measured in the first 20 cm of soil (see Konopatzky and Freyer, this volume, Konopatzky and Freyer, 1996).

- Biomonitoring data (defoliation, discoloration) of different measurement pogrammes (see also Ende *et al.*, this volume)

- Topography (elevation, slope and aspect) and depth of groundwater level.

Table 1. The available data sets and their temporal and spatial resolution

	Parameters	Temporal and spatial resolution
Forest inventory data	Stand age, height, diameter, density, structure, canopy closure, yield class, N- fertilization	1970, 1992 per year and forest stand
Biomonitoring	Defoliation, needle yellowing, necrosity, needle biomass, leaf area index	1965, 1985 per year and forest stand
Soil conditions	Soil texture, field capacity, soil depth, pH, humus type, C/N ratio, nutrient concentrations, groundwater distance from surface	1967/1977, 1988–1994 per year, irregular point data
Meteorology	Temperature (air / soil), precipitation, humidity, snow cover, wind velocity and direction, radiation / cloudiness	1967–1994, daily values, point data
Immission	SO_2, NO_x, dust	1970, 1974, 1982, 1989, daily values, 500 m × 500 m raster
Deposition	Ca, Mg, K, Na, SO_4, NO_3, NH_4, Cl, dust	1977–1980, 1985–1989, per time period, 1993–1994 per year, 1969–1981 per year, point data
Emission	SO_2, NO_x, NH_3, dust	1970–1992 daily values, point sources and 5 km × 5 km raster for non-point sources

2.3. The driving forces

The concentration of SO_2, the total amount of nitrogen and alkaline dusts and meteorological data are the driving forces of FORSANA. For the retrospective validation between 1970 and 1989, meteorological data were available from meteorological stations situated near the study area (DWD German weather service).

Because the spatial and temporal resolution of immission/deposition data from monitoring programmes were too coarse for the appropriate resolution of the landscape model, a model cascade was performed to supply the model with data for the whole validation time period (Figure 3).

An emission inventory data base for SO_2, NO_x, NH_3, CO and VOCs was created by the Institut für Energiewirtschaft und Rationelle Energieanwendung (IER, University Stuttgart), considering the whole area of East Germany (see Bellmann *et al.*, this volume) to calculate daily emissions of point (e.g. power plants and industrial complexes) and non-point sources (e.g. traffic, private heating in 5 km × 5 km spatial resolution) for the complete validation time period (Friedrich *et al.*, 1996). Daily values were calculated, taking into account production schemes, sequences of working days, weather and other factors.

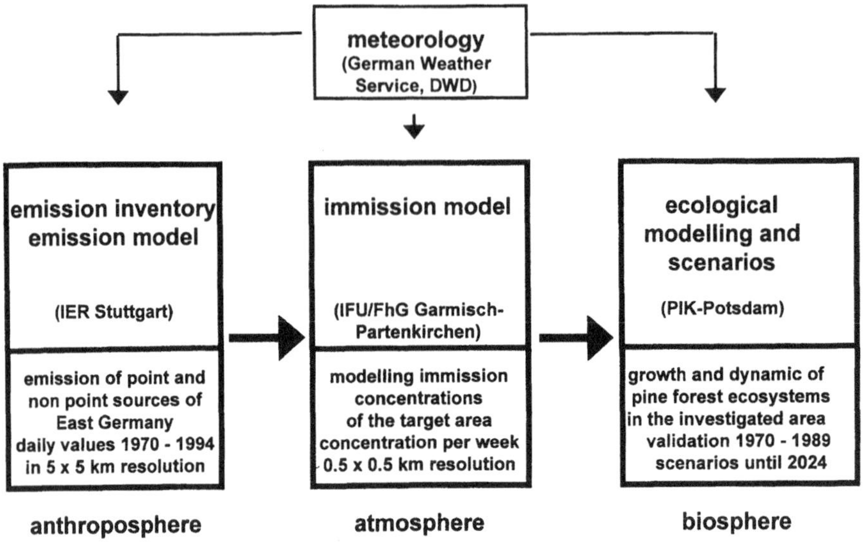

Figure 3. Modelling cascade for simulating long term immission concentrations with high temporal and spatial resolution

These model outputs were used to model daily immissions of SO_2, NO_x and dust (aerosols with diameter <20 μm) for the target area Dübener Heide. The calculations were carried out by the IFU/FhG (Fraunhofer-Institut für Atmosphärische Umweltforschung, Garmisch-Partenkirchen) using the Gaussian distribution model. Emission–immission matrices were calculated for every substance and emission source (point and non-point sources) depending on daily wind fields and atmospheric stability classes within an emission source area, which had been determined previously by sensitivity analysis (Schaller, 1996). Calculations of daily immissions within the target area are based on superposition of the weighted matrices and a 'background immission' from outside the main source area. At the present time model results are available for the years 1989, 1974 and 1982 at a spatial resolution of 500 m x 500 m.

To obtain information about deposition rates, monitoring programmes were evaluated and the data collected. Different data sets of wet deposition of cations and anions and total amount of fly ash deposition are available (Tables 4, 5a–c).

2.4. *Scenarios*

For the development of scenarios, neither information about the further trends in emission nor about climate trends was available in adequate spatial and temporal resolution. Emission trends were therefore estimated, making some simple assumptions on the basis of the environmental protection laws of Germany concerning the regulation of emissions. The year 1989 was selected as representative of the emission situation of the former GDR, 1992 for the time of economic transition and 1996 (estimated values) as representative for future trends.

Besides the immission/deposition load, meteorological conditions are dominating forest ecosystem dynamics. Climate scenario construction was carried out using the meteorological data of the validation period. Meteorological years were classified into normal, wet and dry years with respect to the vegetation period, using cluster analysis (Gerstengarbe and Werner, 1997). A representative year was selected from each class ('normal', 'dry' and 'wet' years). Climate scenarios were constructed as a sequence of these selected years (Table 2 and Figure 4). An important reason for choosing such an approach was that reliable model outputs of GCM's (general circulation models) are still not available for central Europe (Machenhauer, 1996) and also were not scaled down to an appropriate temporal and spatial resolution.

Table 2. Precipitation of the 'dry', 'normal' and 'wet' years selected from the meteorological data of 1967–1992 using Cluster analysis. In this analysis saisonality was also be considered

Scenario year	Meterological year	Precipitation (mm) Total sum	Veg. period	%
'Dry'	1982	364	167	46
'Normal'	1990	551	327	60
'Wet'	1974	761	346	45

The following emission scenarios were constructed (see Figure 5):

- a 'worst case' or 'GDR' scenario, extrapolating the emission situation of 1989 until 2024.
- a 'reunification' scenario, extrapolating the emission situation of 1992 until 2024. In 1992, emission reached in total approximately 50% of the year 1989.
- an 'emission reduction' scenario, reducing emissions of 1992 by approximately 40% which is estimated as the trend due to execution of current emission reduction regulations.

Different combinations of normal, wet and dry years with the emissions of 1989, 1992 and 1996 were created as driving forces for FORSANA.

Since lateral flows were not considered in the model, it was possible to perform complex emission and climate scenarios of the whole area subsequently for all simulated patches.

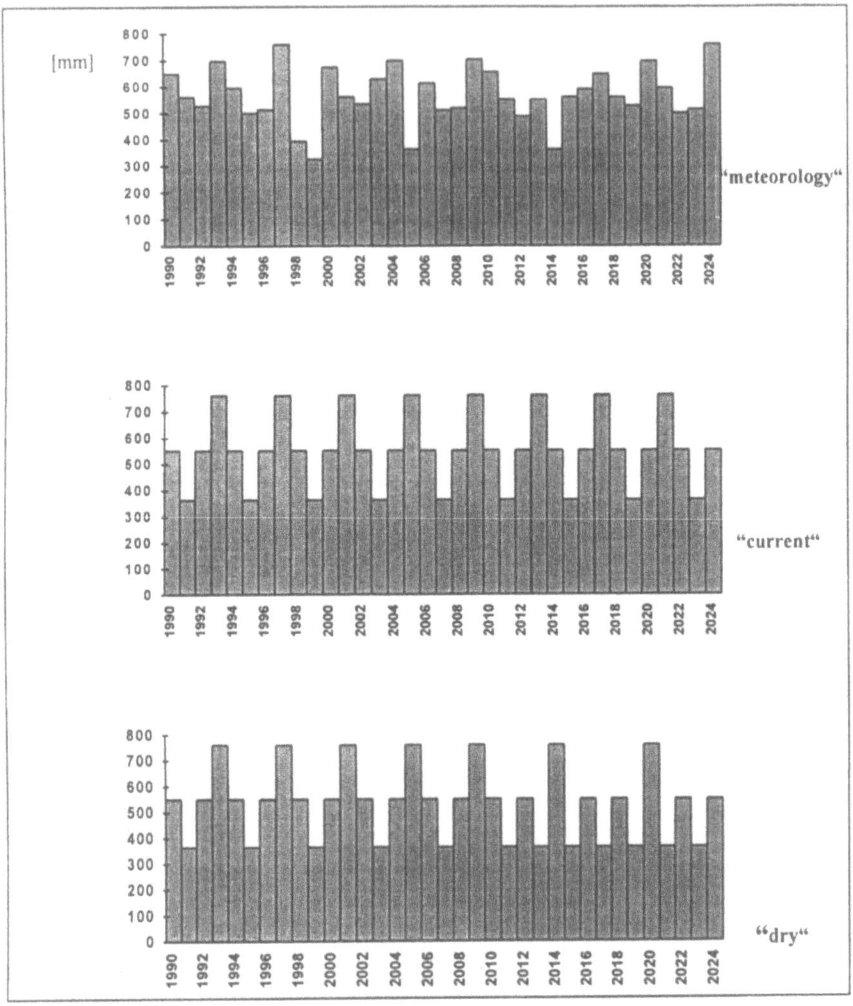

Figure 4. Climate scenarios 'current' and 'dry' as compositions of the representative years 'dry', 'normal' and 'wet' (see Table 2). The scenario 'meteorology' represents the climate of the validation period 1967–1991 (data: DWD)

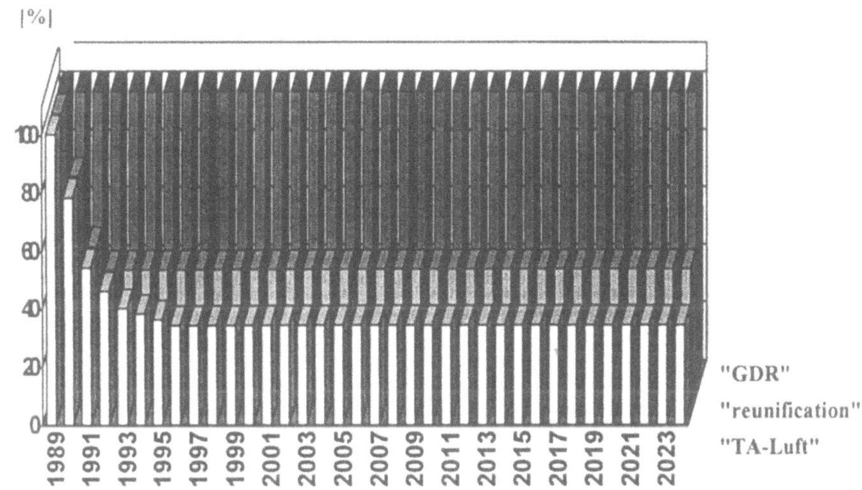

Figure 5. Emissions of the years 1992 ('reunification' scenario) and 1996 ('best' scenario) relative to 1989 (=100% 'GDR' or 'worst case' scenario) (Data: IER, Stuttgart)

Therefore the parallel simulation environment SPRINT-S could be used to perform the parallelization of the scenario runs on the scalable parallel computer system SP2 without modifying the model itself. SPRINT-S manages the organisation of data supply, distribution of the model runs on the different processors and the transfer of the results to the GIS for mapping purposes (Flechsig, 1998). With a 55 processor system it was possible to reduce the model run time from about one week to three hours by simulating the growth of 5000 forests stands for 22 years with daily resolution.

3. Results and discussion

3.1. *The changes of the environmental conditions in the area*

Forest damage has been observed in the Dübener Heide since about 1950 and was mapped systematically for the first time in the sixties using the method of crown condition assessment, where defoliation and needle discoloration are the most important indicators (Lux, 1965). The appropriate classification is shown in Figure 6a. The gradient of decreasing damage from west to east seems to be determined by the location of the main emittents and the dominating westerly winds. A second monitoring carried out 20 years later shows a quite similar pattern (Figure 6b). The areas of severe to low damage had expanded, but also the most severe damage (zone 1a) was no longer found. This dominating west–east gradient seems to be modified by local effects like ammonia immissions (stock-farming) and the lowering of the ground water table for drinking water

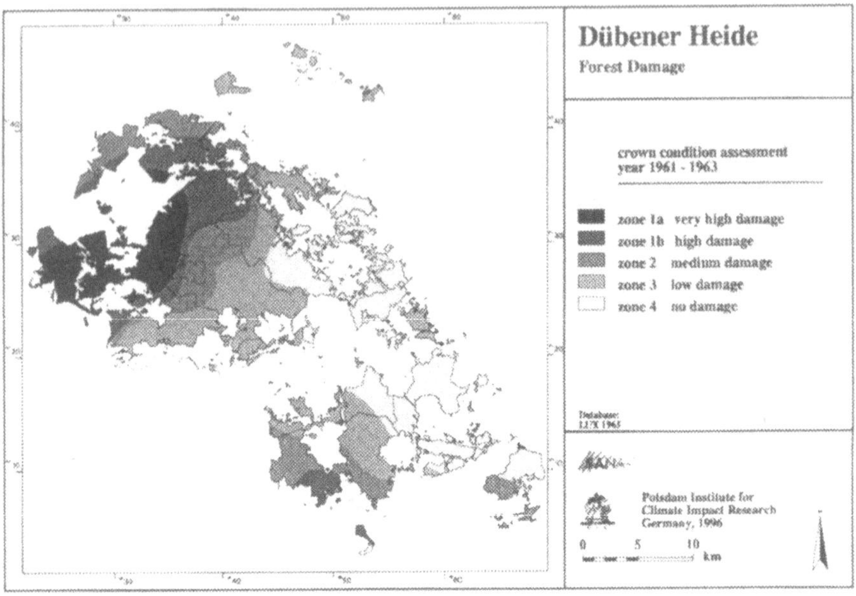

a

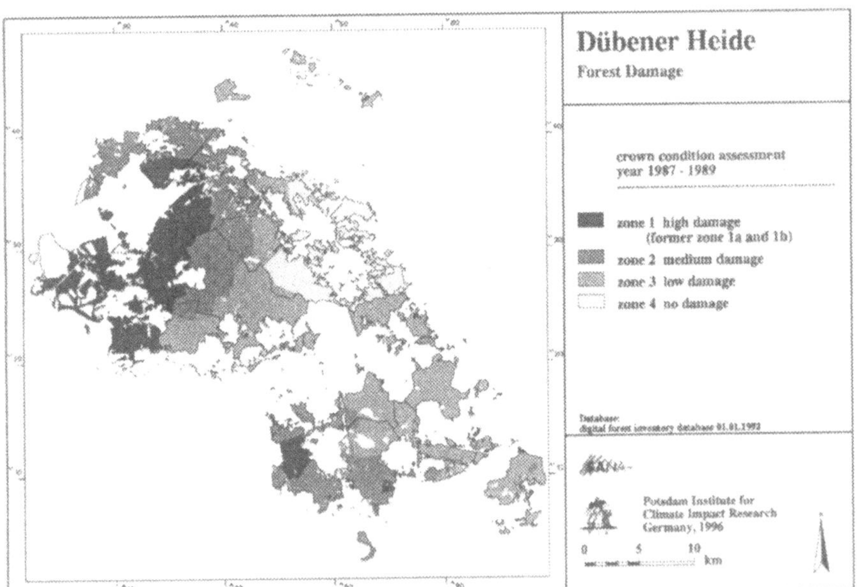

b

Figure 6. (a) Forest damage zones of the years 1961–1963 as described in Lux (1965). (b) Forest damage zones of the years 1987–1989 as described in the digital forest inventory data base of 1992

usage, which led to further patches with severe forest damages mostly in the southeastern part of the area.

The damages following the west-east gradient seem to be caused mainly by high ambient air concentrations of SO_2, which led to direct inhibitions of photosynthesis in the needles (Mohren *et al.*, 1992). The results of the immission model for 1989, the year with the highest SO_2 emissions (see Figure 7), show a gradient similar to the damage pattern. Average SO_2 concentrations reached about 150 μg m^{-3} in the western part and about 50 μg m^{-3} in the eastern part of the area. Similar values were also found by Flemming (1969). Figure 8 shows the variation within the year on the basis of average values per week at the locations of the measurement sites 'Rösa' and 'Taura' as examples (see also Figure 2). The most interesting aspect is that peak values of more than 200 μg m^{-3} were not only found in winter when high values can be expected due to meterological inversion situations in combination with enhanced energy demand (heating) and when photosynthesis is low or zero, but also within the growing season (weeks 33–37).

Modelling average SO_2 concentration using meteorological data of extremely dry and wet years shows no significant changes in the spatial pattern. Therefore it can be assumed that average SO_2 concentration for every location was strongly correlated with the SO_2 emission trend of the emission sources within the whole validation period (1970–1989).

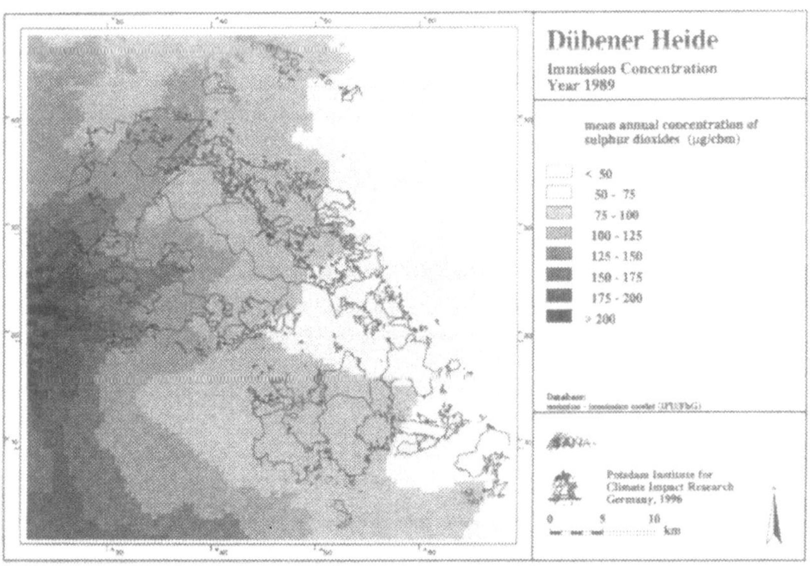

Figure 7. Mean annual immission concentration of SO_2 of the year 1989

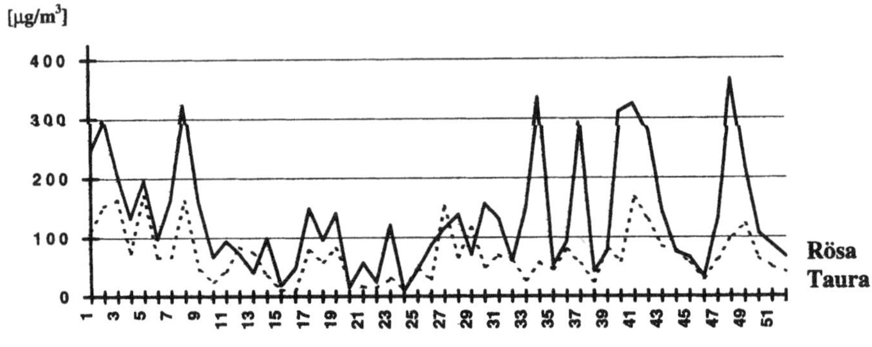

Figure 8. Simulated weekly SO_2-concentrations of 1989 (Data from IFU/FhG) at the SANA experimental sites 'Rösa' and 'Taura' (see Figure 3)

Deposition measurements of SO_4^{2-}-S show with 130 kg ha^{-1} yr^{-1} highest values at the site of Bitterfeld for the years 1977–1980. The gradient between Bitterfeld in the west and Torgau in the east of the area, as well as the trend of decreasing inputs in the eighties can be found in several measured deposition data (Table 5a–c).

The immission of alkaline dusts (fly ashes) shows a similar distribution pattern. Only the fine fraction of these dusts (particle <20 µm diameter) could be simulated with the emission–immission model. The concentrations ranged from 80 µg m^{-3} in the western part to 35 µg m^{-3} in the eastern part of the area. The chemical composition of these dusts is dominated by Ca^{2+} (mainly CaO) and Fe-oxides, but contains also high amounts of heavy metals such as Pb and Cu (Table 3). As can be seen from the measurements of dust in bulk and wet deposition data sets (Table 4), this gradient in small particle dust deposition mentioned above was enhanced by the sedimentation of fly ash of coarser size fractions (>20 µm diameter) in the west, which was transported only over short distances.

Table 3. Chemical composition of dust (in ppm) after Möller and Lux (1992)

Location	(n)	Na	K	Ca	Mg	Fe	Mn	Cu	Pb	Cd	Zn
Bitterfeld	15	3 150	1 760	61 800	7 950	32 100	240	340	250	6.6	510
Torgau	8	7 860	3 540	55 800	6 860	23 900	330	160	220	4.1	860

n= number of mean values per month

Table 4. Dust deposition (kg ha^{-1} yr^{-1}) at measurement sites (see Figure 3) after Niehus *et al.* (1982)

Year	Bitterfeld Dust dep.	Bitterfeld SD (n = 9)	Site 1 Dust dep.	Site 1 SD (n = 7)	Site 2 Dust. dep (n = 1)	Site 3 Dust. dep (n = 1)	Site 4 Dust dep. (n = 1)
1969	17190	10929	7690	85	3430	1900	–
1970	19114	12292	8737	2133	3240	1670	–
1971	23049	13485	9583	3214	3530	1520	2840
1972	29469	21547	9796	3821	3480	1830	2540
1973	18560	12784	8167	3650	3920	1750	1520
1974	12623	8564	6571	2448	2510	1520	1280
1975	11410	6870	6023	1361	2170	2540	1380
1976	6620	4843	4693	1564	2700	1870	760
1977	7443	6614	4800	1603	1820	1370	1230
1978	5933	3989	3533	958	2040	1200	1170
1979	5638	5162	3866	1713	1960	1170	1180
1980	5380	3880	3451	1920	1410	800	910
1981	5581	3584	3074	1126	1130	870	80

SD = standard deviation

Measurements near industrial sites in the early seventies show maximum rates of dust deposition up to 30 tons ha^{-1} yr^{-1} with a great variety within short distances (Table 4, Figure 2). After 1975, dust deposition decreased due to the installation of filters, and appears to decrease continuously until the year 1989. The measured deposition rates are higher in summer than in winter time. The measurements seem to be influenced by wind erosion (Niehus and Brüggemann, 1995). Therefore, the amount of industrial dust emission may be overestimated.

Ca deposition was estimated using these dust deposition measurements (Table 4) and analyses of their chemical composition (Möller and Lux, 1992). Deposition is therefore linked to the dust deposition trend (Figure 9). This trend of decreasing deposition can also be seen from several other measurements (Table 5a–c). Ca deposition reached values of 187 kg ha^{-1} yr^{-1} in Bitterfeld in 1977–1980, which is only less than 50% of the value calculated from dust deposition data. This and the calculations of total deposition in Möller and Lux (1992) (Table 5c) shows how much bulk deposition measurements underestimate total deposition. For 1984–1989 59 kg ha^{-1} yr^{-1} Ca-deposition was reported, which shows the general trend of decreasing deposition and which is still three times higher than the values for Torgau and ten times higher than the deposition at the background site.

The spatial pattern of dust deposition can be verified by mapping soil pH and the portion of Ca saturation of the soils (Figure 10 a,b and Figure 11a,b). Comparing the two chemical parameters of the soil confirms this gradient in deposition of fly ash. Both pH-values and Ca saturation of CEC (cation exchange capacity) decreased between the years 1967/1977 and 1989. The area

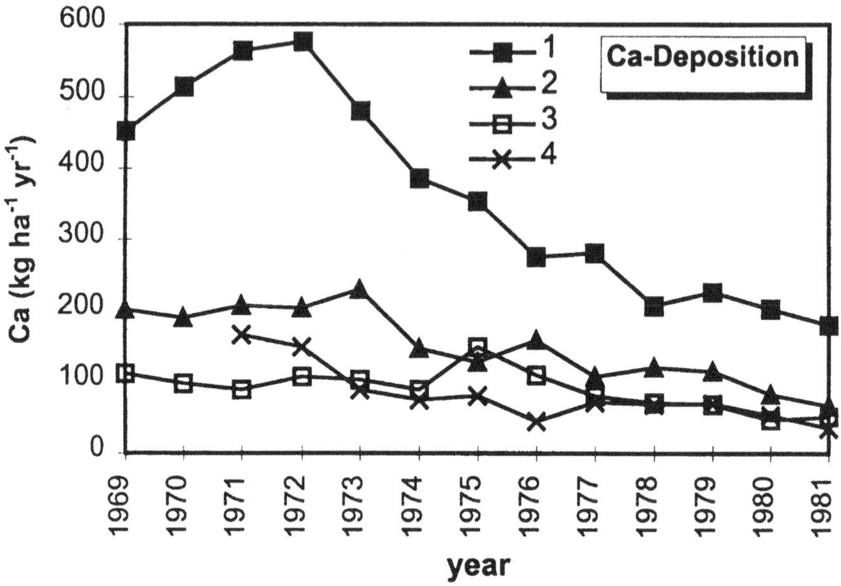

Figure 9. Ca-depositon at measurement sites 1–4 (see Figure 3) estimated from total dust deposition (Niehus *et al.*, 1982) and their chemical composition (Table 3)

with pH values below pH 4.2, increased from about 8500 to 16 600 ha in the east of the investigation area. Below this value, an enhanced mobilisation of Aluminium has to be expected, which reduces the Ca/Al ratio in the soil and may lead to nutrient imbalances and root damages (Magistad, 1925, Ulrich, 1987, Schulze *et al.*, 1989). At present the Ca/Al ratio is one of the most important indicators for forest damage caused by acidification, (Sverdrup and Warfvinge, 1993, Posch *et al.*, 1995). Because base cation supply in the humic layers of the soils is still high (see Weisdorfer *et al.* and Konopatzky and Freyer, this volume) these results should only be interpreted as a first hint of where acidification may cause such problems in future.

Mapping the Ca-storage of the soil this gradient cannot be found. In Figure 12a and 12b one maximum in the western and one in the eastern part of the area can be seen, which cannot be explained by soil chemical properties. Although the results of the interpolation may be influenced by the location of the sampling points as well as by random effects of the sampling itself, these peaks can be found in both data sets. On analysing the results of SO_2 immission modelling and mapping of nitrogen load (see Figures 7 and 13a), the area in the west seems to be polluted directly by one of the main emittents located in or near Bitterfeld. But no explanation has yet been found for the peak in the east.

The nitrogen status of the soils was mapped in 1967/1977 and 1992 by using humus type, C/N ratio and ground vegetation species as indicators. The information was quantified by empirical functions as described in Kopp and Kirschner (1992) and verified for the investigation area by Konopatzky and Freyer (this volume and Konopatzky and Freyer, 1996). In the seventies most of the sites had 'low' or 'optimum' nitrogen status whereas the 'optimum' status is to be considered as 'typical' in Pine forests (Kopp and Kirschner, 1992). Only some small areas in the north (caused by emissions of a nitrogen fertilizer plant near Wittenberg see also Figure 2 and Table 5a) and some patches at the edge of the forests (polluted by stock-farming) were well supplied with N at this time. The information about nitrogen deposition (Table 5a–5c) show rates of about

Table 5a. Bulk deposition data (kg ha^{-1} yr^{-1}) normed with mean annual rainfall during 1967–1990 to analyse the trend of deposition (data from Möller and Lux, 1992). Data of the SANA background measurement site 'Neuglobsow' are shown for comparisions

	Time	pH	Na	K	Ca	Mg	NO_3 -N	NH_4 -N	SO_4 -S	Cl	PO_4 -P	Fe	Pb	Cd
B	1977–1980	6.50	6.58	4.93	186.9	8.22	6.31	7.23	129.9	39.46	0.27	10.2	0.14	0.01
B	1984–1989	5.80	2.62	0.95	59.02	3.67	4.09	6.65	52.43	15.71	0.33	0.38	0.05	0.02
W	1985–1989	5.44	6.50	3.55	47.40	3.84	5.11	15.97	59.39	11.11	–	–	–	–
T	1975–1979	5.35	5.20	3.12	36.40	3.64	4.11	6.46	38.18	16.64	0.25	5.20	0.10	0.01
T	1984–1989	4.70	2.03	0.81	18.29	1.47	4.25	6.31	23.74	16.76	0.30	0.66	0.02	0.02
N	1985–1989	4.30	3.73	1.19	5.71	0.85	4.30	5.05	13.37	7.80	–	–	–	–

B = Bitterfeld, T = Torgau, N = Neuglobsow, W = Wittenberg

Table 5b. Bulk (b) deposition and throughfall (t) (kg ha^{-1} yr^{-1}) in 1985–1989 (data: Simon and Westendorff, 1991) and 1993–1994 (data: Niehus and Brüggemann, 1995) at a forest measurement site (see Figure 3)

	Time	Precip (mm/yr)	pH	Na	K	Ca	Mg	NO_3 -N	NH_4 -N	SO_4 -S	Cl	PO_4 -P	Fe	Al
b	1985–1989	609	4.86	5.41	4.82	68.28	6.41	7.72	6.17	52.24	25.1	0.04	0.32	0.61
t	1985–1989	394	4.65	8.18	16.70	119.9	12.20	8.19	12.61	112.10	34.8	0.03	0.81	3.50
b	1993–1994	811	4.50	–	0.94	9.34	0.06	5.44	6.56	14.40	7.14	–	0.12	0.36
t	1993–1994	645	3.80	–	20.80	37.00	3.97	12.60	12.2	58.30	15.50	–	1.40	2.98

Table 5c. Total deposition (wet and dry in kg ha^{-1} yr^{-1}) in Möller and Lux (1992) for the period 10/1984–10/1989

Station	Na	K	Ca	Mg	N	S	Cl	P	Fe	Pb	Cd	Zn	F
Bitterfeld	15.7	8.3	320.0	36.5	38.0	190.0	54.0	0.32	125.0	1.03	0.084	2.31	9.82
Torgau	6.4	3.8	52.0	5.0	23.0	61.0	34.5	0.22	10.9	0.13	0.033	0.81	2.52

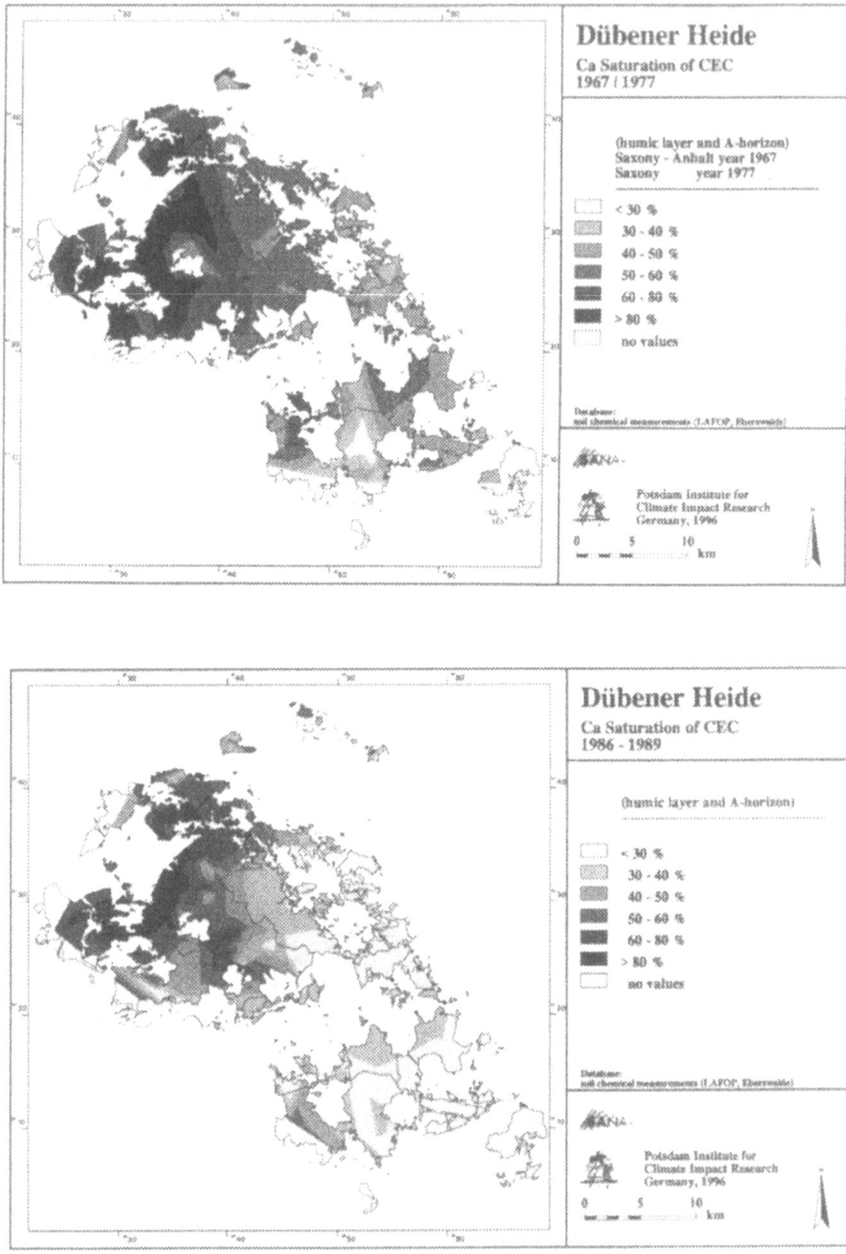

a

b

Figure 10. (a) pH values of the year 1967/1977 of the upper soil (20 cm depth) as interpolated from soil chemical measurements (data: LAFOP, Eberswalde). (b) pH values of the years 1986–1989 of the upper soil (20 cm depth) as interpolated from soil chemical measurements (data: LAFOP, Eberswalde)

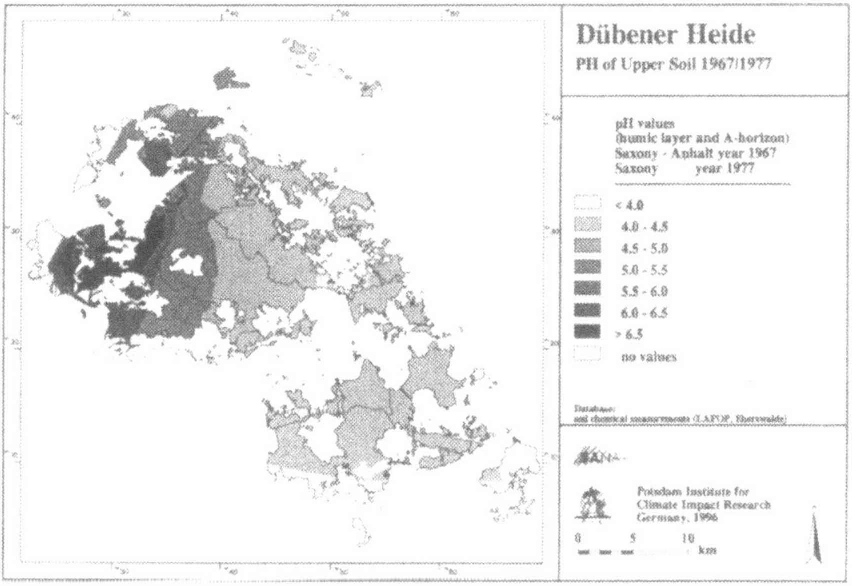

a

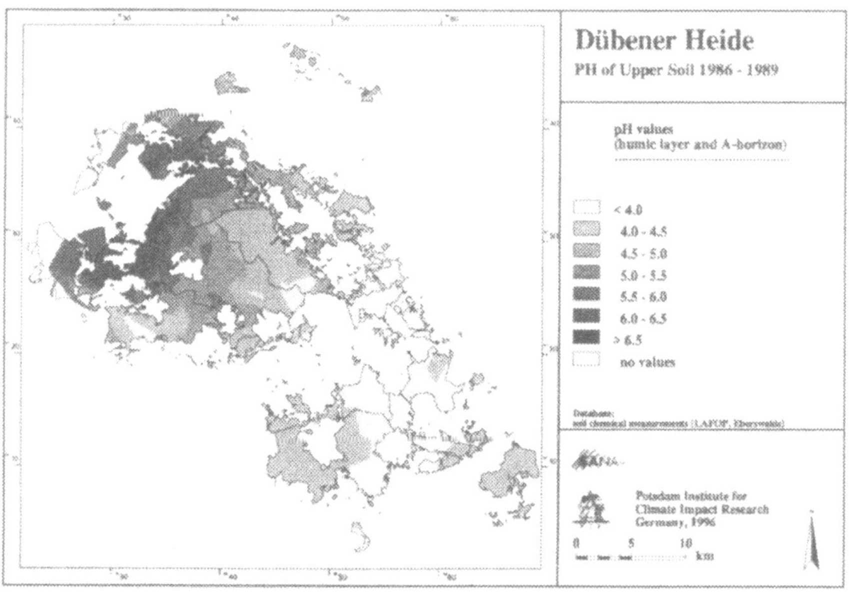

b

Figure 11. (a) Calcium saturation of the effective cation exchange capacity (CEC) of the upper soils for the years 1967/1977 (data: LAFOP, Eberswalde). (b) Calcium saturation of the effective cation exchange capacity (CEC) of the upper soils for the years 1986–1989 (data: LAFOP, Eberswalde)

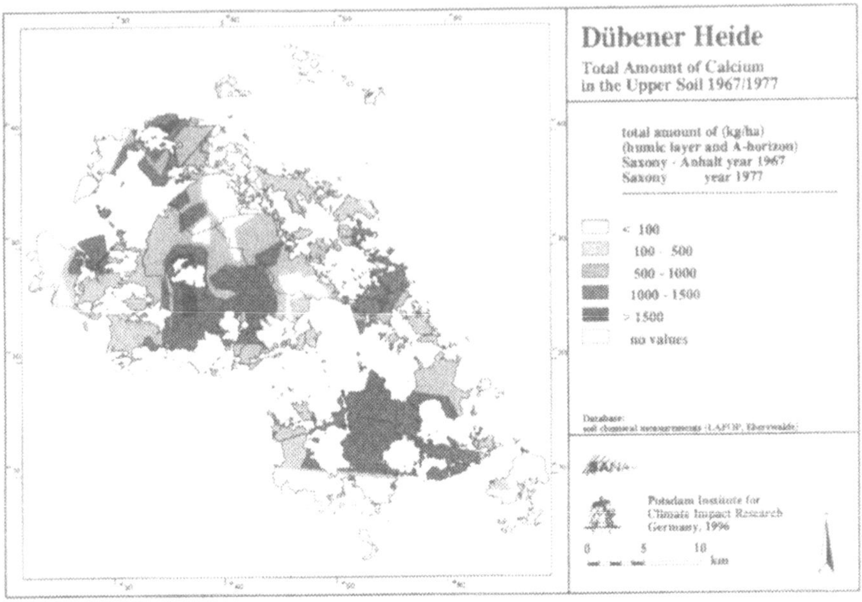

a

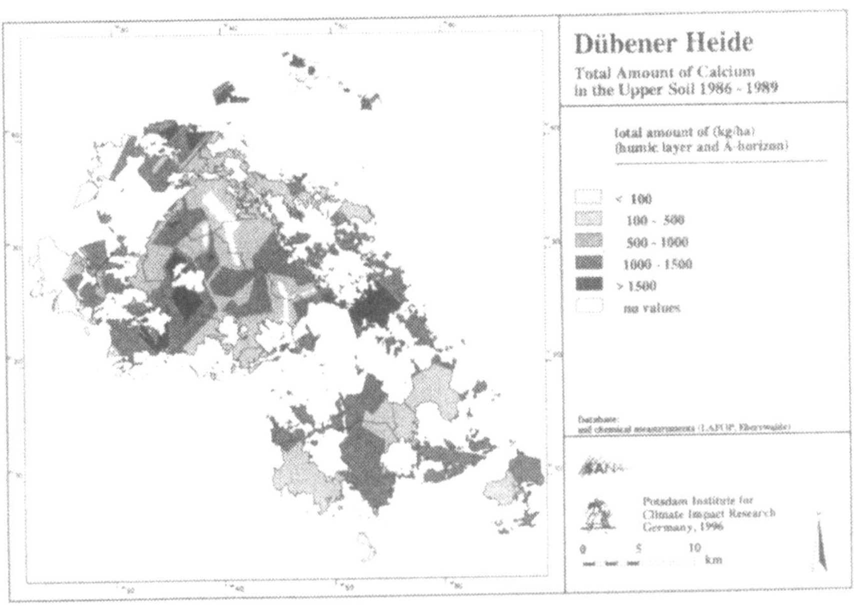

b

Figure 12. (a) Amount of calcium in the upper soils in kg ha^{-1} (until 20 cm depth) of the years 1967/1977 (data: LAFOP, Eberswalde). (b) Amount of calcium in the upper soils in kg ha^{-1} (until 20 cm depth) of the years 1986–1989 (data: LAFOP, Eberswalde)

10–15 kg ha^{-1} yr^{-1} Nitrogen (NO_3^- and NH_4^+-N) bulk deposition, which means about 20–25 kg ha^{-1} yr^{-1} N-input into the forest stands (throughfall). These rates do not seem to have changed very much since 1989. Niehus and Brüggemann (1995) compared measurements of two periods at the site 'Dübener Heide' (see Figure 2) and found bulk deposition rates of 12 kg ha^{-1} yr^{-1} N-deposition for the year 1993/1994 vs. 14.4 kg ha^{-1} yr^{-1} in 1985/1989. For throughfall values of 25 kg ha^{-1} yr^{-1} N vs. 21 kg ha^{-1} yr^{-1} N were described. Mean values from the same authors from 6 sites in the west of the investigation area lead to values of 13.5 kg ha^{-1} yr^{-1} N-bulk deposition and 20.5 kg ha^{-1} yr^{-1} throughfall for the years 1993/1994.

To improve the growth conditions of the forests, N-fertilization programmes were started in 1970. Amounts of about 300–360 kg ha^{-1} of nitrogen were applied over periods of 3 to 4 years, with rates of 100–120 kg ha^{-1} N per year as urea fertilizer. In most cases, these applications were repeated up to three times (that means 900 to 1000 kg ha^{-1} N-input) in the damage zone 1, twice in zone 2 (600–700 kg ha^{-1} yr^{-1} N) and once in zone 3 (300–350 kg ha^{-1} yr^{-1}). N-net uptake by tree growth is estimated as about 4–5 kg ha^{-1} yr^{-1} (Grote *et al.*, this volume). Most of the N-surplus was incorporated in the humic layer and the A-horizon of the soils until a status of 'maximum' N-storage (N-saturation) was reached. If N-input into these soils continues, then the amount of N in the soils is reduced again (Table 6). In this case an increased outflow of nitrate into the groundwater has to be assumed (Konopatzky and Freyer, 1996). Gaseous nitrogen losses are estimated at about 5 kg ha^{-1} yr^{-1} (Brumme, 1995). Estimating maximum N-uptake and gaseous losses as being about 10 kg ha^{-1} yr^{-1}, an N-surplus of 10 to 15 kg ha^{-1} yr^{-1} will be left in the forest stands, under present immission conditions. Comparing bulk deposition values with total deposition as shown in Table 5a–c these values have to be multiplied by factor of 2–3 to obtain total deposition. Therefore the amount of N-surplus will increase to 30–45 kg ha^{-1} yr^{-1}.

Mapping the changes of soil N-status (Figure 13b), another deposition pattern can be seen than that shown for SO_2 and dust immissions (Figures 7 and 11 a,b). N-deposition increases from the middle to the edge of the area, with the exception of a small strip just leewards of the location of the main emittents (Figure 13a). The N-content of the soils increased in almost all of the stands (Table 6). No correlation was found between the N-increase of the soil and the fertilization rate and/or with the N-deposition pattern. One of the reasons for this may be that the amount of fertilizer applied together with the N-deposition rate is within the range of the different humus classes (Table 6). Furthermore a certain amount of nitrogen may be incorporated in the biomass of the ground vegetation as long as this biomass increases. Because no information is available about N-losses during and shortly after fertilizer application, the effectiveness of these applications cannot be evaluated.

As the result of fertilization and deposition, the area of nitrogen saturated soils in Pine forests (nitrogen status 'maximum' see Table 6 and Figure 13b) increased from 1300 ha to 5800 ha, the area where N-losses have to be assumed

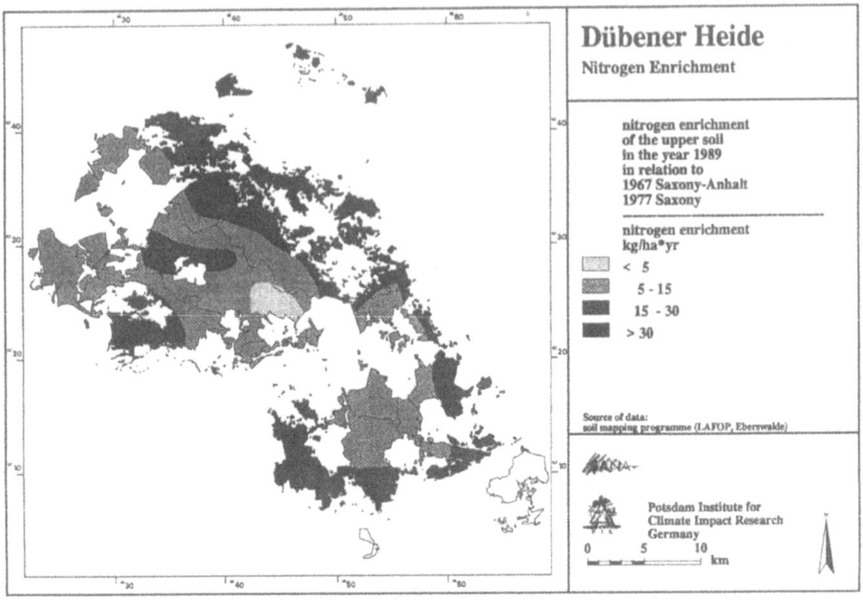

a

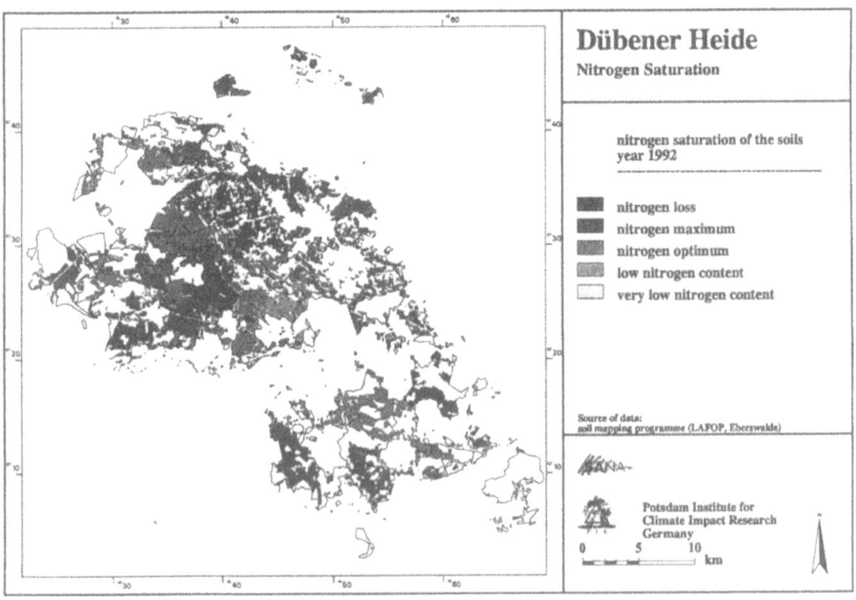

Figure 13. (a) Nitrogen increase of the humus layer and the A-horizon as mapped 1967/1977 and 1992 using humus type, C/N ratio and soil vegetation species (data: LAFOP, Eberswalde). The values are estimated for soils with 'low nitrogen content' (see also Table 6). (b) Nitrogen status of the soils in the year 1992, calculated on the basis of the data of 1967/1977 and the information as shown in Figure 13a

from 33 ha to 1050 ha within the validation period. The actual rates of nitrogen deposition indicate that this process will continue in future and may lead to an enhanced output of nitrate into the ground water.

Comparing the stem diameter, height and yield of the two forest inventory data sets (1970 and 1992), a general improvement of growth conditions in every damage zone within this time period is shown (Table 7). Although SO_2 immission increased within the validation period, the nitrogen input by

Table 6. Changes in soil nitrogen content and nitrogen storage capacity depending on the degree of saturation (after Konopatzky and Freyer, 1996)

Nitrogen content of humus	Area 1967/ 1977 (%)	Area 1992 (%)	N % of C_{total}	Humus amount (t/ha)	N-amount (ka/ha)	N storage capacity (ka/ha)
Very low nitrogen content	0.001	0.001	2.3	73.1	1423	
Low nitrogen content	71.11	0.03	2.9	98.4	2008	585
Nitrogen optimum	17.86	67.91	3.8	108.3	2637	629
Nitrogen maximum	4.09	17.82	4.9	91.9	2831	194
Nitrogen loss	0.001	3.24	6.2	63.8	2559	–272
No informations	6.68	6.68	–	–	–	

Total area = 32635 ha

Table 7. Mean values and standard deviations of stem diameter, stem height and yield of the forest inventory data in 1970 and 1992 (for damage zones see Figure 6a)

Damage Zone	1970 Mean	1970 SD	1970 n	1992 Mean	1992 SD	1992 n
			Stem diameter (m^2 ha^{-1})			
1a	14.506	5.954	178	18.154	5.767	642
1b	15.170	5.273	718	19.137	6.306	1779
2	16.533	5.886	460	19.345	6.612	1158
3	17.087	6.007	565	20.164	6.830	1452
4	15.629	5.240	650	19.652	6.999	1587
			Stem height (m)			
1a	11.226	3.442	177	14.020	3.175	656
1b	12.368	3.152	718	14.969	3.522	1804
2	13.567	3.738	460	15.422	3.817	1168
3	14.101	3.864	566	15.965	3.990	1466
4	13.438	3.365	650	15.730	3.978	1602
			Yield (m^3 ha^{-1})			
1a	82.424	34.987	177	118.555	43.844	633
1b	103.742	34.789	698	134.367	50.910	1771
2	121.733	40.552	449	147.745	54.139	1152
3	132.623	49.725	560	158.612	59.493	1447
4	132.130	46.770	644	161.547	74.240	1585

SD = standard deviation n = number of forest stands

deposition and fertilization and eventually the high amounts of base cations deposited in this time have improved the growth conditions. In the low polluted areas (zone 4) tree growth may be limited by nitrogen.

Multiple linear regression shows only weak relationships between site conditions, pollution zones and tree growth ($r^2 < 0.50$). The reason for this could be that the classification of zones of different pollution types is still not sensitive enough for regression analysis because the gradients of pollution are different for the different substances (see nitrogen vs. SO_2 gradients) or the relationships are nonlinear. Further investigations have to be carried out to analyse this.

3.2. *Model validation*

The forest growth model FORSANA was validated at 50 different sites with different levels of pollution. The number of sites was restricted to forest stands where measurements of soil chemical data were available, which of course increases confidence in data accuracy as compared to interpolated data. Modelling was performed using forest inventory data of 1970 as initial values and comparing the results at 1992 with forest inventory data of the same year. As can be seen in Figure 14 the current version of FORSANA seems to underestimate the stem diameter and stem height of younger trees and in the same way overestimate both if old stands are simulated. Variations in the initialisation data as well as differences between real and simulated management seem to be the reason for this. Otherwise total stand growth is calculated quite realistically instead of the different pollution patterns. One explanation for this is the compensation of differences between measured and simulated stem size by stand density.

3.3. *Scenario analyses*

Scenario calculations were carried out considering climate, nitrogen budget and SO_2 effects on forest stand growth. For the analysis of the results, the state variables total wood growth, stem growth and amount of needles were considered.

Needle loss is one parameter of the yearly monitoring programme of forest condition in Germany (WZB, 1995; EC-UN/ECE, 1995) and can be considered as an indicator of stability of the stands and their resistance to stress factors such as droughts and insects.

As shown in Table 8, increase of needle biomass is directly related to the reduction of pollution and seems only be limited by water supply. So the former highly polluted areas will profit mostly from pollution reduction. Further reductions of SO_2 emission will only have a small effect on this parameter. An increase in needle biomass may also reduce the biomass of ground vegetation and therefore improve the water supply of the trees, if this effect is not compensated by increased interception.

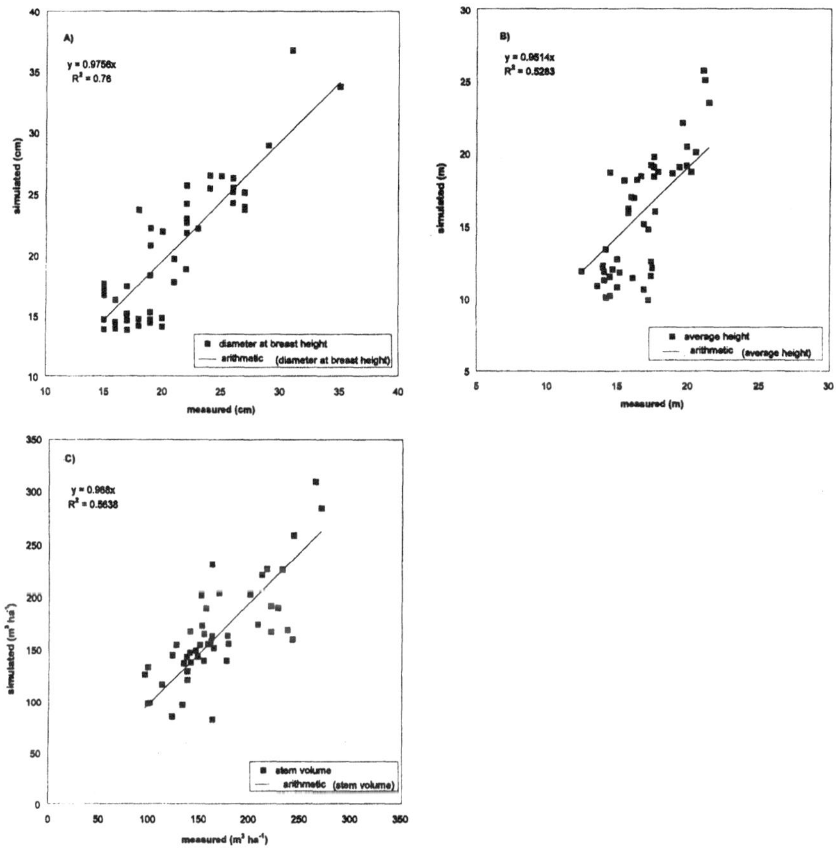

Figure 14. Relationship between measured and simulated stand dimensions (A: stem diameter; B: stand height; C: stemwood volume)

Table 8. Average needle percentage (%) in the damage zones and needle development under different emission scenarios with 'recent' climate

Damage zone	1961–63 Lux 1965	1987–89 Forest inventory data 1992	Emission scenarios 2024 'GDR'	'Reunification'	'TA Luft'
1a	25–50	54	60.0	84.4	93.0
1b	50–60	61	70.3	90.0	96.0
2	60–67	67	73.0	91.5	97.4
3	67–75	69	74.1	90.0	96.3
4	75–90	76	79.3	91.1	96.2

The development of total wood growth, an indicator for the productivity of the forests, calculated as yield (2024) – yield (1990) + [harvest (1990 – 2024) by 'normal management'] shows the same trend. Comparing the 'worst case' with the 'best' emission scenario, total wood growth will increase at an average rate of 6–15% in the western part of the area, above all in the former damage zones 1 and 2 (see also Figure 6a,b). In the other zones growth will remain more or less constant.

The spatial pattern of the growth trends is shown in Figure 15. The heterogeneity of production changes within short distances are primarily

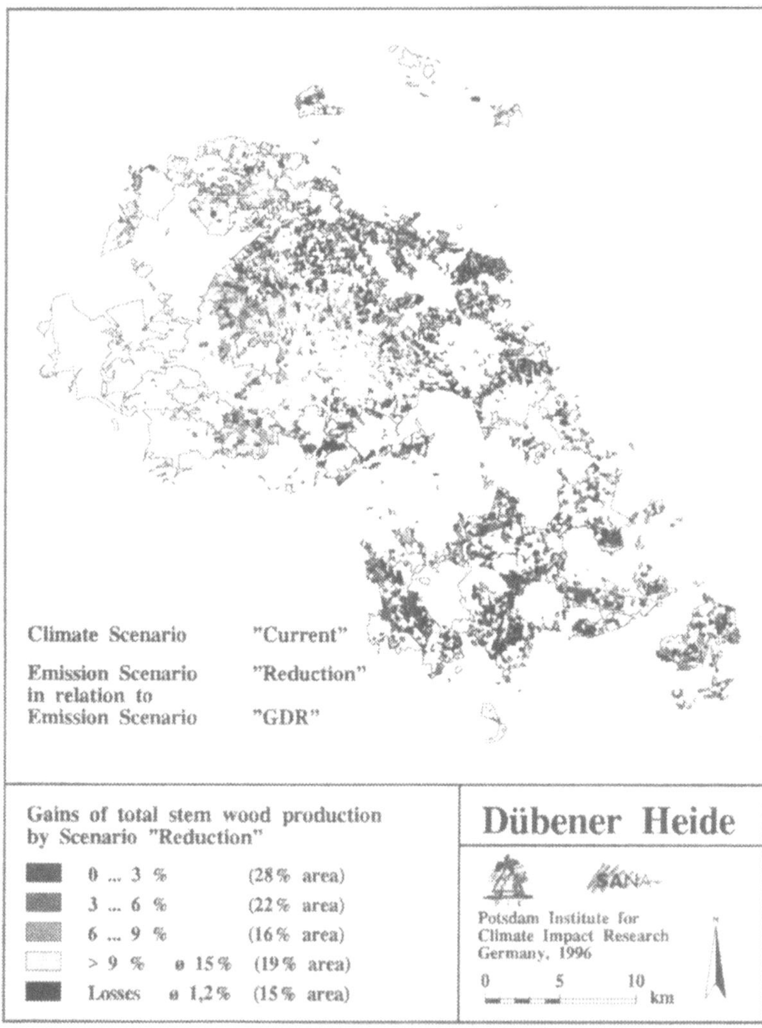

Figure 15. Spatial pattern of changes in total wood growth as difference between the scenario 'worst' and the scenario 'best' under 'recent' climate conditions

caused by the heterogeneity of the initial conditions, which are normally a consequence of management. In most cases the year of the last management may be the most important factor. Of course this effect becomes relatively more important if changes in growth rate are only small ($\pm$3%).

As a result of these trends, the production of stem wood (yield 2024 – yield 1990) which is available for usage will increase significantly on 79% of the investigated area as the consequence of pollution reduction. It seems, that this compartment will mostly profit from emission reduction. So the highest increase rates can be expected in the former highly polluted western part of the region, but areas with distinct increases of stem wood production are also found in the eastern parts of the area.

All emission scenarios are based on the spatial pattern of emission of 1989. Immission model results based on emission data sets of 1992 or later are still not available. But because the industrial structure has changed considerably within the last six years, the spatial pattern of immission concentrations has changed. This will modify the results of scenarios. Present deposition data indicates that there is no longer such a steep gradient of SO_2 concentration, as it is shown in Figure 7. In this case, growth increase is still underestimated in the western part of the area.

A moderate reduction of water availability implemented in the climate scenario 'dry' by increasing the number of dry years has only a weak effect on stand growth (Table 9) under 'best' emission scenario conditions. Transpiration is reduced by about 25% in the dry years, but this does not affect assimilation to the same order of magnitude. That means an increase in the water use efficiency. Furthermore, periods of water stress are observed in almost all the years, so that the reduction of growth is less than may be expected from the precipitation data (Grote, personal communication, 1996).

Table 9. Changes in wood production under different scenarios. The spatial distribution of changes in growth are shown in Figure 15

Total wood growth emission scenario 'best' – 'worst' climate 'present'		Stem wood growth emission scenario 'best' – 'worst' climate 'present'		Total wood growth emission scenario 'best', climate 'present' – 'dry'	
Growth in % (average value)	% of investigated areea	Growth in m^3 (average value)	% of investigated area	Reduction in % (average value)	% of investigated area
<0.0 (1.2)	15	<0 (4.5)	8	>-3.0 (3.5)	40
0.0 – 3.0	28	0 – 5	13	-3.0 – -2.0	19
3.0 – 6.0	22	5 – 10	18	-2.0 – -1.0	8
6.0 – 9.0	16	10 – 20	31	-1.0 – 0.0	7
>9.0 (15)	19	>20 (35)	30	>0.0 (0.8)	26

4. Conclusions

The coupling of emission, atmospheric transport and impact models is an important task for the investigation of the impact of pollution abatement strategies on the environment. The model approaches of the different scientific disciplines involved in such a study are based on different temporal and spatial scales according to the response time of the sub-system under investigation. For the linking of the different approaches a compromise has to be found. To tackle this problem at the landscape scale, an area in the order of magnitude of 100 km^2 to 1000 km^2 was selected and a time period of years to decades, the response time of forest ecosystems to air pollution which allows investigation of concentration and accumulation effects, was chosen. The results presented here give an overview about the potentials and limitations of such an approach.

For modelling immission concentrations over decades only quite simple approaches can be used. Highly sophisticated models (so-called Langrange models) are available, but cannot be applied in this case because of still insufficient computing capacity and the high demand for model calibration data (Schaller, personal communication, 1996).

Another important point is that the coarser fraction of the alkaline dusts could not be simulated with the immission model because the transport of these particles is strongly influenced by gravity, which means that another modelling approach (as developed for wind erosion investigations) has to be used. Although these substances are only transported over short distances (10 to 20 km), they are very important for the nutrient budget of the ecosystems within this area. The spatial pattern of dust deposition can be measured quite easily by soil chemical analysis, but as the results demonstrate, it is difficult to quantify such measurements for modelling purposes.

Dynamic process models seem to be the appropriate tool to map ecosystem dynamics in areas with different gradients of pollution. The comparision between simulated and measured data demonstrates that the dynamic model reproduces the spatial distribution of forest growth within a wide range of different site conditions. The deviations between measured and simulated data seem to be systematic, so that model results may be improved by modifying the management module. This may improve the simulation of stem height and diameter but will probably have no effects on the simulation of stand growth. The lack of sufficient immission data since 1992 is the most limiting factor for obtaining more realistic scenario results.

The concept of modelling forest ecosystem dynamics with process-based models for larger areas is limited by two factors. One is the difficulty in deriving model validation and calibration data from ecological monitoring programmes, as already mentioned above. The other is that some relationships between tree growth and environmental parameters are only known as empirical relationships. The physiological processes themselves have not finally been investigated and are still under discussion. For example, the relationship between acidification and destabilisation of forest ecosystems is described in Ulrich (1987). In the concept of Critical Loads/Levels the Ca/Al ratio in the soil solution is used as an indicator to estimate the risks for further forest development (UBA, 1996). Such empirical growth reduction functions as described in Sverdrup and Warfvinge (1993) were not yet implemented in the process-based model FORSANA; the extent to which this is possible and how much of the flexibility of the model will be lost still have to be discussed. Consequently, only qualitative or semi-quantitative risk assessments can be carried out currently in this case.

Further investigations by analysing the available data will improve the assessment of forest dynamics in the investigated area. In this case the statistical interrelationships between base cation supply, nitrogen input, SO_2 immission, the natural site conditions and stand growth are of special interest. This will also give some hints for an optimal investigation of regions with similar pollution loads, which may be quite common near the industrial areas of eastern Europe.

The present results show, on the one hand, that forest growth conditions will be improved generally by emission reduction, especially in former highly polluted areas. The production of stem wood seems to increase relatively more than total growth rate. The increase of growth rates may be underestimated in the western part of the area because present deposition data indicates that the gradient of SO_2 concentrations (see Figure 7) no longer exists. On the other hand, there are some hints that ecosystem stability and function can be affected by acidification and nitrogen saturation in the future.

Due to the deposition of alkaline dusts the western part of the area is well provided with Ca^{2+} and Mg^{2+} (Weisdorfer *et al.*, this volume), but the rate of base cation loss is high due to the low exchange capacity of the soils (Konopatzky and Freyer, this volume) and the reduction of dust emissions after the year 1989. Tree growth may be limited by low phosphate supply (Ende *et al.*, this volume) and also affected by the mobilisation of heavy metals in the future if the acidification of the soils continues. In the eastern part of the area, pH values of the upper horizons have already reached levels of below 4.2. Therefore an enhanced Al-mobilisation rate is observed but is still compensated by relatively high concentrations of basic cations in the soil solution (Weisdorfer *et al.*, this volume).

The high rates of nitrogen input by deposition and fertilization have improved the forest growth conditions in the past even under increasing SO_2 immissions (see Table 7). According to the empirical relationship between humus type and nitrogen storage capacity (Table 6) the area where soils are nitrogen saturated has increased significantly. Combined with relatively high N-deposition rates it has to be expected that nitrate pollution of the groundwater will increase (see Table 6).

The observations concerning the soil processes can only be interpreted as general risk assessments. To obtain reliable information about the further development in this area, measurements should be carried out to observe the soil processes under acidification in the former highly polluted areas. Nitrogen dynamics at N-saturated sites should be investigated for the same reasons.

Acknowledgement

This project could only be realized with the goodwill and support of many other working groups and institutions.

The authors thank Dr. Friedrich from the University of Stuttgart, Institut für Rationelle Energieanwendung, for his investigations in the emission inventory, Prof. Schaller from the Fraunhofer-Institut für Atmosphärische Umweltforschung Garmisch-Partenkirchen (now Technische Universität Cottbus) for emission–immission modelling, Dr. Niehus (Institut für Troposphärenforschung, Leipzig), Dr. Hasselbach (forest administration of Saxony-Anhalt) and Prof. Braun (forest administration of Saxony) for their data support, and last but not least the ecology groups of the SANA project who have provided us with the results of their investigations.

This study was supported by the German Federal Ministry for Education, Science, Research and Technology under grant No. 07 VLP 02.

References

Bellmann K *et al.* 1996. Untersuchung der immissionsbedingten Dynamik von Ökosystemen unter dem Einfluß sich verändernder Schadstoffemissionen durch Emissionssenkungsmaßnahmen – ökologische Wirkungen; Abschlußbericht zum Teilprojekt E 2.1 des BMBF – Verbundforschungsvorhabens SANA; Band IV; GSF, München.

Brumme R. 1995. Mechanisms of carbon and nutrient release and retention in beech forest gaps; III. Environmental regulation of soil respiration and nitrous oxide emissions along a microclimatic gradient. Plant Soil. 168–169, 593–600.

EC-UN/ECE 1995. Forest Condition in Europe. Results of the 1994 Survey. 1995 Report, Brussels, Geneva, 106 pp.

Flechsig M. 1998. A parallelization tool for experiments with simulation models. PIK Report No. 47, Potsdam, 66 pp.

Flemming G. 1969. Rechnerische Kartierung von SO_2-Relativwerten im Rauchschadensgebiet Dübener Heide. Angewandte Meteorologie. 5, H (112), 44–49.

Friedrich R. *et al.* 1996. Ermittlung von Luftschademissionen in den neuen Bundesländern. Abschlußbericht zum Teilprojekt A 1.2 des BMBF – Verbundforschungsvorhabens SANA. Band I. GSF, München, 41 S.

Gerstengarbe F-W, Werner P. 1997. A new quality criterion to separate clusters. Theor Appl Climatol. 57, 103–110.

Grote R. 1998. Integrating dynamic morphological properties into forest growth modeling II. Allocation and mortality. For. Ecol. Manage. 111, 193–210.

Kartschall T *et al.* 1990. Simulation of nitrogen, water and temperature dynamics in soil. Syst Anal Model Simul. 7(6), 33–40.

Konopatzky A, Freyer C. 1996. Erfassung und Analyse des Wandels der Oberbodenzustände von Kiefern-Ökosystemen des Eintragsmodellgebietes Dübener Heide. Abschlußbericht zum Teilprojekt E 1.2 des BMBF – Verbundforschungsvorhabens SANA; Band IV; GSF, München.

Kopp D, Kirschner G. 1992. Fremdstoffbedingter Standortswandel aus periodischer Kartierung des Standortszustandes in den Wäldern des nordostdeutschen Tieflandes nach Ergebnissen der Standortserkundung. Beiträge für Forstwirtschaft und Landschaftsökologie. 26(2/4), S. 62–71.

Leeuwen van EP *et al.* 1996. Mapping dry deposition of acidifying components and base cations on a small scale in Germany. RIVM Report No. 722108012, 47 pp.

Lux H. 1965. Die großräumige Abgrenzung von Rauchschadenszonen im Einflußbereich des Industriegebietes um Bitterfeld. Wiss Z Techn Univ Dresden. 14(2), 433–442.

Machenhauer B. 1996. The Regionalization Project: A Synthesis of the Final Report. EEC Environment Program Contract No. EV5V-CT92-0126 and PECO Suppl. Agreement No.1 CIPD CT93 0005.

Magistad OC. 1925. The aluminium content of the soil solution and its relation to soil reaction and plant growth. Soil Sci. 20, 181–226.

Mohren GMJ. 1987. Simulation of forest growth, applied to Douglas Fir stands in the Netherlands. Agricultural University, Wageningen, The Netherlands, 184 p.

Mohren GMJ *et al.* 1992. Quantifying the direct effects of SO_2 and O_3 on forest growth. Forest Ecol Managem. 51, 137–150.

Möller D, Lux H. Eds. 1992. Deposition atmosphärischer Spurenstoffe in der ehemaligen DDR bis 1990. Methoden und Ergebnisse. Schriftenreihe Kommission Reinhaltung der Luft im VDI und DIN Bd. 18, 308 pp.

Niehus B, Brüggemann L. 1995. Untersuchungen zur Deposition luftgetragender Stoffe in der Dübener Heide. Beitr. Forstwirtsch. u. Landsch.ökol. 29(4), 160–163.

Niehus B *et al.* 1982. Zur Einschätzung der Immissionssituation in einem industriellen Ballungsgebiet. Unpublished manuscript of the Institut für Geographie und Geoökologie Leipzig. 46 pp.

Posch M *et al.* Eds. 1995. Calculation and Mapping of Critical Thresholds in Europe. Status Report 1995; RIVM Report No. 259101004; 198 pp.

Schaller E. 1996. Berechnung saisonaler Konzentrationsfelder und Depositionsraten als Eingangsdynamik in ökologische Modelle; Abschlußbericht zum Teilprojekt D 2.1 des BMBF – Verbundforschungsvorhabens SANA. Band III. GSF, München

Schulze E-D *et al.* Eds. 1989. Forest decline and air pollution, a study of spruce (*Picea abies*) on acid soils. Springer-Verlag Berlin, 469 pp.

Simon K-H, Westendorff K. 1991. Stoffeinträge mit dem Niederschlag in Kiefernbestände des nordostdeutschen Tieflandes in den Jahren 1985–1989. Beitr. Forstwirtschaft, Berlin 25(4), 177–180.

Sverdrup H, Warfvinge P. 1993. The effect of soil acidification on the growth of trees, grass and herbs as expressed by the (Ca+Mg+K)/Al ratio. Reports in ecology and environmental engineering 1993:2, Lund University, 2nd ed., 177 pp.

UBA (German Federal Environment Agency). Ed. 1996. Manual on methodologies for mapping critical loads/levels and geographical areas where they are exceeded. Final Draft. UBA Texte. 71/96, 136 S.

Ulrich B. 1987. Stability, elasticity and resilience of terrestrial ecosystems with respect of matter balance. In: Potentials and Limitations of Ecosystem Analysis. Eds. E-D Schulze, H Zwölfer. Ecological Studies Vol. 61. Springer, New York. 11–49.

WZB. 1995. Waldzustandsbericht der Bundesregierung (forest condition report of the German government) 1995. Bundesministerium für Ernährung, Landwirtschaft und Forsten, Bonn, 104 pp.

16
Synopsis of the ecological SANA investigations. An integrated analysis of the results

K. BELLMANN and R. GROTE

1. Tree and stand response

The radial growth of forest trees is an important indicator for stand development because it provides integrated information about environmental conditions over the entire lifetime of the stand. If it is used as an indicator for forest productivity, however, several uncertainties have to be considered. Firstly, the diameter increase is only proportional to the stem wood production if the tree mortality and the height:diameter-relation of the trees are constant. These conditions depend heavily on stand age and management, but also on the occurrence of environmental impacts. Secondly, there is a considerable variability in tree growth even within a homogenous plantation that affects the size of the sample needed to produce reliable results. Furthermore, it has to be considered that the measurements do not quite reflect the actual average growth, because it is only possible to investigate the trees that have survived, while many of the slow growing individuals have already died.

Despite these uncertainties in mind, the information obtained about diameter growth from the SANA investigations was nevertheless used here to define certain time frames, which are characterised by specific growth patterns of the stands. Furthermore, the information about the general behaviour of the ecosystem can serve to identify the dominant processes that determine these responses.

Figure 1 illustrates the four growth phases at the three investigated sites, which represent high (Rösa), medium (Taura) and low (Neuglobsow) air pollution load.

Until 1970 (i.e. phase 1), there were no major differences in the diameter increase of the stands. During the next 10 years (phase 2), however, a general growth increase in all three stands occurred, which was less pronounced at the reference site Neuglobsow, where the growth declined to the previous values after a few years. In the third phase, approximately between 1980 and 1988, the diameter growth at the polluted sites decreased below that of the stand at the references site and reached extremely small values, particularly at the highly impacted site Rösa. During the last years of the investigation (phase 4), diameter growth at the polluted sites sharply increased, whereas at the background site no change or even a decline in growth was observed.

R.F. Hüttl and K. Bellmann, Changes of Atmospheric Chemistry and Effects on Forest Ecosystems, 313–322.

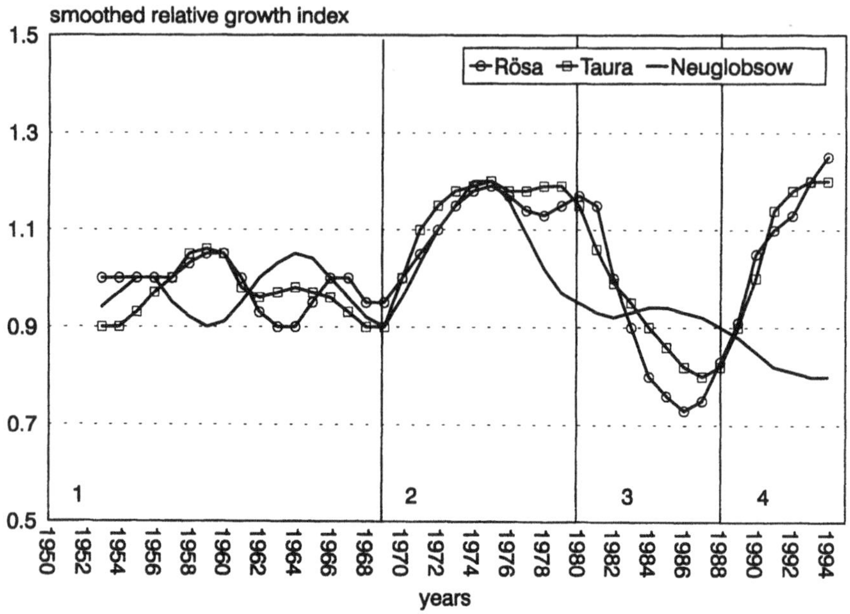

Figure 1. Simplified index of stand growth for the test sites, derived from stand investigations by Neumann (this volume) and separation into four growth phases

1.1. The early years

Because only very little data was available until the late 1960s, it is not quite clear whether the anthropogenic impacts in this first period were insufficient to affect the growth of the stands, or whether different influences, for example higher SO_2 air pollution and higher nitrogen deposition, compensated each other. There is some evidence, however, that air pollution increased more or less steadily during this period, leading to considerable foliage damage (Table 1).

1.2. The period of increased growth

During the 1970s and the early 1980s, forest managers responded to the symptoms of foliage loss fertilising the forest plantations with huge amounts of urea (e.g. Niefnecker, 1985). Even at relatively unpolluted sites, some fertilisation was applied; in this context also the test stand in Neuglobsow received a total of 200 kg N applied in 1973 and 1974. Apparently, the stands responded to this additional nitrogen input by an increased diameter growth (Figure 1). The long lasting positive growth effect at Rösa and the early growth decline at the backgroundsite is well correlated with the amount and duration

Table 1. Indices of foliage health* for highly polluted sites (from Ende *et al.*, in Anonymous, 1997)

	BENA*	% of unpolluted site
1950–1967	185	70
1968–1975	160	60
1976–1983	175	66
1984–1989	150	56
1990–1995	180	67

*Benadelungsgrad = sum of relative needle abundance of each needle age class

of fertiliser application at these sites. In the literature, numerous examples can be found for this effect (e.g. Aronsson and Elowson, 1980; De Visser, 1995; Niefnecker, 1982), generally relating the response to an increased net photosynthesis (Makino and Osmond, 1991) or a changed carbon allocation pattern be(Beets and Whitehead, 1996), particularly leeding to an increase of leaf area index (Fife and Nambiar, 1997).

However, to a large extent this generally observed response is presumably also related to the indirect effect of an increased thinning prior to fertilisation providing better resource availability and, thus, improved growth conditions for the remaining trees. Of course, natural impacts can also change the stand density by means of one or a few single but disastrous effects or by increasing the natural mortality rate of the stands.

1.3. The time of highest pollution

The 1980s, particularly the years between 1985 and 1989, were the years with the highest emission rates of SO_2 in the investigated area. At the same time fertilisation did not result in a further increase of forest stand productivity and, therefore, was not applied any longer. In this time period the average diameter growth at the high and moderately polluted sites declined sharply (Figure 1). Total volume production was supposed to have responded even more, as tree mortality was assumed to be higher during this time period. Therefore, it is hypothesized, that the depression of growth in this period is mainly due to the peaking SO_2 load and was further reinforced by the stopped fertilisation practice. This is in accordance with results for the experimental area of the Dübener Heide of Lux (1974) and is further supported by similar findings of Legge *et al.* (1996) and McLeod *et al.* (1992).

Various mechanisms may have been involved in this process such as reduced photosynthetic rate (Meng *et al.*, 1994), a changed tree morphology (Huttunen *et al.*, 1983) and the enhanced formation of stress enzymes well known to increase foliage mortality and necrosis (Table 1; Taylor *et al.*, 1994; Matschke, 1988; Shaw *et al.*, 1993).

In addition the competition for water of a spreading ground vegetation may impact the vitality state of polluted stands (Tölle and Hofmann, 1970). This is especially true for nutrient rich and sparsely covered stands (Perry *et al.*, 1994). The assumption that SO_2 played the dominant role in the observed growth decrease is also supported by the finding that the diameter development at the background site remained almost unchanged.

1.4. The period of recovering growth rates

On the polluted sites a restoration of growth started at the end of the 1980s and still continues. The current diameter growth rates in these stands clearly exceed the growth rates of the early years or those of the background site in former times (Figure 1, phase 3+4). In contrast, the radial increase at the background site keeps on declining probably indicating a gradual overall decrease of the nitrogen availability and may also be related to a higher stand density of this pine stand.

The major reason for the restoration of growth on the impacted sites can be seen in the decrease of the SO_2 air pollution in synergy with an improved use of available light and nitrogen by the sparse stand structure. Based on a simulation of the pine forest ecosystem model it was calculated that the decrease of the SO_2 pollution accounts for up to 40% to 50% of the current increase in diameter growth.

If a linear response of photosynthesis to nitrogen supply is assumed, and the increase in photosynthetic production per unit of increased nitrogen concentration in the foliage according to Aber (1996) is applied, the increase in net photosynthetic production due to the better nitrogen supply is approximately 30%. Considering the change in carbon allocation as estimated from the biomass measurements at the three investigation sites, the contribution of nitrogen supply to the observed total growth recovery, is around 25%. This result is very well in the range of growth increases resulting from similar kinds of fertilisation in the Pacific Northwest, as documented by Cole (1995).

It has to be considered that a huge amount of nitrogen (and other nutrients) is stored in either organic surface layer or in the forest floor vegetation itself. Thus, the overall nutrient availability has been maintained although the actual deposition showed a gradual decline (see Schaaf *et al.*, in this volume). It has been shown that ground vegetation plays an important role in the conservation of nutrients (Fischer *et al.*, in Anonymous, 1997). Model-based estimates indicate that more than twice as much nitrogen is cycled within the ground vegetation at Rösa and Taura than at the background site, whereas the cycling of nitrogen within the trees don't vary very much (Grote *et al.*, this volume).

The better nutrient supply at the impacted sites is supposed to be have reduced the fine root production (Strubelt *et al.*, this volume). At the impacted sites the relatively high nutrient availability is correlated with lower fine root density, indicating a higher lifespan and a lower turnover of roots. Thus, less carbon is required for fine root formation, and can be used for growth of other tree compartments, e.g. wood biomass.

However, not only the total fine root biomass, but also the distribution pattern within the soil is different. Whereas fine roots at Neuglobsow are concentrated rather in the upper soil layers fine roots at the other sites are more evenly distributed throughout the soil profile (Strubelt *et al.*, this volume). In general, root distribution with a relatively large proportion of the total fine root biomass located in deeper soil layers has advantages during drought periods, because the available soil water can be used more efficiently. This may positively influence the water availability of stands at the impacted sites since a higher shore of fine roots in the deeper soil horizons may facilitate the maintainance of the water supply in drought period. Thus, in general, a smaller drought stress develops during the dry summer period at the impacted sites (Lüttschwager *et al.*, this volume).

Model-based estimations indicate that on experimental sites drought stress leads to a reduced net primary production of about 7%. Assuming that the fine root distribution is the only difference between sites. A shallow root system like in Neuglobsow would cause a decrease of stem wood production by 5% at Rösa and Taura.

All impacts discussed above explain between 75 and 90% of the total increase in radial growth between 1987 and 1994, based on the investigations of the last two years in this period. However, it has to be considered that the investigation period gives only a short-term picture of the processes, which are not linearly related to environmental conditions.

In return, based on the simulation studies weather conditions, physiological and morphological stand properties and soil conditions are also dynamic variables, which may alter the sensitivity of the pine forest ecosystems to changing environmental conditions. Some of these changes will probably have increased the sensitivity of the stand to respond to the more favourable conditions of recent years. In this context, measurements indicate that the leaf area index may have increased over several years (Ende *et al.*, in Anonymous, 1997). In addition, the hydraulic stem properties seem to need some years to recover (and possibly have not yet recovered at all; see Rust, this volume). However, a mechanistic and quantitative understanding of these processes has not been achieved yet.

2. Soil response

The deposition of basic cations (e.g. calcium, potassium and magnesium) and nitrogen (ammonium as well as nitrate) over several decades led to considerable changes in the soil element status and budget. Particularly the pH value of the soils was found to be unusually high, despite large sulphate deposition rates (see also Schaaf *et al.*, this volume). However, the installation of dust filters since the mid 1980s and the breakdown of large sections of industry since 1990 decreased the input of cations much more than the input of sulphate. Thus, a re-acidification can already be observed in the upper soil horizons in large areas of the Dübener Heide (Konopatzki, 1995).

Despite this development, the sites within the investigation, which were exposed to high air pollution were well supplied with nutrients. However, output rates of basic cations are already much greater than the input rates into the soil, indicating that site conditions are changing. At Taura, where cation deposition but not sulphate deposition generally has been smaller in the past, Al^{3+} ions have already been released in the upper soil horizons. In contrast to the basic cations, total output of nitrogen (loss with percolation, fixation in plant biomass including export with harvest) is smaller than the input by deposition. Thus, even at the most affected sites, the soils seem to be still not quite saturated with nitrogen. However, the development in pH and in nitrogen dynamics are closely linked together as shown in Figure 2. The decreasing pH value has a negative effect on mineralisation and nitrification (Fischer *et al.*, in Anonymous, 1997), which decreases the plant available nitrogen and increases percolation losses.

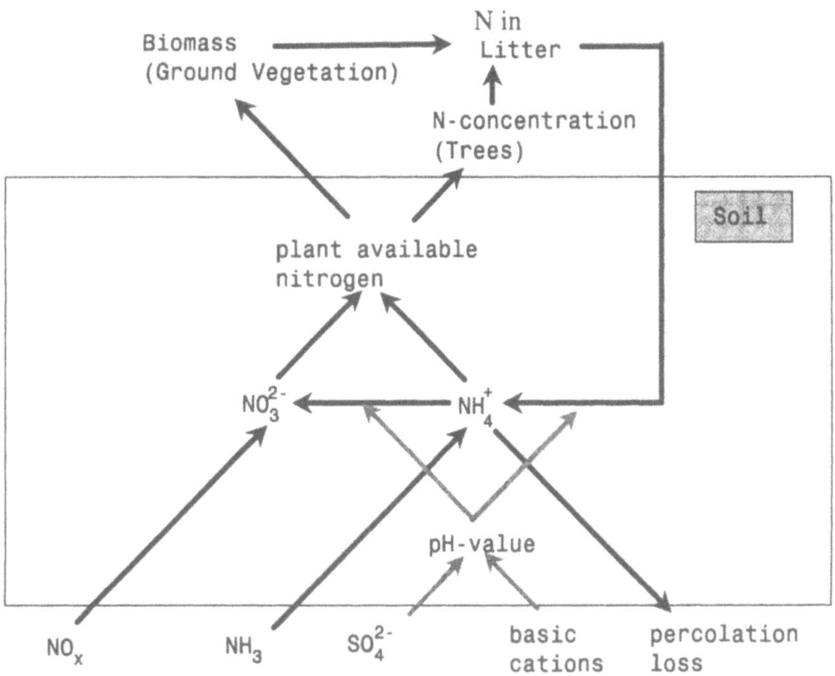

Figure 2. Effect of deposition on vegetation. (Black lines = mass fluxes, grey lines = effect on processes; words in parentheses = main impacts)

3. Modelling and regional assessment

In SANA, the modelling activities formed an integrated part of the ecological investigations right from the beginning of the project. The results obtained by the working groups were used to develop a mechanistic model of the ecosystem, including the qualitative and quantitative interrelationships within and between stand climate, trees, ground vegetation and the soil (Grote *et al.*, this volume). Although the period of evaluation of the physiological processes was only two years, the new model successfully describes the development of a large number of stands of the Dübener Heide area during the last three decades (Erhard and Flechsig, this volume).

For the differentiation of the pollution impacts within the Dübener Heide area, and for the assessment of different pollution scenarios, a new GIS-based approach (stand and soil data) for simulating the time- and space-related variables, which determine forest growth dynamic, was applied. This was carried out with spatially distributed air pollution and deposition records for each pine stand (>25 years) within the Dübener Heide area, which refers to approximately 400 km^2 (Erhard and Flechsig, this volume). In the simulation of 30 years of forest development, different scenarios showed that the present pollution load, i.e. a reduction by 60% of the 1989 SO_2-emission (Figure) would result in a long-term positive effect on stemwood production between 0 and 10%. The greatest increase in growth is expected in the areas close to former main industrial centres, i.e. the high loaded test site at Rösa and Taura (Erhard and Flechsig, this volume).

4. Conclusion

The impacted sites under study are by no means in an equilibrium stage. Further developments can still be expected even if environmental conditions will be kept stable. This development, however, will probably not be driven by the changes in SO_2 concentration, which certainly played the dominant role during the last 5 years. Even if the current SO_2 concentration is decreased further, as can be expected from the technical development and legislation, the effect on tree and stand growth will be only small. The reason lies in the shape of the response curve of the respective physiological processes, particularly with respect of foliage mortality, indicating a steep reaction only if certain boundary limits are exceeded (Gluch, 1988).

The nutrition of the vegetation, however, is expected to be exposed to a long-lasting change, which is driven mainly by the change in the element budget of the soils and the resulting decrease in the pH value (Figure 2). This process is expected to decrease the nitrogen availability by means of its decreasing effect on mineralisation. Direct losses of nitrogen with percolation will also be increased. However, it is not expected that output rates will increase essentially above deposition input (see also Skeffington and Wilson, 1988).

It is expected that this development will be enhanced by stand development processes, because the decreasing air pollution obviously has increased the vitality of the trees at the affected sites. Thus, the stand density will increase compared to the current conditions, while it is supposed that ground vegetation will decrease with decreasing light conditions at the forest floor. This development, however, will depend on the forest management and the deer grazing intensity at the sites, which are not investigated in this study. Less ground vegetation will decrease the amount of nitrogen, which is exchanged annually, and thus the nitrogen losses due to percolation will increase.

In response to the alterations in nutrition, the allocation of the trees will change in order to avoid nutrient stress. Thus, the fine root biomass will increase and the distribution of fine roots within the soil will become more concentrated in the upper soil layers, which contain more nutrients. However, this will increase the susceptibility to water stress and thus the turnover rate, although the water competition from ground vegetation may be decreased.

The duration of this process cannot be estimated with the information gained by the current investigations, because the current speed of the development is widely influenced by the specific climatic conditions within the two test years. From the large pools of elements, as found at Rösa, however, it can be assumed that the development at such sites will last at least until a new generation of trees becomes established (see also Erhard and Flechsig, this volume). This may be different at sites with a smaller pool of basic ions like Taura, where the fast changes of the pH value may lead to conditions which are less favourable as currently observed at the background sites within a few years.

The origin and the problems due to enhanced acidification related to these conditions are typical for sites in western Europe and have been intensively investigated (e.g. Heij *et al.*, 1991). Nevertheless, it is not clear, whether the same mechanisms can be assumed, because there is little well founded information about the amount of heavy metals, the processes of their release, and their damage potential at the sites within the target area. Furthermore, it cannot be fully excluded that considerable environmental effects have been overlooked or underestimated in this study. For example, it can be assumed that the importance of ozone to plant physiological processes has increased (in relative and absolute terms) since the SO_2 concentration has decreased while NO_X concentrations remained unchanged or even increased (Schulz, this volume). Also the relation between plant development and other nutrients than nitrogen, e.g. phosphorus, at the investigated sites remains unclear, although it is known that such elements are possible limitations to photosynthetic responses(e.g. Conroy *et al.*, 1988; Küppers, 1993).

References

Aber JD, Reich PB, Goulden L. 1996. Extrapolating leaf CO_2 exchange to the canopy: a generalized model of forest photosynthesis compared with measurements by eddy correlation. Oecologia. 106, 257–265.

Anonymous. 1997. SANA – Wissenschaftliches Begleitprogramm zur Sanierung der Atmosphäre über den neuen Bundesländern. BMBF, Bonn.

Aronsson A, Elowson S. 1980. Effects of irrigation and fertilization on mineral nutrients in Scots pine needles. In: Structure and function of northern Coniferous forests – An ecosystem Study. Ed. T Persson. Ecol Bull, Stockholm. p. 219–228.

Barbour RJ, Chauret G, Cook J, Karsh MB, Ran S. 1994. Breast-height relative density and radial growth in mature jack pine (*Pinus banksiana*) for 38 years after thinning. Can J For Res. 24, 2439–2447.

Beets PN, Whitehead D. 1996. Carbon partitioning in *Pinus radiata* stands in relation to foliage nitrogen status. Tree Physiol. 16, 131–138.

Cole DW. 1995. Soil nutrient supply in natural and managed forests. Plant Soil, 168–169, 43–53.

Conroy JP, Küppers M, Küppers B, Virgona J, Barlow EWR. 1988. The influence of CO_2 enrichment, phosphorus deficiency and water stress on the growth, conductance and water use of *Pinus radiata* D. Don Plant Cell Environ. 11, 91–98.

De Visser PHB. 1995. Effects of irrigation and balanced fertilization on nutrient cycling in a Douglas fir stand. Plant Soil. 168–169, 353–363.

Fife DN, Nambiar EKS. 1997. Changes in the canopy and growth of *Pinus radiata* in response to nitrogen supply. Forest Ecol Management. 93, 137–152.

Gluch W. 1988. Zur Benadelung von Kiefern (*Pinus silvestris* L.) in Abhängigkeit vom Immissionsdruck. Flora. 181, 395–407.

Heij GJ, De Vries W, Posthumus AC, Mohren GMJ. 1991. Effects of air pollution and acid deposition on forests and forest soils. In: Acidification Research in The Netherlands. Final report of the Dutch Priority Programme on Acidification. Studies in Environmental Science. Eds. GJ Heij, T Schneider. Elsevier, Amsterdam. p. 97–137.

Huttunen S, Karhu M, Laine K. 1983. Air pollution induced stress and its effects on the photosynthesis on *Pinus sylvestris* L. in Oulu. Aquilo Ser Bot. 19, 275–282.

Konopatzki A. 1995. Untersuchungen zum langjährigen Oberbodenzustandswandel in den Waldökosystemen der Dübener Heide. In: Atmosphärensanierung und Waldökosysteme. Eds. RF Hüttl, K. Bellmann and W Seiler. Blottner-Verlag, Taunusstein, Umweltwissenschaften. 4, 238 S.

Küppers M, Conroy JP, Barlow EWR. 1993. Wirkung erhöhten CO2-Angebotes, der Phosphat- und Wasserverfügbarkeit auf Gasaustausch und Wachstum von Monterey-Kiefern (*Pinus radiata*). Verhandlungen der Gesellschaft für Ökologie. 22, 1–4.

Legge AH, Nosal M, Krupa SV. 1996. Modeling the numerical relationships between chronic ambient sulphur dioxide exposures and tree growth. Can J For Res. 26, 689–695.

Lux H. 1974. Diagnose der Rauchschäden in den Forsten der Niederlausitz, Sektion Forstwirtschaft, Tharandt.

Makino A, Osmond B. 1991. Effects of nitrogen nutrition on nitrogen partitioning between chloroplasts and mitochondria in pea and wheat. Plant Physiol. 96, 355–362.

Matschke J, Hertel H, Ewald C, Nöring E. 1988. Vitalitätsbestimmung immissionsbeeinflußter Koniferen durch die Analyse von Defensivenzymen. Beitr Forstwirtschaft. 22(3), 125–133.

McLeod AR, Shaw PJA, Holland MR. 1992. The Liphook Forest Fumigation Project: studies of sulphur dioxide and ozone effects on coniferous trees. For Ecol Management. 51, 121–127.

Meng F-R, Cox RM, Arp PA. 1994. Fumigating mature spruce branches with SO_2: effects on net photosynthesis and stomatal conductance. Can J For Res. 24, 1464–1471.

Niefnecker W. 1982. Volumenmehrzuwachs nach aviotechnischer Großflächendüngung in Kiefernreinbeständen. Soz. Forstwirtschaft. 32(8), 233–235.

Niefnecker W. 1985. Ertragskundliche Untersuchungsergebnisse zur stabilisierenden Wirkung der mineralischen Großflächendüngung in immisionsgeschädigten Kiefernwaldgebieten. Beitr Forstwirtschaft. 19(3), 120–124.

Perry MA, Mitchell RJ, Zutter BR, Glover GR, Gjierstad DH. 1994. Seasonal variation in competitive effect on water stress and pine responses. Can J For Res. 24, 1440–1449.

Shaw PJA, Holland MR, Darrall NM, McLeod AR. 1993. The occurrence of SO_2-related foliar symptoms on Scots pine (*Pinus sylvestris* L.) in an open-air forest fumigation experiment. New Phytol. 123, 143–152.

Skeffington RA, Wilson EJ. 1988. Excess nitrogen deposition: Issues for consideration. Environ Pollut. 54, 159–184.

Taylor GE, Johnson DW, Andersen CP. 1994. Air pollution and forest ecosystems: A regional to global perspective. Ecol Applic. 4(4), 662–689.

Tölle H, Hofmann G. 1970. Beziehungen zwischen Bodenvegetation, Ernährung und Wachstum mittelalter Kiefernbestände im nordostdeutschen Tiefland. Archiv für Forstwesen. 19(4), 385–400.

Tzschacksch O. 1987. Zur Labor- und Feldresistenz der Kiefer (*Pinus sylvestris* L.) gegenüber phytotoxischen Stoffen und Schlußfolgerungen für die Anbauwürdigkeit von Kiefernarten in den Immisssionsschadgebieten des oberen Erzgebirges. Beitr Forstwirtschaft. 21, 97–102.

17
Concluding remarks

K. BELLMANN and R.F. HÜTTL

In conclusion, the following major findings can be stated:

- High deposition loads of air pollutants in the past caused high accumulation rates of the respective substances in the soil; locally base saturation and pH values of the soils were improved, also Ca, S, and N soil stores were increased; as a consequence, humus forms changed.
- In relation to former deposition loads, high output rates still occur from the rooted solum, particularly related to SO_4^{2-}, Ca^{2+}, and at Taura – because of subsoil acidification – also Al^{3+}.
- Despite regionally high historic N deposition rates and N input via fertilization no N-losses with the seepage water could be detected.
- The strong reduction of the alkaline deposition loads after reunification of Germany and the still relatively high acid deposition rates locally lead to re-acidification of soils.
- Reduced concentrations of air pollutants in the atmosphere after reunification of Germany led to an improvement of canopy density and foliar tissue of the investigated Scots pine stands.
- The improvement of the photosynthetic potential of the foliage is related to an enhanced growth performance of the trees.
- Previously high deposition loads and the subsequent change of the chemical soil status in combination with sufficient sunlight radiation to the forest floor caused a massive expansion of forest floor vegetation; this development leads to the immobilization of significant amounts of N and at least partly explains that no N-leaching losses could be observed at the respective sites.
- The forest floor vegetation at the heavy deposition site Rösa makes up for a remarkably large proportion of the leaf area index (LAI) and hence is also responsible for a relatively large proportion of the transpiration amount of the ecosystem (Table 1).

R.F. Hüttl and K. Bellmann, Changes of Atmospheric Chemistry and Effects on Forest Ecosystems, 323–324.

- The total transpiration rate does not differ much between the high and the moderate deposition site irrespective of the proportion of the deposition related differences in forest floor versus forest stand vegetation (Table 1).

Table 1. Water balance of the experimental Scots pine stands with previously high (Rösa) and previously moderate (Taura) atmogenic deposition load (mean percentage values of the years 1993 and 1994)

Water balance parameter		High deposition (%)	Moderate deposition (%)
LAI	Scots pine	54	84
	Forest floor vegetation	46	16
Transpiration	Scots pine	64	77
	Forest floor vegetation	36	23
Total ecosystem	Interception	39	30
	Transpiration	30	28
	Soil evaporation	1	1
	Seepage loss	30	41

With regard to the condition of reduced deposition load of air pollutants the following future developments in Scots pine stands on sandy substrates in north-eastern Germany may be expected:

- decrease of the base saturation of the soil
- reduction of the acid neutralization capacity of the soil
- continued re-acidification of the top-soil and eventually enhanced subsoil acidification
- decreasing mineralization and nitrification rates
- continued high nitrogen storage in soil and vegetation
- increased adsorption capacity for sulfate in the mineral soil
- gradual reduction of matter output rates via seepage water
- continuously improved vitality of the needle apparatus

From these predictions it could be expected that the nitrogen supply of the trees and, hence, the growth rates of the stands will remain at a relatively high level. Whether changes in the forest floor vegetation and subsequently also in the tree species will occur, remains to be seen.

Nutrients in Ecosystems

1. R.F. Hüttl and W. Schaaf (eds.): *Magnesium Deficiency in Forest Ecosystems*. 1997 ISBN 0-7923-4220-8
2. E. Schnug (ed.): *Sulphur in Agroecosystems*. 1998 ISBN 0-7923-5123-1
3. R.F. Hüttl and K. Bellmann (eds.): *Changes of Atmospheric Chemistry and Effects on Forest Ecosystems*. A Roof Experiment without Roof. 1999 ISBN 0-7923-5713-2

KLUWER ACADEMIC PUBLISHERS – DORDRECHT / BOSTON / LONDON

SPRINGER NATURE

GPSR Compliance

The European Union's (EU) General Product Safety Regulation (GPSR) is a set of rules that requires consumer products to be safe and our obligations to ensure this.

If you have any concerns about our products, you can contact us on ProductSafety@springernature.com

In case Publisher is established outside the EU, the EU authorized representative is:

Springer Nature Customer Service Center GmbH
Europaplatz 3
69115 Heidelberg, Germany

FSC
www.fsc.org
MIX
Papier | Fördert gute Waldnutzung
FSC® C083411

Kolibri 360 GmbH
Ferdinand-Jühlke-Straße 7
99095 Erfurt, Deutschland
produktsicherheit@kolibri360.de